# COMPUTATIONAL AUCTION MECHANISMS FOR RESTRUCTURED POWER INDUSTRY OPERATION

# THE KLUWER INTERNATIONAL SERIES IN ENGINEERING AND COMPUTER SCIENCE

## Power Electronics and Power Systems

*Consulting Editors*

**Thomas A. Lipo and M. A. Pai**

*Other books in the series:*


ANALYSIS OF SUBSYNCHRONOUS RESONANCE IN POWER SYSTEMS
K.R. Padiyar, ISBN: 0-7923-8319-2
POWER SYSTEMS RESTRUCTURING: *Engineering and Economics*
Marija Ilic, Francisco Galiana, and Lester Fink, ISBN: 0-7923-8163-7
CRYOGENIC OPERATION OF SILICON POWER DEVICES
Ranbir Singh and B. Jayant Baliga, ISBN: 0-7923-8157-2
VOLTAGE STABILITY OF ELECTRIC POWER SYSTEMS, Thierry Van Cutsem and Costas Vournas, ISBN: 0-7923-8139-4
AUTOMATIC LEARNING TECHNIQUES IN POWER SYSTEMS, Louis A. Wehenkel, ISBN: 0-7923-8068-1
ENERGY FUNCTION ANALYSIS FOR POWER SYSTEM STABILITY, M. A. Pai, ISBN: 0-7923-9035-0
ELECTROMAGNETIC MODELLING OF POWER ELECTRONIC CONVERTERS, J. A. Ferreira, ISBN: 0-7923-9034-2
MODERN POWER SYSTEMS CONTROL AND OPERATION, A. S. Debs, ISBN: 0-89838-265-3
RELIABILITY ASSESSMENT OF LARGE ELECTRIC POWER SYSTEMS, R. Billington, R. N. Allan, ISBN: 0-89838-266-1
SPOT PRICING OF ELECTRICITY, F. C. Schweppe, M. C. Caramanis, R. D. Tabors, R. E. Bohn, ISBN: 0-89838-260-2
INDUSTRIAL ENERGY MANAGEMENT: *Principles and Applications*, Giovanni Petrecca, ISBN: 0-7923-9305-8
THE FIELD ORIENTATION PRINCIPLE IN CONTROL OF INDUCTION MOTORS, Andrzej M. Trzynadlowski, ISBN: 0-7923-9420-8
FINITE ELEMENT ANALYSIS OF ELECTRICAL MACHINES, S. J. Salon, ISBN: 0-7923-9594-8

# COMPUTATIONAL AUCTION MECHANISMS FOR RESTRUCTURED POWER INDUSTRY OPERATION

**Gerald B. Sheblé**

Springer Science+Business Media, LLC

**Library of Congress Cataloging-in-Publication Data**

Sheblé, Gerald B.
Computational auction mechanisms for restructured power industry operation / Gerald B. Sheblé.
p. cm. -- (The Kluwer international series in engineering and computer science ; SECS 500. Power electronics and power systems)
Includes bibliographical references.
ISBN 978-1-4613-7348-3 ISBN 978-1-4615-5157-7 (eBook)
DOI 10.1007/978-1-4615-5157-7
1. Electric utilities--Finance--Computer simulation. 2. Commodity exchanges--Computer simulation. 3. Auctions--Computer simulation. I. Title. II. Series: Kluwer international series in engineering and computer science ; SECS 500. III. Series : Kluwer international series in engineering and computer science. Power electronics & power systems.
HG6047.E43S54 1999
333.793'2--dc21

99-13571
CIP



*Printed on acid-free paper*

This printing is a digital duplication of the original edition.

Dedicated to my family: Jason, Laura,
Judy, Ron, Mike, Cathy, & Walter

# CONTENTS

# Preface

The most interesting times are those including massive changes. The electric power industry is in the most turbulent era except for the era when the industry was going from competitive to regulated. The society viewpoint of electricity has fundamentally changed. The nuclear crises as well as the lack of consistent pricing around the world have awakened a sense of distrust of the electric energy monopolies. Schweppe [1988] stated the paradigm precisely: "Electric energy must be treated as commodity which can be bought, sold, and traded, taking into account its time- and space-varying values and costs." Schweppe's work on the application of spatial (spot) prices to the electric power industry, as has been done in other industries [Thompson, 1992; Hillier 1996], is a classic work upon which this work is deeply indebted. However, that work did not truly represent what the spot price of electricity is based. The spot price of any commodity in a free market is based on the perceived value as negotiated through millions of contracts between the consumers and the suppliers. It is the need for a free market that this work builds various auction techniques and bidding techniques to find the best price. The best price is based not only on the supplier costs but also the consumers valuation of that commodity. As with any commodity with special handling and transportation, the price is a function of time and location. Under very restrictive conditions, the spot price will gravitate to the same level in all locations and to the marginal cost. However, any technological changes, any unfair market play, any social or industrial welfare added would drive the price away from the ideal market price. This must be avoided during the transition from a regulated industry to a competitive industry and during the eventual operation as a competitive industry.

This work is written for electric power engineers, economists, and financial planners to include all aspects of competition into the industry. Detailed optimization formulation is provided for those who wish to reproduce the results obtained.

Commodity trading of electricity can be of immense benefit to society and to the industry only if the markets are properly defined and implemented. The history of commodity markets over time (since 400 BC) has shown many market and society crises due to improper market implementation. The benefits of proper price signals are well documented for the expansion of the industry, for the research and development of new technology and processes, and for the continued improvement in the quality of life for our society. The primary advantage I foresee is the joint venture agreements to spread the risk of system outages and of fuel supply uncertainty between the suppliers and the consumers. The proper establishment of the market structure and of the market procedures will require regulation by federal and by state government agencies. Markets are a means to focus controlled greed to the improvement of society. There is much to be accomplished during this transition period. The markets are presently chaotic in nature. The future supply of electricity is not assured. It is only through the simulation and analysis of market structure that the industry can mature into a positive weather vane for democratic societies to succeed.

## Organization of Book

This work deals with many topics published by this author with many students: auction basics, auction by transportation, optimal power flow (OPF) linear programming and auctions, comparison with economic dispatch calculation (EDC), auction market simulation for single hour bidding, unit commitment and auctions, ancillary services, auctions and generation bidding, auctions and supportive services bidding, transmission auctions, wholesale market access, and interaction with automatic network control (ANC). Other topics published by this author are not included due to time a space constraints: hydro thermal coordination and multiple auctions, transmission planning and auctions for network expansion, generation planning and auctions for franchises, energy services and auctions for demand expansion. A very interesting topic not covered is valuation of exotic derivatives for risk management. The interested reader will have to search the journal and conference papers for these topics.

Chapter 1 reviews the present state of the industry and the direction this author recommends for auction market implementation.

Chapter 2 on auction basics outlines the different type of auctions available. This chapter should provide an interesting start for those who wish to compare the theory of auctions with the auctions that have actually evolved throughout history.

Chapter 3 presents auctions as transportation and assignment problems based on Linear Programming optimization. Auctions are also posed as price decomposition problems. The auction as an optimization technique is compared with the classical economic dispatch calculation and Dantzig Wolfe equations. Auction techniques can be compared with optimizing a vertically integrated company using optimization through decentralized production. A complete solution with multiple buyers and sellers are then presented with assignment to find a partial equilibrium solution.

Chapter 4 develops economic dispatch, unit commitment, and optimal power flow by linear programming as auction mechanisms. This work is based on LaGrangian Relaxation and Linear Programming as explained in most operations research texts. This chapter outlines the markets for supportive services and the need to bundle such services. This chapter reviews the needs for ancillary services, spinning reserve vs contingency or capacity options. Auctions and generation bidding for multiple contracts, list types and clearing problems are outlined.

Chapter 5 presents the basic tools needed by suppliers and by buyers to function in a competitive marketplace. The use of control theory concepts for modeling bidding strategies is outlined as well as the modeling of competitors based on real-time bids. The basic algorithms for managing cash flow, transaction risks, fuel market scheduling, transaction selection, as well as contract allocation are included. Auction market simulation for single hour bidding by multiple players by control simulation and classical optimization is detailed. The interaction with automatic network control (ANC), formerly automatic generation control (AGC), for on-line market simulation and training is defined.

Chapter 6 defines the standard operational problems encountered by both GENCOs. The need for multiple hour auctions and relationships similar to a decentralized unit commitment problem is discussed. Auctions for spot prices and forward prices for the next two weeks are minimal information sources. Generation models for fully evolved decentralized markets are segmented and defined. Hedging, comparative analysis, scheduling and capacity options are discussed. Several bidding building techniques are defined.

Chapter 7 reviews the software needs for ESCOs to provide competitive services. Auction implications and generation bidding for multiple contracts, various contract types and clearing problems are outlined. Auctions and energy services bidding, transformation of auction markets to customer contracts, price volatility, forecasting and demand side management by composite market mechanism is provided. Transformation of auction markets to customer contracts, price volatility, forecasting and risk management is outlined. The software needs for ESCOs to provide customer services, spinning reserve, contingency or capacity options are outlined. Auction implications and general bidding for multiple contracts, contract types and clearing problems are outlined. Auctions and energy services bidding to transform auction markets to customer contracts, especially with demand side management, are outlined.

There are many topics not covered. Transmission auctions as an extension to commodity markets have been attempted. The release of information and market power, market share, wholesale market access, probabilistic access while maintaining loss of load probability (LOLP) and expected unserved energy (EUE) are topics central to long term operation and planning of the overall system. Complete operational planning including hydro-thermal coordination and maintenance scheduling are also important. The comparison of classical techniques to multiple auctions, such as water and electricity interdependence, is not mentioned. This is based on the suggested markets proposed by this author for the Brazilian reregulation. The essence of these markets is the equivalent estimation of pseudo prices for fuel and hydro as done in the past for fuel and hydro scheduling. Complete system planing concepts of capital budgeting, portfolio analysis, transmission planning and industrial expansion for multiple or for a single product, generation planning and auctions for franchises, capacity tariffs, auctions for network expansion are left for future work.

This material has been used for a graduate level course in electrical engineering at Iowa State University. The student is assumed to be aware of material from classical optimization for power system operation as found in [Wood, 1996]. The student should also have a firm basis in unconstrained and constrained optimization techniques. This material and material on risk management and financial derivatives are used in a one semester three credit hour course. This is a graduate level course dealing with the economic analysis of power system operation and planning in a de-regulated environment. The objective is to define the alternatives for the new business environment. Algorithmic equivalents are then defined for select alternatives. Then, algorithmic refinements to alleviate market deficiencies for each of the power system environments will be sought. As with any economic solution, many optimization methods will be included to find solutions. Embedded optimization methods include Linear Programming, Constrained Nonlinear Programming, Network Flow Programming, Integer Programming, and Dynamic Programming. The use of artificial life techniques is very beneficial. I have had the benefit of a course in artificial life techniques designed by Dr. Daniel Ashlock offered for students before taking this course. Background in fuzzy logic is also an additional benefit as most pricing techniques are, at best, fuzzy.

The graduate level course has used the case study technique instead of traditional engineering homework. Students are assigned to a specific company based on their interests. Typically, three companies are sufficient to show the equilibrium solutions. I have limited the company types to GENCOs, ESCOs, and ICAs. The major lectures per topic for this course are shown in the following table.

| *Lectures* | *Topic* |
|---|---|
| 1-2 | Review of Corporate Modeling |
| 3 | Review of Engineering Economics |
| 4-5 | Review of Linear Optimization Fundamentals |
| 6-7 | Review of Non-Linear Optimization Fundamentals |
| 8-10 | Overview of Micro-economics (theory of the firm) |
| 10-14 | Overview of Finance |
| 15-16 | Overview of Financial Management |
| 17 | Electric Power Industry Structure - Chapter 1 |
| 18-19 | Industrial infrastructure – Chapter 2 |
| 20 | Auction by Linear Programming – Chapter 3 |
| 21-24 | Economic Dispatch, Unit commitment, Optimum Power Flow as Auction - Chapter 4 |
| 25 | Supportive Services, Bundled or Separate - Chapter 4 |
| 26 | Forecasting markets, Production Frontiers - Chapter 5 |
| 27 | Decentralization of production, Vertically Integrated Company Production - Chapter 5 |
| 28 | Financial Management - Chapter 5 |
| 29 | Decision Analysis, Value of Information, Value of Perfect Information, and Bidding Possibilities - Chapters 6 & 7 plus handouts |
| 30 | Inventory Management & Decisions - Chapters 6 & 7 plus handouts |
| 31 | Production Costing, Risk of Production - Chapters 6 & 7 plus handouts |
| 32 | Activity Analysis & Multiple Markets - Chapters 6 & 7 plus handouts |
| 33 | Input-Output Analysis and Cash Management - Chapters 6 & 7 plus handouts |
| 34 | Contracts And Incentives/Yield Management - Chapters 6 & 7 plus handouts |
| 35 | Options, Portfolio Analysis - Chapters 6 & 7 plus handouts |
| 36 | Capital Budgeting & Investment Programs - handouts |

# *Acknowledgments*

The title page of this work has one name on it but the bibliography shows who really contributed to this work. After 15+ years in the electric power industry, I was reluctant to enter the academic world in 1986. Now, after 12+ years, I have found that both worlds are excellent career choices. However, the gratification of working with many excellent graduate and undergraduate students is very rewarding. The experience at Purdue University working with Professors Gerald Heydt and Ahmed El-Abiad has given me a very lofty goal to attain. This research started in 1986 when Leo Grigsby recommended that I become an expert at something worthwhile. Leo, like many others, was convinced that deregulation would not happen. Leo felt that I did not make a sane choice since deregulation should not happen. Leo thoroughly supported my choice and provided enormous support during my tenure at Auburn University. This work at Auburn enabled me to work with many gifted students, most especially George Fahd. I also extend thanks to Professors Charles Gross and Mark Nelms for their comments. After moving to Iowa State University, I was very ably assisted by George Fahd as a post-doctoral student. I have been very fortunate to have Jayant Kumar and Douglas Post as students, as well as Darwin Anwar. At the time of this writing, Chuck Richter has become a very valuable collaborator. I have also been very fortunate to have Somgiat Dekrajangpetch and Kah-Hoe Ng as graduate students. I am grateful to the students with whom I have worked in this area and those students who have helped me in other areas of research. As it has been with my own children, just when they get interesting, they leave to attain their own goals. I am especially indebted to the efforts of Dr. Richter and Mr. Dekrajangpetch in putting this document together. Kah-Hoe Ng assisted with the ESCO chapter. The comments at the beginning of each chapter were the famous last words of several of these students.

I am also indebted to Ms. Gloria Oberender for her patience in typing my scribbles and changing the text so many times. I extend thanks to the support of Ms. V. Williams for general comments. I extend thanks to Professor M. A. Pai and Mr. A. Greene for their valuable comments on how to assemble this work. Finally, I must thank all of my associates in the industry who have provided valuable constructive criticism during the last decade. I especially thank Mr. John Pope, Dr. A. Phadke, Dr. G. T. Heydt, Dr. S. Rahman, Dr. C. C. Liu, Dr. R. Thomas, and Dr. K. Clements. The recent input provide by Mr. Dale Stevens is most appreciated. I hope that all of you will forgive me for the things I did not include, it is hard to keep up with the input you have provided. The support by Georgia Power, MidAmerican, National Science Foundation, ISU Electric Power Research Center, and the Electric Power Research Institute is deeply appreciated.

Finally, I extend deepest appreciation to my colleagues at Iowa State University S. S. Venkata, V. Vittal, J. McCalley, V. Ajjarapu, J. Lamont, K. Kruempel, G. Hillesland and A. Fouad. Your input, collaboration, and constructive comments have provided value to my research. As always, I remain responsible for all errors and omissions.

# 1 INTRODUCTION

*Engineers thought it the best of times, economists thought it the worst of times.*

Power system interconnections have increased over the past several decades. The two main reasons for this are increased need for security and improved economic operation and planning. Interconnections increase the ability of automatic generation control to maintain system stability. They also increase the possibility of economic gains through power interchanges. When marginal generation costs between two or more areas differ and extra capacity is available, the areas can negotiate an economic energy transaction and share the resulting cost savings.

Unfortunately, the price difference exceeded customer acceptance between states within the United States. The onset of the global economy requires increased pressure on reducing the cost of basic industrial infrastructure. Electric energy is clearly a major component of the industrial complex. Indeed, it is even considered ring three among the military experts. Such a central structure is considered crucial to the survival of any country. Thus, it is easy to understand why the pricing of electricity has become the focal point of many economists and financial analysts since the first U.S. oil crisis in the early '70s. Thus started the reregulation, not deregulation, of the electric energy industry.

Recently, interest in competitive pricing of transactions has increased because of imminent re-regulation of the electric power industry throughout the world. This work considers the application of auction mechanisms to the pricing of electric power. Auction methods are analyzed and the approach that best lends itself to the pricing of electric power is solved by linear and nonlinear programming techniques.

## 1.1 History of the Electric Industry

The electric utility industry which most of us take for granted began over 100 years ago with the electrical pioneers of the late 1800s. The electric light bulb had just been invented, but wasn't going to be a big hit until people had a place to plug it in. According to humorist Dave Barry [Barry, 1985]:

> *The greatest Electrical Pioneer of them all was Thomas Edison, who was a brilliant inventor despite the fact that he had little formal education and lived in New Jersey. Edison's greatest achievement came in 1879 when he invented the electric company. His design was a brilliant adaptation of the simple electrical circuit. The electric company sends electricity through a wire to a customer, then immediately gets the electricity back through another wire, then, (this is the brilliant part) sends it right back to the customer again.*
>
> *This means that an electric company can sell a customer the same batch of electricity thousands of times a day and never get caught, since very few customers take the time to examine their electricity closely. In fact, the last year any new electricity was generated was 1937. The electric companies have been merely reselling it ever since, which is why they have so much time to apply for rate increases.*

Most people in the electric utility industry would agree that this system is more complex than Dave Barry's humor describes it. Many brilliant people have been working throughout the past century to make power systems reliable and as economically efficient as possible. The era when electric companies had so much time to apply for rate increases is gone, and the supply and demand forces are now setting U.S. electric rates. Indeed, many countries are now fully aware of the consumer forces and electricity prices.

For decades, electric consumers had only their local, vertically integrated utility as a source of electricity. In exchange for the guarantee to be the only electricity provider within a given service territory, the electric utility had the obligation to serve everyone within its territory. The electric utility was sole producer, transporter, and distributor of electric energy. Rates were subject to review by responsible government regulatory bodies (e.g., public utilities commissions or government agencies) to prevent price gouging. Public Utility Commissions (PUC) allowed the rates to cover the utility's cost plus a reasonable return on investment. This system of vertically integrated monopolies was established by political intrigue within the United States. Other countries realized that electricity was an essential infrastructure and managed it in the same manner as water, sewage, etc. They then avoided expensive duplication of transmission and distribution by competing companies. Economies of scale meant that a single large power plant operated by the monopolistic utility could produce electricity more efficiently than two smaller power plants operated by competing utilities. The efficiencies gained from economies of scale outweighed the deadweight losses associated with monopoly operation and system over-building. This is especially true in the area of distribution. Earlier in this century, before regulation, there were two or three sets of lines to serve customers down every street! For the rest of the 20th century, the electric industry was viewed as a natural monopoly.

## 1.2 The Shift Toward Deregulation

The high energy prices during the 1970s oil embargoes focused attention on the need to find more efficient methods of using electricity. Improved appliance performance was dictated by consumer demand. Government regulations required efficiency labeling. Better prices for end user services are the desired outcome. The industry needs to shift from the selling of electrons to selling services. Since electrons were never *made*, this would be more appropriate. The need to decouple the industrial economy from the electric expense became apparent. The political winds within the United States and around the world wanted to remove the monopolistic pricing and achieve better prices that might come with competition. The United States began to taking small steps toward a competitive electricity system. Independent power producers were given the right to produce and sell power to the local electric utility at a price which represented the costs that were avoided by not having to produce that electrical energy themselves. For the most part, the electric energy industry was still hanging on to its monopolistic structure, although other U.S. industries (e.g., natural gas, airline, and communications) were being deregulated. Countries worldwide with government-owned industries began to see the benefits of privatization. Especially when the government coffers were insufficient to maintain the production and the expansion of electricity while satisfying all of the union demands, investor demands, and customer demands.

The impact of monopolistic inefficiencies because of high fuel prices was not the only reason deregulation surfaced in the late 20th century. The nuclear storage problem and other politically demanding obstacles have increased public distrust of the power industry. It is apparent that the electric industry was not sufficiently mature to meet the prerequisites to competition. Researcher [Clayton, 1996] listed the following intuitive prerequisites to successful competition:

- Mature physical system
- Stable national economy
- Trust in the sanctity of contracts
- Relatively high and/or diverse prices
- Regulated market imperfection

The United States Congress decided it was time for increased competition in the U.S. electrical system. By increasing competition via re-regulation of the electrical system, they would increase power system efficiencies and see benefits for electric consumers. Legislation introduced in 1989 threatened the regulated marketplace with the concept of deregulation. The notions of deregulation and *avoided cost* were introduced by the Public Utilities Regulatory Policies Act (PURPA) in 1978. PURPA encouraged the purchase of energy by utilities from independent power producers (IPPs). The intent was to lessen dependence on foreign oil. The price of the energy would be at the utility's "avoided cost." It is the cost the utility would incur if it uses its own generation. The implementation of "avoided cost" has varied without taking into consideration the reliability of power systems.

The Federal Energy Regulatory Commission (FERC) is entrusted with converting congressional acts into workable regulations through a series of proposed rules. With the passage of the EPAct in 1992, entities that do not own transmission-lines were granted the right to use the transmission system. This was termed *open access* and U.S. electric utilities began to see limited competition in power production. Countries outside of North America (e.g., United Kingdom, Norway, Chile) had already changed, or were in the process of changing, their regulations, or deregulating, to allow a more competitive electric marketplace. Many foreign governments used the electrical infrastructure as an alternative to tax collection. FERC, in various Notices of Proposed Regulation (NOPRs), announced an evolutionary sequence of events intended to expand competition in the U.S. electric marketplace.

The main stated purpose of the new Public Utility Holding Company Act (PUHCA) is to reduce the cost of electricity. Previous attempts by electric utilities to solve such cost problems led to the formation of power pools. Power pools operate as if all companies were a single company. The recent mergers of utilities are attempts to permanently reduce costs beyond that achieved by pool operation.. The merging of utilities effectively invented power pools within another legal framework.

An alternate way to reduce costs is to form energy brokerage systems. Such brokerage systems do not include all the financial and commodity brokerage aspects and attributes. Commodity brokering operations occur in a regulated environment where the procedures are set by commissions. The present paradigm is shown in Figure 1 as depicted by common terms. Energy brokerage systems have functioned very well in the past (Florida Coordination Group). Such successes would lead most to believe that a more general brokerage would function at least as well.

Energy, as a commodity, can be analyzed in a financial framework. Unlike other commodities, energy cannot be stored in warehouses. It has to be consumed when generated. Electricity has demand and supply that must be carefully balanced. This creates a problem: instant demand requires instant generation and instant generation requires instant consumption. Additionally, a path cannot be chosen to move energy from one point to another. The origin and delivery points can be fixed but the delivery path is variable based on physical laws. The path followed by energy can cause problems when wheeling power across intermediate systems because of resulting problems: voltage dipping, reactive power flow, increased losses, reliability problems, etc. Such problems have to be defined technically. Additionally, such problems must be handled in a consistent manner such that trading can occur with a known outcome for each standard violation. Specifically, the user must be aware of what is being bought, where it is being bought, who is responsible for providing the commodity, when it will be delivered, and why it may not be delivered.

When treated as a commodity, energy can be traded in a free market or exchange. A free market exists when traders recognize opportunities and take advantage of them. The major difference between a free market and a centrally directed market is that the free market is controlled by consumer demands and relative costs of production. Supreme Court Justice Louis D. Brandeis commented that "the organization of an exchange is society's attempt to capture the economists' concept of perfect competition." The broad objectives of such exchanges are: (1) greater equality of opportunity; (2) greater efficiency in markets; and (3) improvement in the information flow. There are regulated markets and deregulated markets. Schweppe, et al., identified three main operators in the deregulated energy marketplace: (1) a single regulated transmission and distribution (T&D) company; (2) independent power producers; and (3)consumers. This early work did not foresee the massive divestitures and mergers presently occurring.

The regulated T&D company acts as the agent between the producers and the consumers. It handles monetary transactions for the sale and purchase of energy is rewarded by a margin of profit from such operations. The T&D owns and operates the transmission network. It may be centrally-owned or independently-owned with a regulated rate-of-return. The producers have responsibilities that include: (1) build, maintain, and operate generating units; (2) provide energy to consumers via the T&D company; (3) meet environmental and zoning restrictions; (4) operate within antitrust laws; and (5) operate for profit. The producers do not fall under the jurisdiction of public service commissions. Some of the other participants in the deregulated energy marketplace, as proposed by Schweppe, include market coordinators, consultants, energy brokers, and consumers. Market coordinators collect data from utilities and determine generators to be dispatched. Information consultants forecast future energy prices. Energy brokers are catalysts arranging transactions between the producers and the consumers. The present state of this evolution has gone beyond the vision of Schweppe and others to a more common commodity framework.

The new environment within this work assumes that the vertically integrated utility has been segmented into a horizontally integrated system. Specifically, GENCOs, DISTCOs, and TRANSCOs exist in place of the old. This work does not assume that separate companies have been formed. It is only necessary that comparable services are available for anyone connected to the transmission grid.

As can be concluded, this description of a deregulated marketplace is an amplified version of the commodity market. It needs polishing and expanding. The change in the electric utility business environment is depicted generically below. The functions shown are the emerging paradigm. This work outlines the market organization for this new paradigm.

Attitudes toward reregulation still vary from state to state. Many electric utilities in the U.S. have been reluctant to change the status quo. Electric utilities with high rates are very reluctant to reregulate since the customer is expected to leave for the lower prices. Electric utility companies in regions with low prices are more receptive to change since they expect to pick up more customers. In 1998, California became the first state in the United States to adopt a competitive structure, and other states are observing the outcome. Several states on the eastern coast of the United States have also reregulated. Some offer customer selection of supplier. Some offer markets similar to those established in the United Kingdom. Several countries have gone to the extreme competitive position of treating electricity as a commodity. As these markets continue to evolve, governments in all areas of the world will continue to form their opinions on what market and operational structures will suit them best.

## 1.3 Choosing a Competitive Framework

There are many market frameworks that can be used to introduce competition between electric utilities. Almost every country embracing competitive markets for its electric system has done so in a different manner. The research described here assumes an electric marketplace derived from commodities exchanges like the Chicago Mercantile Exchange, Chicago Board of Trade, and New York Mercantile Exchange (NYMEX) where commodities (other than electricity) have been traded for many years. NYMEX added electricity futures to their offerings in 1996, supporting this author's previous predictions [Sheblé, 1991, 1992, 1993, 1994] regarding the framework of the coming competitive environment. The framework proposed has similarities to the Norwegian-Sweden electric systems. The proposed structure is partially implemented in New Zealand, Australia, and Spain. This new framework is not being implemented because this author performed research to prove that it would work. However, the framework is already implemented in other industries. Thus, it would be extremely expensive to not copy the treatment of other commodities. The details of this framework and some of its major differences from the emerging power markets/pools will be described.

Many believe the ultimate competitive electric industry environment is one in which retail consumers have the ability to choose their own electric supplier. Often referred to as retail access, this is quite a contrast to the vertical integrated monopoly of the past. This company served the typical electricity consumer as decided by the government agency. Already, telemarketers will be contacting consumers asking to speak to the person in charge of making decisions about the electricity service. Depending on consumer preference and the installed technology, it may be possible to do this on an almost real-time basis as a debit card bought at the local grocery store or gas station. Real-time pricing, where electricity is priced as it is used, is getting closer to becoming a reality as information technology advances. Presently, customers in most regions lack the sophisticated metering equipment necessary to implement retail access at this level.

Charging rates that were deemed fair by the government agency, the average monopolistic electric utility of the old environment met all consumer demand while attempting to minimize their costs. During natural or manmade disasters, neighboring utilities cooperated without competitively charging for their assistance. The costs were always passed on to the ratepayers. The electric companies in a country or continent were all members of one big happy family. The new companies of the future competitive environment will also be happy to help out in times of disaster, but each offer of assistance will be priced recognizing that the competitor's loss is gain for every one else. No longer guaranteed a rate of return, the entities participating in the competitive electric utility industry of tomorrow will be profit driven.

## 1.4 Preparing for Competition

Electric energy prices recently rose to more than $7500/MWh in the Midwest (1998) due to a combination of high demand and the forced outage of several units. Many Midwestern electric utilities bought energy at that high price, and then sold it to consumers for the normal rate. Unless these companies thought they were going to be heavily fined, or lose all customers for a very long time, it may have been more fiscally responsible to have terminated services.

Under highly competitive scenarios, the successful supplier will recover its incremental costs as well as the fixed costs through the prices it charges. For a short time, producers may sell below their costs, but will need to make up the losses during another time period. Economic theory shows that eventually, under perfect competition, all companies will arrive at a point where their profit is zero. This is the point at which the company can break even, assuming the average cost is greater than the incremental cost. At this ideal point, the best any producer can do in a competitive framework, ignoring fixed costs, is to bid at the incremental cost. Perfect competition is not often found in the real world, for many reasons. The prevalent reason is *technology change*. Fortunately, there are things that the competitive producer can do to increase the odds of surviving and remaining profitable.

The operational tools used and decisions made by companies operating in a competitive environment are dependent on the structure and rules of the power system operation. In each of the various market structures, the company goal is to maximize profit. Entities such as commodity exchanges are responsible for ensuring that the industry operates in a secure manner. The rules of operation should be designed by regulators prior to implementation to be complete and "fair." Fairness in this work is defined to include non-collusion, open market information, open transmission and distribution access, and proper price signals. It could call for maximization of social welfare (i.e., maximize the total happiness of everyone) or perhaps maximization of consumer surplus (i.e., make customers happy).

Changing regulations are affecting each company's way of doing business and to remain profitable, new tools are needed to help companies make the transition from the old environment to the competitive world of the future. This work describes and develops methods and tools that are designed for the competitive component of the electric industry. Some of these tools include software to generate bidding strategies, software to infer the bidding strategies of other competitors, and updated common tools like economic dispatch and unit commitment to maximize profit.

## 1.5 Present Overall Problem

This work is motivated by the recent changes in regulatory policies of inter utility-power interchange practices. Economists believe that electric pricing must be regulated by free market forces rather than by public utilities commissions. A major focus of the changing policies is "competition" as a replacement for "regulation" to achieve economic efficiency. A number of changes will be needed as competition replaces regulation. The coordination arrangements presently existing among the different players of the electric market would change operational, planning, and organizational behaviors.

Government agencies are entrusted to encourage an open market system to create a competitive environment where generation and supportive services are bought and sold under demand and supply market conditions. The open market system will consist of generation companies (GENCOs), distribution companies (DISTCOs), transmission companies (TRANSCOs), a central coordinator to provide independent system operation (ISO), and brokers to match buyers and sellers (BROCOs). The interconnection between these groups is shown in Figure 1.1.

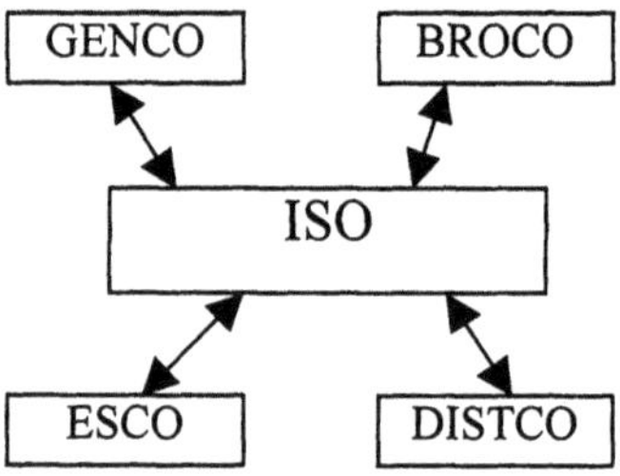


Figure 1.1. New Organizational Structure.

The ISO is independent and a disassociated agent for market participants. The roles and responsibilities of the ISO in the new marketplace are yet not clear. This work assumes that the ISO is responsible for coordinating the market players (GENCOs, DISTCOs, and TRANSCOs) to provide a reliable power system operation. Under this assumption, the ISO would require a new class of optimization algorithms to perform price-based operation. Efficient tools are needed to verify that the system remains in operation with all contracts in place. This work proposes an energy brokerage model for all services as a novel framework for price-based optimization. The proposed foundation is used to develop analysis and simulation tools to study the implementation aspects of various contracts in a deregulated environment.

Although it is conceptually clean to have separate functionals for the GENCOs, DISTCOs, TRANSCOs, and the ISO, the overall mode of real time operation is still evolving. Presently, two possible versions of market operations are debated in the industry. One version is based on the traditional power pool concept (POOLCO). The other is based on transactions and bilateral transactions as presently handled by commodity exchanges in other industries. Both versions are based on the premise of price-based operation and market-driven demand. This work presents analytical tools to compare the two approaches. Especially with the developed auction market simulator, POOLCO, multilateral, and bilateral agreements can be studied.

Working toward the goal of economic efficiency, one should not forget that the reliability of the electric services is of the utmost importance the electric utility industry in North America, in the words of the North American Electric Reliability Council (NERC), uses reliability in a bulk electric system to indicate " *the degree to which the performance of the elements of that system results in electricity being delivered to customers within accepted standards and in the amount desired. The degree of reliability may be measured by the frequency, duration, and magnitude of adverse effects on the electric supply*". The council also suggests that reliability can be addressed by considering the two basic and functional aspects of the bulk electric system - adequacy and security. In this work, the discussion is focused on the adequacy aspect of power system reliability, which is defined as the static evaluation of the system's ability to satisfy the system load requirements. In the context of the new business environment, market demand is interpreted as the system load. However, a secure implementation of electric power transaction concerns power system operation and stability issues:

1. Stability issue: The electric power system is a nonlinear dynamic system comprised of numerous machines synchronized with each other. Stable operation of these machines following disturbances or major changes in the network often requires limitations on various operating conditions, such as generation levels, load levels, and power transmission changes. Due to various inertial forces, these machines together with other system components require extra energy (reserve margins and load following capability) to safely and continuously actuate electric power transfer.
2. Thermal overload issue: Electric power transmission is limited by electrical network capacity and losses. Capacity may include real-time weather conditions as well as congestion management. The impact of transmission losses on market power is yet to be understood.
3. Operating voltage issues: Enough reactive power support must accompany the real power transfer to maintain the transfer capacity at the specified levels of open access.

In the new organizational structure, the services used for supporting a reliable delivery of electric energy (e.g., various reserve margins, load following capability, congestion management, transmission losses, reactive power support, among others) are termed supportive services. These have been called "ancillary services" in the past. In this context, the term "ancillary services" is misleading since the referred services are not ancillary but *closely bundled* with the electric power transfer as described earlier. The open market system should consider all of these supportive services as an integral part of power transaction.

This work proposes that supportive services become a competitive component in the energy market. It is embedded so that no matter what reasonable conditions occur, the (operationally) centralized service will have the obligation and the authority to deliver and keep the system responding according to adopted operating constraints. As such, although competitive, it is burdened by additional goals of ensuring reliability rather than open access only. The proposed pricing framework attempts to become economically efficient by moving from cost-based to price-based operation introduces a mathematical framework to enable all players to be sufficiently informed in decision making when serving other competitive energy market players, including customers.

## 1.6 Economic Evolution

Some economists speculate that regional commodity exchanges within the U.S.A. would be oligopolistic in nature (having a limited numbers of sellers) due to the configuration of the transmission system. Some postulate that the number of sellers will be sufficient to achieve near perfect competition. Other countries have established exchanges with as few as three players. However, such experiments have reinforced the notion that collusion is all too tempting. Such experiments have shown that market power is the key issue to price determination as in any other market. Regardless of the actual level of competition, companies that wish to survive in the deregulated marketplace must change the way they do business. They will need to develop bidding strategies for trading electricity via an exchange.

Economists have developed theoretical results of how variably competitive markets are supposed to behave under varying numbers of sellers or buyers. The economic results are often valid only when aggregated across an entire industry and frequently require unrealistic assumptions. While considered sound in a macroscopic sense, these results may be less than helpful to a particular company not fitting the industry profile that is trying to develop a strategy that will allow it to remain competitive.

Generation companies (GENCOs), energy service companies (ESCOs), and distribution companies (DISTCOs) that participate in an energy commodity exchange must learn to place effective bids in order to win energy contracts. Microeconomic theory states that in the long term, a hypothetical firm selling in a competitive market should price its product at its marginal cost of production. The theory is based on several assumptions (e.g., all market players will behave rationally, all market players have perfect information) which may tend to be true industry-wide, but might not be true for a particular region or a particular firm. As shown in this work, the normal price offerings are based on average prices. Markets are very seldom perfect or in equilibrium.

There is no doubt that deregulation in the power industry will have many far-reaching effects on the strategic planning of firms within the industry. One of the most interesting effects will be the optimal pricing and output strategies generator companies (GENCOs) will employ in order to be competitive while maximizing profits. This case study presents two very basic, yet effective means for a single generator company (GENCO) to determine the optimal output and price of their electrical power output for maximum profits.

The first assumption made is that switching from a government regulated, monopolistic industry to a deregulated competitive industry will result in numerous geographic regions of oligopolies. The market will behave more like an oligopoly than a purely competitive market due to the increasing physical restrictions of transferring power over distances. This makes it practical for only a small number of GENCOs to service a given geographic region.

## 1.7. Market Structure

Although nobody knows the exact structure of the emerging deregulated industry, this research predicts that regional exchanges (i.e., electricity mercantile associations [EMAs]) will play an important role. Electricity trading of the future will be accomplished through bilateral contracts and EMAs where traders bid for contracts via a double auction. The electric marketplace used in this chapter has been refined and described in various chapters. Fahd and Sheblé [1992a] demonstrated an auction mechanism. Sheblé [1994b] described the different types of commodity markets and their operation, outlining how each could be applied in the evolved electric energy marketplace. Sheblé and McCalley [1994e] outlined how spot, forward, future, planning, and swap markets can handle real-time control of the system (e.g., automatic generation control) and risk management. Work by Kumar and Sheblé [1996b] brought the above ideas together and demonstrated a power system auction game designed to be a training tool. That game used the double auction mechanism in combination with classical optimization techniques.

In several references, [Kumar, 1996a, 1996b; Sheblé 1996b; Richter 1997a] a framework is described in which electric energy is only sold to distribution companies (DISTCOs), and electricity is only generated by generation companies (GENCOs) (see Figure 1-2). North American Electric Reliability Council (NERC) sets the reliability standards. Along with DISTCOs and GENCOs, energy services companies (ESCOs), ancillary services companies (ANCILCOs), and transmission companies (TRANSCOs) interact via contracts. The contract prices are determined through a double auction. Buyers and sellers of electricity make bids and offers that are matched subject to approval of the independent contract administrator (ICA), who ensures that the contracts will result in a system operating safely within limits. The ICA submits information to an independent system operator (ISO) for implementation. The ISO is responsible for physically controlling the system to maintain its security and reliability.

## 1.8 Fully Evolved Marketplace

The following sections outline the role of a horizontally integrated industry. Many curious acronyms have described generation companies (IPP, QF, Cogen, etc.), transmission companies (IOUTS, NUTS, etc.), and distribution companies (IOUDC, COOPS, MUNIES, etc.). The acronyms used in this work are described in the following sections.

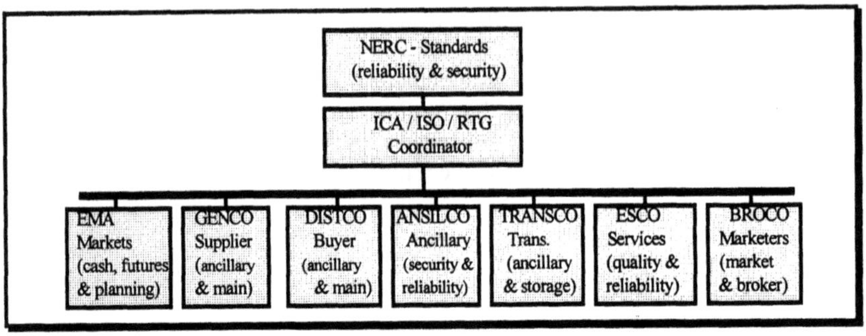


Figure 1.2. Business Environment Model.

### *Horizontally Integrated*

The reregulation of the electric power industry is most easily visualized as a horizontally integrated marketplace. This implies that interrelationships exist between generation (GENCO), transmission (TRANSCO), companies (DISTCO) as separate entities. Note that independent power producers (IPP), qualifying facilities (QF), etc., may be considered as equivalent generation companies. Non-utility transmission systems (NUTS) may be considered as equivalent transmission companies. Cooperatives and municipal utilities may be considered as equivalent distribution companies. All companies are assumed to be coordinated through a regional transmission corporation (or regional transmission group).

### *Generation Company (GENCO)*

The goal for a generation company, which has to fill contracts for the cash and futures markets, is to package production at an attractive price and time schedule. One proposed method is similar to the classic decentralization techniques used by a vertically integrated company. The traditional power system approach is to use Dantzig Wolfe decomposition. Such a proposed method may be compared with traditional operational research methods used by commercial market companies for â "make or buy" decision.

### *Transmission Company (TRANSCO)*

The goal for transmission companies, which have to provide services by contracts, is to package the availability and the cost of the integrated transportation network to facilitate transportation from suppliers (GENCOs) to buyer (ESCOs). One proposed method is similar to oil pipeline networks and energy modeling. Such a proposed method can be compared to traditional network approaches using optimal power flow programs.

### *Distribution Company (DISTCO)*

The goal for distribution companies, which have to provide services by contracts, is to package the availability and the cost of the radial transportation network to facilitate transportation from suppliers (GENCOs) to buyers (ESCOs). One proposed method is similar to distribution outlets. Such proposed methods can be compared to traditional network approaches using optimal power flow programs. The disaggregation of the transmission and the distribution system may not be necessary as both are expected to be regulated as a monopoly for the present time.

### *Energy Service Company (ESCO)*

The goal for energy service companies, which may be large industrial customers or customer pools, is to purchase power at the least cost when needed by consumers. One proposed method is similar to the decision of a retailer to select the brand names for products being offered to the public. Such a proposed method may be compared to other retail outlet shops.

### *Independent System Operator (ISO)*

The primary concern is the management of operations. Real-time control (or nearly real-time) must be completely secure if any amount of scheduling is to be implemented by markets. The present business environment uses a fixed combination of units for a given load level and then performs extensive analysis of the operation of the system. If schedules are determined by markets, then the unit schedules may not be fixed sufficiently ahead of real-time for all of the proper analysis to be completed by the ISO.

### *Regional Transmission Group (RTG)*

The goal for a regional transmission group, which must coordinate all contracts and bids among the three major types of players, is to facilitate transactions while maintaining system planning. One proposed method is based on discrete analysis of a Dutch auction. Other auction mechanisms may be suggested. Such proposed methods are similar to the warehousing decision on how much to inventory for a future period. As shown later in this work, the functions of the RTG and the ISO could be merged. Indeed this should be the case based on organizational behavior.

### *Electric Markets*

Competition may be enhanced through the various markets: cash, futures, planning, and swap. The cash market facilitates trading in spot and forward contracts. This work assumes that such trading would be on an hourly basis. Functionally, this is equivalent to the interchange brokerage systems implemented in several states. The distinction is that future time period interchange (forward contracts) is also traded.

The futures market facilitates trading of futures and options. These are financially derived contracts used to spread risk. Such contracts are a means of risk management. The planning market facilitates trading of contracts for system expansion. Such a market has been proposed by a West Coast electric utility. The swap market facilitates trading between all markets when conversion from one type of contract to another is desired. It should be noted that multiple markets are required to enable competition between markets.

The structure of any spot market auction must include the ability to schedule as far into the future just as the industrial practice did before re-regulation. This would require extending the spot into the future for at least six months as proposed by this author [Sheblé, 1994]. Future month production should be traded for actual delivery in forward markets. Future contracts should be implemented at least 18 months into the future if not three years. Planning contracts must be implemented for at least 20 years into the future, as recently offered by TVA, to provide an orderly, predictable expansion of the generation and transmission systems. Only then can timely addition of generation and transmission be assured. Finally, a swap market must be established to enable the transfer of contracts from one period (market) to another.

To minimize risk, the use of option contracts for each market should be implemented. Essentially, all of the players share the risk. This is why all markets should be open to the public for general trading and subject to all rules and regulations of a commodity exchange. Private exchanges, not subject to such regulations, do not encourage competition and open price discovery.

The described framework [Sheblé, 1996b] allows for cash (spot and forward), futures, and planning markets as shown in Figure 1-3. The *spot market* is most familiar within the electric industry [Schweppe, 1988]. A seller and a buyer agree (either bilaterally or through an exchange) upon a price for a certain amount of power (MW) to be delivered sometime in the near future (e.g., 10 MWs from 1:00 p.m. to 4:00 p.m. tomorrow). The buyer needs the electricity, and the seller wants to sell. They arrange for the electrons to flow through the electrical transmission system and they are happy. A *forwards contract* is a binding agreement in which the seller agrees to deliver an amount of a particular product with a specified quality at a specified time to the buyer. The forward contract is further into the future than is the spot market. In both the forwards and spot contracts, the buyer and seller want physical goods (e.g., the electrons). A *futures contract* is primarily a financial instrument that allows traders to lock in a price for a commodity in some future month. This helps traders manage their risk by limiting potential losses or gains. Futures contracts exist for commodities in which there is sufficient interest and in which the goods are generic enough that it is not possible to tell one unit of the good from another (e.g., 1 MW of electricity of a certain quality, voltage level, etc.). A futures *options contract* is a form of insurance that gives the option purchaser the right, but not the obligation, to buy (sell) a futures contract at a given price. For each options contract, there is someone "writing" the contract who, in return for a premium, is obligated to sell (buy) at the strike price (see Figure 1.3). Both the options and the futures contracts are financial instruments designed to minimize risk. Although provisions for delivery exist, they are not convenient (e.g., the delivery point is not located where you want it to be located). The trader ultimately cancels his position in the futures market either with a gain or loss. The physicals are then purchased on the spot market to meet demand with the profit or loss having been locked in via the futures contract. A *swap* is a customized agreement in which one firm agrees to trade its coupon payment for one held by other firm involved in the swap. Finally, a *planning market* is needed to establish a basis for financing long-term projects like transmission lines and power plants [Sheblé, 1993].

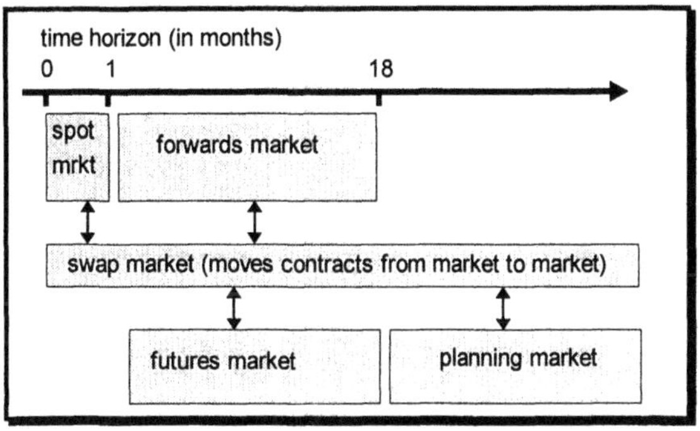


Figure 1.3. Interconnection Between the Markets.

## 1.9 Computerized Auction Market Structure

Auction market structure is a computerized market, as shown in Figure 1.4. Each of the agents has a terminal (PC, workstation, etc) connected to an auctioneer (auction mechanism) and a contract evaluator. Players generate bids (buy and sell) and submit the quotation to the auctioneer. A bid is a specified amount of electricity at a given price. The auctioneer binds bids (matching buyers and sellers) subject to approval of the contract evaluation. This is equivalent to the pool operating convention used in the vertically integrated business environment.

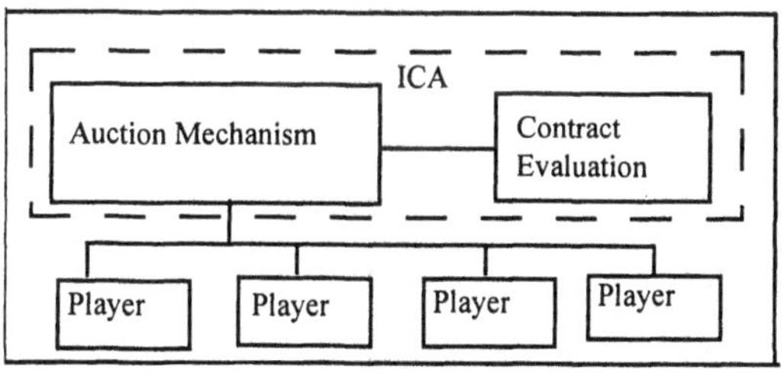


Figure 1.4. Computerized Market.

The contract evaluator verifies that the network can remain in operation with the new bid in place. If the network can not operate, then the match is denied. The auctioneer processes all bids to determine which matches can be made. However, the primary problem is the complete specification of how the network can operate and how the agents are treated comparably as the network is operated closer to limits. The network model must include all constraints for adequacy and security.

The major trading objectives are hedging, speculation, and arbitrage. Hedging is a defense mechanism against loss and/or supply shortages. Speculation is assuming an investment risk with a chance for profit. Arbitrage is crossing sales (purchases) between markets for riskless profit. This work assumes that there are four markets commonly operated: forward, futures, planning, and swaps (Figure 1.5).

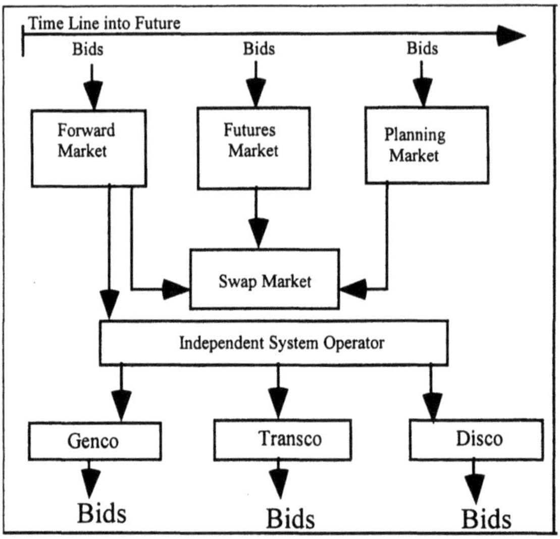


Figure 1.5. Electric Market.

Forward Market: The forward contracts reflect short term future system conditions. In the forward market, prices are determined at the time of the contract but the transactions occur at some future time. Optimization tools for short term scheduling problems can be enhanced to evaluate trading opportunities in forward market. For example, short-term dispatching algorithms, such as economic unit commitment dispatch can be used to estimate and earn profit in forward market.

Futures Market: A futures market creates competition because it unifies diverse and scattered local markets and stabilizes prices. The contracts in futures market are risky because price movements over time can result in large gains or losses. There is a link between forward markets and futures markets which restricts price volatility. *Options* (options contracts) allow the agent to exercise the right to activate a contract or cancel it. Claims to buy are called "call" options. Claims to sell are called "put" options.

A more detailed discussion of an electric futures contract is discussed in [Sheblé, 1994b]. The components include trading unit, trading hours, trading months, price quotation, minimum price fluctuation, maximum daily price fluctuation, last trading day, exercise of options, option strike prices, delivery, delivery period, alternate delivery procedure, exchange of futures for, or in connection with physicals, quality specifications, and customer margin requirements.

Swap Market: In the swap market, contract position can be closed with an exchange of physical or financial substitutions. The trader can find another trader who will accept (make) delivery and end the trader's delivery obligation. The acceptor of the obligation is compensated through a price discount or a premium relative to the market rate.

The financial drain inflicted on traders when hedging their operations in the futures market is slightly higher than the one inflicted through direct placement in the forward market. An optimal mix of options, forward commitments, futures contracts, and physical inventories is difficult to assess and depends on hedging, constraints imposed by different contracts, and cost of different contracts. The exchange of various energy instruments is handled by a clearinghouse such as a swap market.

Planning Market: The growth of transmission grid requires transmission companies to make contracts based on the expected usage to finance projects. The planning market would underwrite equipment usage subject to the long-term commitments to which all companies are bound by the rules of network expansion to maintain a fair marketplace. The network expansion would have to be done to maximize the use of transmission grid for all agents. Collaboration would have to be overseen and prohibited with a sufficiently high financial penalty. The growth of the generation supply similarly requires such markets. However, such a market has been started with the use of franchise rights (options) as established in recent Tennessee Valley Authority connection contracts. This author has published several papers outlining the need for such a market. Such efforts are not documented in this work.

## 1.10 Role of Markets in Society

Consumers maximize use of scarce resources subject to preferences, budgets, and other constraints (mostly physical laws) based on expected future events. An essential concept is stating indexed commodities (SIC). Electricity has one price if a given set of units is available tomorrow and a different price fewer units are available. Thus, electricity is a state-indexed commodity. Indeed, such variability in price valuations is very easily observed in today's spot markets. All customers' have a risk indifference curve to indicate a preference for loss recovery or insurance payment. Many customers want a contingent claim (agreement) to provide a fixed, acceptable electricity price future contingent states. The customer's demand for such contracts has led to an evolution of various markets for not only the basic commodity but also for contracts to smooth the price variations over multiple periods (price volatility). Other commodity trading involves a number of market structures based on the evolution of the market in each industry and country. An elementary block of trade is a claim to a single commodity (j) for a given state (s) if that state occurs. The price of such an elementary block is widely published ($p_{sj}$). Contracts that are more complex are simply assemblies of such elementary contracts.

A combination of contracts to maintain profit (hedging) is critical in many industries to smooth price variations to win and to maintain market share. The combination of contracts to profit from infrastructure instability, local economic instability, global economy instability, or customer preferences is speculation by insurance providers. The goal has always been to provide consumer preferences for selecting the amount of each commodity based on budget and on intangible desires (i.e. green energy). This goal is often expressed in the quality of the markets—specifically, efficiency and robust design. This goal is measured over many periods. Thus, the measure of a market is based on the history of market use and not on the possible response of the market to all potential situations. This goal must be the cornerstone of any market design not only for spot prices but also for forward, future, and planning horizons.

## 1.11 Commodity Contracts

A commodity is defined as anything useful or valuable that can be turned into a commercial or other advantage. Commodity markets provide the setting to trade commodities. Commodity market prices generally have a maximum allowable price move on a given day to provide price stability. It is the element of risk that commodity markets uniquely address. If no market participant can impose a position through a change in actions, then a state of general equilibrium exists. Otherwise, partial equilibrium exists. A stable market process "is one where revised bids and offers lead to a uniform price with uniform quantities offered and demanded at that price." An efficient market is the result of a minimum of waste, expense, and dissatisfaction by the participants. Three prerequisites exist for commodities investment decisions: (1) the correct formulation of the business objectives; (2) an information network; and (3) a decision process to apply the information toward achieving the business objectives.

Commodity contracts must specify price, contract length (horizon), quantity, place of delivery, and quality. The types of commodity contracts traded are:

- Cash contracts traded by direct placement between the buyer and the seller, resembling price-only contracts in the electric energy marketplace
- Option contracts that give the buyer the right but not the obligation to buy or sell a commodity, they are similar to price-quantity contracts in the electric energy marketplace
- Futures contracts that involve a commitment to buy or sell commodities in the future, equivalent to long-term contracts in the electric energy marketplace

The options and futures markets bear a striking resemblance to their energy counterpart. They provide the buyer with the opportunity to reduce the risk involved in trading. They lower the price due to unforeseen events, and contribute to price stability. The distinction between the markets is that the energy market regulates a specific transaction while the options and futures markets regulate the general operations of the market.

The major role of markets for these objectives is to identify the information necessary to maximize profit over a given period. If the market structure reflects only current economic impacts, then all participant decision will be based on short time frames. If the market structure reflects future economic impacts, then the participants can make decisions based on correspondingly longer time duration.

## 1.12 Auction Market Mechanism

There are two kinds of markets: the two-party direct placement market that occurs between a buyer and a seller with or without the intervention of broker, and the open outcry market that allows transactions between a buyer and a seller only through a broker.

Note that most of the author's previous work [Fahd, 1992a, Fahd, 1992b, Anwar, 1994, Post, 1995, Kumar, 1996e, Kumar, 1996f] has dealt with multi-lateral contracts (multiple buyers and sellers) to maximize the number of bid awards for each time period of bid matching. This is equivalent to the pool operating convention used in the vertically integrated business environment. It may also be equivalent to bilateral contracts for incremental exchange of commodity.

## 1.13 The Spot Market-Place

The basic framework for the research described in this work is adopted from [Sheblé, 1991-1997], which has been implemented in New Zealand and is similar to the California power exchange. The referenced research described the different types of commodity markets and their operation. Sheblé outlined how each type of commodity market could be applied in the evolved electric energy marketplace. Under this framework (Shown in Fig. 1.1, which was presented in [Sheblé, 1993]) companies presently having generation, transmission, and distribution facilities would be divided into separate profit and loss centers. Power would be generated by generation companies (GENCOs) and transported via transmission companies (TRANSCOs). Energy service companies (ESCOs) would purchase power from the generator for the customer. It has been proposed that NERC would set the reliability and security standards in the U.S. It is predicted that we'll see energy services companies (ESCOs) replacing the current distribution utilities as the main customer representative. An Independent Contract Administrator (ICA) will review the power transactions to ensure that system security and integrity is maintained. An ICA is a combination of Independent System Operator (ISO) and Regional Transmission Group (RTG). Distribution companies would own and maintain the distribution facilities. Companies providing ancillary services (ANSILCOs) may emerge in this new framework. Energy mercantile associations (EMAs) have emerged in this new framework as power exchanges.

The double auction, the basic method for this research, uses the bids and offers from GENCOs and ESCOs sorted into descending and ascending order respectively, similar to the Florida Coordination Group approach described by Wood and Wollenberg [Wood, 1996]. If the buy bid is higher than the sell offer that is to be matched, then this is a potential valid match. The ICA must determine whether the transaction would endanger system security and whether transmission capacity exists. Specifically, the contract approval will be subject to meeting requirements for maintaining sufficient spinning reserve, ready reserve, reactive support, and area network control (contract-based AGC). If the ICA does approve, the valid offers and bids are matched and the difference in the bids ($/MW) is split to determine the final price. This is termed the *equilibrium price* and is similar to the power pool split-savings approach that many regions have been using for years.

If there is an insufficient number of valid matches, then *price discovery* has not occurred. The auctioneer reports the results of the auction to the market participants. If all bids and offers are collected and insufficient valid bids and offers are found to exist, then the auction has gone through one cycle. The auctioneer reports that price discovery did not occur. The auctioneer asks for bids and offers again. The auctioneer requests buyers and sellers to adjust their bids and offers. To aid in eventually finding a feasible solution, during subsequent cycles within a round, buyers may not decrease their bid and sellers may not increase their offer. The cycles continue until price discovery occurs, or until the auctioneer decides to bind whatever valid matches exist and continue to the next round or hour of bidding

After price discovery, the buyers and sellers whose bids were bound will have a conditional contract. This contract is subject to the approval of the ICA, who verifies that none of the security criteria have been violated. Following the completion of one round of bidding, the auctioneer asks if another round of bidding is requested. If the market participants have more power to sell or buy, they request another round. Allowing multiple rounds of bidding each hour (vs. one-shot bidding) allows the participants the opportunity to use the latest pricing information in forming their present bid. This process is continued until no more requests are received or until the auctioneer determines that enough rounds have taken place (Figure 1.2).

One variant is a one-sided auction that would accept quantity bids from buyers and price and quantity bids from sellers. The one-sided bid is addressed first by this work.

## 1.14 Commodity Markets

A commodity is defined as anything useful or valuable that can be turned into a commercial or other advantage. Commodity markets provide the setting to trade commodities. Commodity market prices generally have a maximum allowable price move in a given day to provide price stability. It is the element of risk that commodity markets uniquely address. If no market participant can impose a position through a change in actions, then a state of general equilibrium exists. Otherwise, partial equilibrium exists. A stable market process "is one where revised bids and offers lead to a uniform price with uniform quantities offered and demanded at that price. An efficient market" is the result of a minimum of waste, expense, and dissatisfaction by the participant.

There are two kinds of markets: the two-party direct placement market that occurs between a buyer and a seller with or without the intervention of broker and the open outcry market that allows transactions between a buyer and a seller only through a broker. The majority of markets are one-sided open auctions. Specifically, only the sellers bid to provide a service. However, one-sided markets do not provide the buyer with the diversity or with the competition, which was originally envisioned by the lawyers and economists. It is not appealing to many of the large purchasers of electricity. There is no differentiation of reliability or of timing of delivery. This is especially true for the largest purchaser, the residential customer. It is difficult for this author to believe that the customers will not demand a truly competitive marketplace where reliability and quality are distinguishable parts of the product provided.

Three prerequisites exist for commodities investment decisions: (a) the correct formulation of the business objectives, (b) an information network, and (c) a decision process to apply the information toward achieving the business objectives.

## 1.15 Commodity and Futures Trading Commission's (CFTC's) Regulations

Markets are regulated to prevent unfair competition, unscrupulous traders, and economic instability. A market without regulations is an invitation to chaos. Regulations are meant to uphold the integrity of market prices and fiduciary responsibility of floor traders and brokers. An individual or a group of investors cannot capture a large proportion of the open contracts and control market prices. The exchanges and CFTC have imposed a limit on the number of open speculative contracts for select commodities. Congress authorized CFTC in the Commodity Exchange Act of 1974. CFTC replaced the Commodity Exchange Authority.

The energy market is presently in its infancy. The majority of brokers operate behind closed doors, setting fees without any overview. Such markets are not beneficial to the customers the suppliers of electricity. Closed competition results in excessive profits for the companies holding the key information.

The operation of markets depends on the steps involved in a trade, the role of participants, the regulations of the trade, and the reports and accounting rules. The participants in the "ring" system of markets include the following players as depicted in Figure 1.3. The buyers and sellers: They are not members of the exchange. They place their order with the exchange Commission Merchants (CMs): Some of them are members of the clearing corporation and some of them are not. They transmit the customers orders to the floor and receive payments. Non-clearing members can not handle the margin responsibility for an open trade. Non-clearing members direct the trade to a clearing member. By definition, an order is an instruction to the CM or broker to take action for an account.

- Associated Persons (APs): They are employees of the CMs. They are also called commodities brokers.

- Floor brokers: They are members of the exchange who handle the orders from CMs and attempt to execute the orders for which they earn commissions.
- Floor traders: They trade for their own accounts and cannot deal with the public.
- Clearing corporation: It will guarantee the customers that the exchange's regulations for price and delivery will be enforced.

CFTC's regulations govern individuals on a procedural basis. Futures contracts between two individuals must take place on an organized exchange. Large traders must report their open positions. The size of a position can be restricted by CFTC depending on the commodity and the intent of the transaction. CFTC may require the identity disclosure of a foreigner. Individuals cannot sell or buy silver and gold leverage contracts. Individuals must satisfy maintenance margin requirements; otherwise the CM can liquidate the open position. If price limit moves are exceeded, then liquidation of a futures contract can be restricted by CFTC. CFTC can force the exchanges to increase margin requirements. CFTC can suspend trading in any contract and require trading for liquidation purposes only. CMs and floor traders are restricted by CFTC from disclosing customers' orders, trading ahead of customers' orders, and trading against customers' orders. The last two rules are known as the "dual trading" rules.

Regulated CMs and APs. APs are carefully screened by CMs. Both CMs and APs must be registered and approved by the CFTC. CMs must analyze the qualifications of customers based on net worth, family structure, employment history, personal traits, and ability to understand and take risk. CMs must notify customers in writing of the risk involved in futures trading due to the inherent leverage. By definition, leverage is the amount of money needed to control a given amount of resources. The customer gives the CM a margin payment which must be segregated from the CM's working capital. When trade is initiated, a margin is posted with the clearing corporation and as the position shifts, maintenance margins must be paid in cash. CFTC requires CMs to report any business position, that might impact trading.

Regulated commodity trading advisers (CTAs). CTAs are individuals or corporations who provide advice on commodities trading. They must be registered with the CFTC if a fee for the advice is anticipated. They are not required to list all their trades or allow clients to examine their books.

Commodity Pool Operators (CPOs): CPOs are individuals who solicit and group funds from small investors to reduce the risk of trading. They must be registered with the CFTC. They are the general partners in a commodity pool. Most pools hire a CTA to make trading decisions. They can manage more than one pool but they have to keep funds separated. They can place orders for different pools together to take advantage of discounts resulting from volume trading.

Exchange Regulation: Self-regulations are imposed to avoid racketeering, to control the exchange operations, and to preserve the integrity of trading. The exchanges are "private membership organizations geared to providing the trading arena, setting the professional standards of trading, and maintaining the integrity if the markets."

The settlement price usually reflects the bids and offers currently in the market and not necessarily those that transpired before the close. A common rule is to take the lowest bid as the settlement price if no trades are executed on the close. If several trades occur on the close, the average of the last trade prices is often used as the settlement price. A fixed settlement price and limited trading hours may not apply to the energy marketplace due to the continuous nature of energy flow and the price movement of energy at a given hour, month, and season. Exchanges can limit price movements depending on the previous day's activities. Trading activities can be suspended if price fluctuations are manipulated by traders.

## 1.16 General Exchange Regulations

Self-regulations are imposed to avoid racketeering, to control the exchange operations, and to preserve the integrity of trading. The exchanges are private membership organizations geared to providing the trading arena, setting the professional standards of trading, and maintaining the integrity of the markets.

The settlement price usually reflects the bids and offers currently in the market and not necessarily those that transpired before the close. A common rule is to take the lowest bid as the settlement price if no trades are executed on the close. If several trades occur on the close, the average of the last trade prices is often used as the settlement price. The fixing of the settlement price and the limiting of trading hours may not apply to the energy marketplace due to the continuous nature of energy flow and the price movement of energy depending on the hour, month, and season. Exchanges can limit price movements depending on the previous day's activities. Trading activities can be suspended if price fluctuations are manipulated by traders.

Any trader who fails to provide the contracts held, at the time the commodity is delivered is assisted by the association to provide all services and/or products, even if such actions deplete the trader's financial holdings. The stick held over the traders is liquidation.

## 1.17 The Energy Marketplace

The impact of deregulation on the energy marketplace will be increased competition and hopefully lower energy prices. The objective of investors will still be to make profits. In achieving their goals, investors should not forget that the security and reliability of the electric service are of the utmost importance. Rules and regulations should be instituted to ensure the quality of service is not compromised. Agreements and compromises should be reached between the electric utilities in order to create a fair environment for trading. The inception of an energy exchange, the ENERGY MERCANTILE ASSOCIATION (EMA), with a governing body can help overcome some of the obstacles standing in the way of deregulation. If energy is to be treated as a futures commodity rather than an options commodity, then margin requirements paid by the traders must be set. The exchange to be implemented in California adheres to the majority of the rules needed of any association operating under CFTC's regulations.

### *1.17.1 The Energy Mercantile Association (EMA)*

The functions of this fictitious association, created for the explanation of markets within this work, will be similar to those of any exchange and will fall under the jurisdiction of the CFTC. EMA should have a board of directors elected by members of the exchange. Any power producer or transmitter can become a member of the exchange by applying for membership. The number of members representing any particular utility should be proportional to the size of the produced power and/or transmission network. Members should be grouped into voting and non-voting members. A maximum number of three voting members can represent an individual utility. This will help ensure that a large utility cannot have a large number of voting members and cannot sway decisions in his/her favor. Non-voting members are those who do not own generation or transmission facilities. They are called Electric Power Marketers (EPMs). They resemble floor traders and brokers in the futures and options markets. EPMs facilitate transactions for utilities that do not have access to the exchange. Membership in EMA is not mandatory but can be desirable due to benefits and incentives. Energy Brokers are voting members of the EMA and are employed by a utility. They must register with the EMA and are subject to the regulations of the EMA. Energy Brokers from different utilities cannot enter into individual transactions without prior clearance from the exchange.. This will help ensure fair and uniform trading on the floor of the exchange according to standard contracts. The possibility should exist for independent contracts to be negotiated between utilities through brokers if the transaction cannot be conducted on the floor of the exchange and the transaction is deemed essential to the operation of a utility by the EMA. Otherwise, energy brokers enter into transactions for their respective utility under the broad guidelines of the EMA. Electric Power Marketers (EPMs) are non-voting members of the EMA. They are subject to the regulations of the EMA. They should have the following characteristics: 1) They cannot own generating and transmitting facilities; 2) They only deal with utilities who are not members of EMA; and 3) They cannot exercise monopoly over the market, (i.e., they have no market power, they must file the transactions' rates with the exchange, and the rates must be market-based.)

### *1.17.2 Reliability Corporation*

The Reliability Corporation (RC) has not been defined in the electric market as it has in other markets. The RC underwrites (insures) the operation of the exchange (EMA). The RC verifies that the rules of the exchange are being followed and that the players are adhering to all laws and regulations. The electric power industry has defined several entities which would have some of the RC responsibilities. They are: Regional Transmission Group (RTG), Independent System Operator/Administrator (ISO/ISA) and Independent Contract Administrator (ICA). This work assumes that the RC responsibilities would be covered by a combined RTG and ICA. There is no need to replace existing control centers with new ones.

The owner of the equipment should have the authority and responsibility to remove or to place equipment into service. Transferring the responsibility to another entity only confuses the litigation when something goes wrong. The RC should simply determine if all players are following the rules and that the integrated system is performing as well as possible for the given conditions. The RC would audit all contracts to determine if any players are cheating or infringing on the operating principles. The RC is thus a market coordinator.

To avoid major disturbances by maintaining voltage stability, the market coordinator must perform contingency and state estimation analyses, and compute the amount of available reserves. Disturbances affect the dynamic behavior of the system and can cause instability. Users who aggravate the dynamic behavior of the power system must be penalized and those who help improve the dynamic behavior of the power system must be rewarded. Reactive power flow should be treated in the same manner as the dynamic behavior of the power system because it contributes to additional losses and a rise or fall in the voltage profile of the system. A charge for reactive power flow must be levied/credited depending on the situation. The reactive power flow must be monitored at the boundaries of each power system to determine the contributors to the increase or decrease in flow. The debit/credit system or aggravating the dynamic behavior of the power system and for reactive power flow will be a feature of the deregulated energy marketplace. The responsibility of assessing reliability of the power service can be delegated to the North American Electric Reliability Council (NERC) who should have extended power to oversee the operation of the utilities. NERC will keep records of power systems operation and forward the information to the RC who will have the jurisdiction to act upon the information. NERC should have the responsibility of developing a set of guidelines to be followed in emergencies. The RC would implement the cash market contracts.

### *1.17.3 EMA's Regulations*

Regulations must be imposed to ensure fair trade and to control the operation of the exchange. These regulations are imposed and enforced by the board of directors of the EMA. Following are some of the regulations to be included:

- Setting the standards for trading with fixed rules and procedures;
- Classifying brokers according to qualifications;
- Determining the size of the blocks of energy to be traded;
- Establishing the maximum amount of energy to be traded by any one utility in a given day;
- Collecting pertinent to cost-of-service data from utilities;
- Setting margin requirements to move with market prices and to reflect the risk position;
- Determining the delivery conditions;

- Assessing the impact of transactions on the security and reliability of the power system as determined by NERC.
- Imposing a charge on increased reactive power flow and/or violation of the dynamic behavior of the power system.
- Applying a credit to decreased reactive power flow and/or enhancement of the dynamic behavior of the power system.
- Determining a basis for computing the wheeling charges based on the utility's cost-of-service.
- Providing a mechanism for arbitration of disputes.
- Monitoring private trading by brokers.
- Protecting against unauthorized trading.
- Providing the traders with the clearinghouse function.
- Processing price and commodity information to the members and to the public.
- Setting limits on trading authority.
- Establishing parallel audit trails.
- Selecting a priority order for the transactions.
- Determining which contracts to cancel in the case of an emergency.

This is only a partial listing.

## 1.18 Regulation of Pricing of Ancillary Services

Recently, FERC has issued a notice of proposed rulemaking (NOPR) seeking comments on six ancillary services: reserve margins, transmission losses, load following, reactive power, energy imbalance, and redispatch. The nature of the comments requested indicates that this is an area that is largely new to the FERC, and that it feels itself on rather shaky ground. The NOPR establishes pricing for stage one since this is necessary to enable the FERC to get open access tariffs in place promptly. Specific pricing of two ancillary services, namely transmission losses and energy imbalances, are proposed. The charges for transmission losses are defined as 3% loss factor and 110% of sellers incremental cost. The tariff for energy imbalances is set at 100 mills/kWh for imbalances in excess of ±1.5%. There is no separate charge (included in account 556) for redispatch. Pricing of other ancillary services, such as load following, reserve margins and reactive power is defined by 1 mill/kWh ceiling for the group.

Utilities are free to propose revisions to these rates. Utilities must charge themselves for ancillary services when they make off-system sales, using the same rate they charge others for ancillary services. Thus, a utility that charges high prices for ancillary services reduces its ability to compete for off-system sales. This may cause the FERC to desire a ceiling on. The resolution of such issues demands that adequate study and analysis be invested to gracefully change the industry into a more competitive electric marketplace.

The role of ancillary services is very important in achieving a reliable power system operation. The stage one pricing scheme proposed by the FERC does not take into account the network configurations, system conditions, and the reliability desired by the market participants. Thus, the intent of achieving a true market optimum in price based operation as well as maintaining the reliability of power system is completely defeated. Unbundling of ancillary services should be viewed in the context of dis-aggregated utilities and competitive markets. The end-state should be management of ancillary services through specified requirements for all grid users and competitive bidding. The competitive environment brings a new set of complicated problems in designing an efficient market for ancillary services to provide reliable power system operation. This requires a comprehensive analysis of price-based operation by including ancillary services that are presently embedded within the vertically integrated industry.

## 1.19 Regulatory Questions [Sheblé, 1993]

The energy industry is presently regulated through federal and state agencies that set rates and approve expansion plans. To date, the delivery of energy to consumers has been solely the responsibility of the local electric utility. In this sense, utilities have a monopoly on the energy market. In an effort to open the market to competition, many legislators and consumer advocates are attempting to break the monopoly of electric utilities by deregulating access to the transmission system. Deregulation seems to be an inevitable outcome of a free market society that thrives on competition. The transition to a deregulated market is difficult.

Several unresolved issues must be considered and analyzed prior to the emergence of a fully evolved energy marketplace. Some of these issues are:

1. Wheeling: Should the transmission company be forced to allow any amount of wheeling over its transmission network? What priority should be given to each GENCO? Will utilities break into smaller GENCOs to take advantage of the priority system? What will be the criteria for the priority order? The amount of profits a company can generate from a certain transaction, on how far removed are the wheeling parties the amount of power wheeled the amount of losses incurred, or the amount of increased reactive power flow to support the wheeling transactions?

2. Losses: Who pays for losses? How are losses priced? Who is increasing losses?

3. Emergency situations: In the case of an emergency, who should be dropped first, the distant consumer, the residential consumer, the commercial consumer, or the industrial consumer?

4. Priority of transactions: Should the priority order in which transactions take place and/or are canceled in the case of an emergency depend on the size of the utility, the amount of energy traded the number of utilities affected by the transaction, or the effect of the transaction on security level?

5. Emergency power: Who should supply more power in an emergency? What is the price of emergency power?

6. Transmission network: Who will pay for the upgraded transmission network when the need arises? How will consumers and existing utilities be compensated for the existing transmission facilities that have already been paid for?

Additional questions on unresolved issues of interest to this author are developed in the following paragraphs. Some of these issues are:

1. Fulfilling obligations: Should consumers be allowed to renounce their obligation to be served by their own utility? How far in advance should consumers notify the utility of their decision? What penalty should be imposed on consumers and utilities for violating their obligations?

2. Building new facilities: Should utilities be forced to hold their building operations until deregulation crystallizes? Wouldn't it be more difficult for utilities to accept deregulation if they have built excess generation to serve consumers, cannot get a fair return on their investment, and as a result cannot compete in a free market due to high competition? How should such stranded costs be recovered? How will future stranded costs be handled?

3. Background of board's directors: How technical should the members of the board of directors be? Do all directors need an engineering background with an advanced business degree? Should the exchanged be managed by a hibrid?

4. Generation expansion: Who will pay for the upgraded supplies when the need arises? How are the consumers and existing utilities going to be compensated for the existing facilities that have already been paid for?

5. Changes in software: What changes should be introduced to the existing analytical software for security, reliability, and stability of power systems to take into consideration deregulation?

## 1.20 Engineering Research Problems [McCalley, 1993]

*Maintenance Scheduling.* Maintenance scheduling can be broken into two components: planning and implementation. Timing of maintenance could prohibit wholesale or retail contracts and thus control competition. Previously, NERC and various operating groups coordinated the maintenance of key equipment on a regional basis. It has been common practice to identify the maintenance schedules that would result in improper operation in neighboring transmission systems. When necessary, negotiation occurred to change schedules such that the overall security of the grid was maintained. If a competitive transmission system is required, how will such coordination occur?

Based on the experience in Norway, the transmission grid would remain as a regulated monopoly. However, it is conceivable for the transmission grid to be treated as an open bid resource for the suppliers and buyers to compete for access. It is the implementation of such a highly scheduled resource that is open to question. Indeed, the availability of FACTS devices appears to foster the opinion that the transmission grid could be operated to implement such a transaction-based system. However, the cost of FACTS devices, the systems to provide the necessary accounting, and the control to coordinate such a complex point-to-point transaction scheme has not been costed. The model of a transmission grid as a resource with varying cost structures to represent the increased control and accounting systems has not been identified at this time.

*Operations Scheduling.* Wholesale and retail wheeling imply a very large increase in the number of transactions maintained by the transmission companies. Instead of maintaining hourly records of inter utility transaction, quarter hour records of inter and intra utility transactions might have to be maintained. The increased computer systems and support staff for such systems has not been costed at this time. If more complex control systems are required, such as for proper relaying due to different loading conditions, the interaction of the computer systems and such control systems would have to be designed. Special relay arming programs where the relay targets, the relay type, and the variables sensed change as a function of system configuration have been implemented in several locations. However, the increased complexity of coordinating the control systems based on different matches of supplier and buyer will surely be more complex than present schemes if multiple sources are available to provide each potential buyer.

*Frequency Regulation.* The single function, that will change the most is automatic generation control. This function has not significantly changed since its inception in the early 1930s. Indeed, the need to control generation instead of load is a key philosophical question. Most consumer-based industries use load control as the means of providing operating stability while supplies are controlled on a long-term basis by supply/demand economics. The speed of control is one of the critical issues to be addressed. Presently, the decision variables (tie line flows, generation levels, and frequency) are sampled on a two-second basis. The control commands are issued once every four seconds based on the projected trend of the mismatch between source and demand. Such a fast control cycle is not supported in all countries. Indeed, the cost of such a fast control cycle has been questioned for several decades. A novel approach identified by Schweppe is the FAPER device. The FAPER device is similar to many other demand side management (load control) devices. It was to sense the frequency of the system and change the demand based on the frequency trend. This would result in thousands (millions?) of appliances providing load frequency control. Schweppe proposed that this FAPER device be an integral component of major appliances. It is conceivable that more intelligent appliances, perhaps to provide local voltage control or load relief, could significantly improve the reliability of the overall transmission and distribution grid. Sheblé [1994] proposed that VAPER devices be built into appliances to provide voltage and harmonic assistance to the power system. However, the cost of such devices, the standardization of set points and targets, the replacement of existing appliances, and the overall cost of such a large system design has not been outlined at this time. This author has often recommended that battery backup be provided in all appliances that include a clock display.

*Available Transmission Capacity (ATC).* Transmission owners are now required to sell transmission capacity on request unless the there is insufficient capacity. It is therefore critical that each transmission owner have the ability to efficiently determine ATC so that it does not over-or under-sell the service it is providing. ATC may be defined with respect to a single selling node and a single buying node (i.e., point to point), or with respect to a group of selling and a group of buying nodes (network wheeling). Determination of ATC without regard to security constraints is quite straightforward; simply increase loading until a power flow solution indicates an overloaded circuit. It becomes more difficult when static security constraints (thermal overload and out of limit voltages following contingencies) are introduced, but it is tractable. In addition, optimal power flow technology can be used to maximize ATC by changing dispatch, if that is desired. The problem becomes most difficult, however, when dynamic security constraints are introduced, because of the computational requirements and also because it is difficult to define a boundary between acceptable and unacceptable operation from which ATC can be computed. Establishing such a boundary that includes static, voltage-related, and stability-related systems, and devising probabilistic and deterministic metrics for measuring ATC is a challenge for the future.

*Security Incentives.* Because competitive energy systems will result in less centralized coordination and control and more autonomy for individual participants, and because the energy marketplace is primarily profit-driven, economic incentives may be useful for inducing participants (both suppliers and demanders) to include the effects of those decisions on system security levels. (The same can be said about reliability from a planning perspective and is the focus of new tools such as TRELSS and TPLAN). This point is the thrust of a recently developed transaction selection and pricing algorithm where the "economic incentive" is provided in two distinct ways. First, transactions are accommodated in order of their per MW impact on security levels. The ones having highest impact are accommodated last and therefore risk not being accommodated at all. Second, the transmission service price for each transaction includes existing facility costs (embedded cost) and the incremental costs (due to losses). It also includes a security penalty which is proportional to the per MW security impact, so that those transactions having large security impact pay higher prices for transmission service. The ultimate effect of providing the security incentives is that suppliers will tend to locate closer to the demanders and transactions will tend to be made more often during off-peak and off-season hours.

*Managing Risk.* Participants will have incentive to take risk because their goal is to maximize profits. If the risk is not quantified and well-managed (i.e., taken only when the economic advantages are substantial), such risk-taking on a large scale could have serious effects on system security. This is particularly true for dynamic security because the impact of instability can be costly. Risk management requires integration of probabilistic techniques with system operation. This could be difficult because probabilistic techniques have traditionally been developed and used more by system planners than operators, since the decision horizons and consequently the uncertainty is quite different (the planning horizon is years whereas the operating horizon tends to be months or less). However, operations will have introduced uncertainty because transmission flows, once driven only by normal load growth (which is relatively predictable), is now driven by the economics of long-distance transactions (which may not be predictable at all).

# 2 INDUSTRIAL INFRASTRUCTURE

*Infrastructure is the difference between crossing the river by a bridge or by swimming.*

## 2.1 Review of Deregulation

### 2.1.1 Electric Power Industry

*Wheeling/transmission access*

Wheeling is "the use of the transmission facilities of one system to transmit power of and for another entity or entities [Kelly, 1987]." The concept of wheeling allows buyers and sellers of electricity to maximize their profits by using the lowest generation cost available. It also raises many questions, including how participants should access the transmission grid.

The decrease in regulation desired by economists for effective competition requires open transmission access. However, there are several unsolved relevant issues. Because of the physical laws, current does not follow a specified path. This complicates the costing of electricity transactions. In addition, since there is no inexpensive storage medium, supply and demand must be balanced nearly simultaneously.

Increased power wheeling would probably threaten the security of any transmission network. The consequences would be reduced reliability, increased reserve requirements, and increased production costs. The wheeling of power also results in increased losses and a lower voltage profile. The Federal Energy Regulatory Commission (FERC) has proposed that the wheeling utility be obligated to build new transmission lines as needed to support wheeling transactions [Open Access, 1990]. FERC has also suggested that the wheeling utility become responsible for providing reactive power if wheeling threatens the voltage profile of the power system. These proposals result in additional expenses that will eventually be passed on to the consumer who would not necessarily derive any benefit from the wheeled transaction.

The pricing of wheeling services is an unresolved issue. The transmission cost for a utility is specific to the characteristics of its system, the nature of the transmission service requested, and the location of the buyer and seller. Proper pricing procedures must give wheeling utilities incentive to build new lines and increase capacity when necessary. Opportunity costs of wheeling and transmission-related operating costs, incurred to preserve system reliability, should be reflected. Power system reliability and security, typically assumed unaffected by wheeling, must be preserved.

The ideal objectives of a wheeling scenario according to Caramanis et al. [Caramanis, 1986] are:

- Profits for all involved parties
- Efficient, reliable, and secure operation of power systems
- Multiple transactions
- Easy deployment
- Decentralized decision making
- Wheeling rates independent of political and arbitrary decisions

Current FERC pricing policy was formulated in the era of short-term, nonfirm coordination exchanges, not in present long-term wheeling arrangements. FERC requires pricing to be based on embedded costs and to use average transmission operating and maintenance costs. In addition, average transmission losses are required and "percentage adders" on wheeling transactions are limited [Weiss, 1991]. Recently, FERC has moved away from strict reliance on embedded cost rates to permit wheeling utilities to share the benefits of trade.

### *2.1.2 Economic problems*

Although useful, traditional cost-based pricing has a number of inherent economic problems [Merrill, 1991]:

- Allocation of costs to various customer groups requires arbitrary judgment
- Cost allocation of long-term utility investments involves time. When should it be paid for?
- Cross subsidies occur between customer classes if cost allocations are not fair
- Under recent legislation, utilities question if traditional returns cover the risks

- Traditional rates are generally below marginal costs that are, according to economists, necessary for efficiency
- Firms whose profits are based on their investments, and that in fact make profits, have an incentive to make excessive investment in facilities. In other words, such firms have no incentive to operate efficiently
- Regulation protects utilities against competition by guaranteeing a monopoly in a service territory

*Regulation and a re-regulated competitive environment*

Regulation is a substitute for the lack of competition. Hence, deregulation is proposed to end the monopoly of utilities and to protect consumers. However, increased competition should not happen at the expense of the consumer. It might be more appropriate to name the transition "re-regulation." Competition may be implemented in the form of a brokerage or commodity exchange where rules and regulations govern the transactions of electric power. Competitive bidding would allow the substitution of market pricing for administrative pricing [Phillips, 1990], thus reducing FERC's role. The New York Mercantile Exchange has recently proposed setting up such a market.

Transmission networks can be seen as one of two businesses [Tabors, 1984]:

- Fixed asset based business put in place to guarantee (+/-) sufficient transmission capability to maintain reliability and security during peak load, or
- Transportation business that wants to move the greatest amount of energy at the greatest profit margin.

To date, pricing simulates competition in a regulated monopoly environment. A competitive environment can be simulated by basing price on the marginal cost of delivery. This provides incentive to operate and invest for increased system efficiency. It also provides incentive for both generation and end users to locate efficiently in the network. Most importantly, marginal cost pricing maximizes social welfare and profit.

If regulatory control evolves to a simulated competitive environment, market information will play a greater role in network expansion planning. Network utilities will minimize costs independent of the supply service. The concept of wheeling will disappear into a simultaneous buy/sell structure, in which total benefits of transactions will be maximized when least cost exchanges take place [Tabors, 1994].

Like any other approach to selling and buying electricity, the competitive approach has advantages and disadvantages. Regulatory efficiency is a poor substitute for competition, but the present utility industry works well and any restructure may introduce new types of inefficiency. Joseph Swidler [Swidler, 1991] states: "The relations between government and industry are a complex problem and do not lend themselves to simple or formulaic solutions. But the faults of regulation should move us to better regulation—rather than to a leap in the dark toward the creation of new entities, free of regulation, which pose a threat to the stability of the most important element in the infrastructure of our society." Swidler believes real competition is infeasible in the electric power industry because large investments for generation capacity require the assurance that ratepayers assume all the risks.

Competition in generation has been tested with three different approaches [Merrill, 1991]:

- Qualifying facilities (QFs) under PURPA are paid the utility's avoided cost. One of the problems is defining and measuring avoided cost
- Bidding for new capacity
- Privatization/disintegration of vertically integrated utilities

However, competition, to the extent applicable in generation, is not possible in transmission. Thus, the regulatory issues of access terms and access prices must be addressed.

Discarding regulatory rules will change the justification procedures used by utilities for further investments in the electric industry. An unstable business environment could force a gradual withdrawal of utilities within their market areas and, thus, could become detrimental to consumers.

Allocating rate-based pricing to wheeling transactions will probably have to be further modified. It might be appropriate to have various levels of rates dependent on the priority of access and based on the cost-of-service [Brown, 1991][1].

*Market-based pricing*

Recent interest in marketplace pricing that bases transmission services on the value of wheeling to all benefiting parties.

Marketplace pricing can take several forms. Three of these forms are presented here [Weiss, 1991]:

- An auction can be held to ration capacity when wheeling demands exceed power transfer capabilities

1 Also see the paper by Chao et al. [Chao, 19986] which presents a detailed study of reliability differentiated electricity pricing.

- Rates can be based on the profits of the generating parties
- Each utility can be allowed to charge what the market will bear. Such flexible pricing usually has a price cap associated with it.

Marketplace pricing of transactions encourages utilities to offer access and to build new lines when necessary. It allocates scarce resources efficiently, rewards innovators who contribute to increasing the value of industry to society, and searches out methods to meet consumer's lowest cost [Outhred, 1993]. Its flexible pricing is thought to increase system efficiency and overcome some perceived problems with current regulation. This method allows wheeling utilities to set rates to protect native customers.

Opponents contend that value-based pricing is inappropriate for a monopolistic system. Inappropriate market signals result when line owners charge rates based on service when plenty of capacity is available. Rewards for incremental changes to equipment and behavior are advantageous for the short-term but may not be for the long term. Markets do not address issues of equity—participants with many resources can influence market outcome [Outhred, 1993]. Finally, markets ignore external impacts not reflected by costs and benefits of participants. These externalities include effects on the public and the environment [Merrill, 1991, Outhred, 1993]. FERC has allowed some experimentation provided that the seller does not have monopolistic power.

Electricity markets should possess, among other attributes, the following [Outhred, 1993]:

- Contestability—All participants should be able to influence the outcome of a market process. All participants should be able to enter and leave the market without cost. Projects with large sunk costs, such as generation and transmission, do not allow this, thus they should be considered against more flexible options.
- High quality information—This may include cost reflective prices, forward markets, and information on the implications and risk factors.
- Meaningful choices—All participants must have realistic options to choose from, adequate resources, and access to unbiased advice.
- Effective markets in related goods and services, such as coal, etc.
- Consideration of externalities.
- Effective regulatory procedures with extensive public participation to keep participants from distorting markets to their favor.

A comparison of the various pricing/costing approaches for electric power transactions are summarized in Table 2.1.

Given the trend toward a competitive electric power industry, I believe a market-based approach is the best electricity pricing option. If properly used, a market-based approach would force efficient operation of the electric power industry. In addition, most economic problems caused by current costing methods would be minimized. Implementation of a market-based approach would, of course, require its disadvantages to be minimized. Further research must address its weaknesses, such as improper market signals, equity issues, and external impacts.

## 2.2 Auction Mechanisms

### *2.2.1 The Study of Auctions*

The question might well be asked: "Why consider auctions for pricing electric power?" A strong motive is given by the assumption that the electric power industry will move from regulated rate of return pricing to market-based pricing in the near future. This requires consideration of various pricing mechanisms. An additional reason is that the natural gas industry spent much time and effort researching auction mechanisms for pricing natural gas when their industry underwent deregulation. The electric power industry is quite similar to the natural gas industry. Both industries produce, transport, and sell their respective commodities. The need for a pricing mechanism coupled with the example of the natural gas industry is sufficient reason for considering auctions in the electric power pricing arena.

Economists offer us a wealth of information concerning auctions. Already in 1945, F. A. Hayek argued, "it [the economic problem of society] is a problem of the utilization of knowledge which is not given to anyone in its totality [McAfee, 1987]." One party participating in an exchange often has relevant information the other party does not have. Thus, Hayek argued, omitting the imperfection of information ignores a price system's chief advantage. First, a rational buyer and seller need only know the vector of prices, and not the supply and demand underlying them. Secondly, constraints caused by informational asymmetries can be as significant as resource constraints.

"An auction is a market institution with an explicit set of rules determining resource allocation and prices on the basis of bids from market participants [McAfee, 1987]." Auctions are used for products that have no standard value; for example, the price of electricity depends on the supply and demand conditions at a specific moment in time.

Table 2.1. Comparison of Pricing Methodologies.

| BASIS | ADVANTAGES | DISADVANTAGES |
|---|---|---|
| **Contract path**<br>• Embedded capital costs and average annual operating costs of wheeling utility<br>• Wheeling costs based on specified path | • Use of utility's historic allocation costs | • Economically unattractive<br>• Incorrect pricing signals<br>• Discourages new transmission line construction |
| **Rolled-in**<br>• Embedded capital costs and average annual operating costs of wheeling utility<br>• Wheeling cost is fixed regardless of location or distance | • Use of utility's historic allocation costs<br>• Identifies all affected parties | • Economically unattractive<br>• Incorrect pricing signals<br>• Discourages new transmission line construction<br>• Inefficient siting of new generation |
| **Flow-changing**<br>• Costs are assigned in proportion to change in power flow<br>• Changes in tie-line flow (boundary-flow method)<br>• MW-miles of transmission lines (line-by-line method) | • Allocates embedded costs by power-flow models<br>• Reflects wheeling line losses<br>• Identifies all affected parties<br>• Encourages construction of new transmission | • Transaction distance not reflected<br>• Complex simulations |
| **SRMC**<br>• Rate of change in operating cost, given change in wheeled energy | • Efficient where excess capacity exists<br>• Proper pricing signals<br>• Cost rate changes with respect to amount wheeled<br>• Identifies all affected parties<br>• Reflects line losses | • Capital costs ignored<br>• Complex calculation models<br>• Volatile results |
| **LRIC**<br>• Change in total cost of providing unit of wheeling service<br>• Wheeling costs obtained by comparing capacity of base case and wheeling cases (SI)<br>• Minimizes use of long-term forecasts (LRFIC) | • Estimates investment costs for new facilities because of wheeling<br>• Estimates change in cost because of wheeling<br>• Identifies all affected parties<br>• Encourages construction of new transmission lines<br>• Reflects wheeling line losses | • Inappropriate price signals during excess capacity<br>• Computationally complex and time consuming |
| **Marketplace pricing**<br>• Transmission service based on value to parties benefiting from wheel | • Encourages construction of new transmission lines<br>• Flexible pricing increases efficiency<br>• Protects native customers | • Extra capacity combined with service-based rates causes inappropriate market signals<br>• Value-based pricing inappropriate for monopoly |

Auctions can be considered games with incomplete information. Richard Engelbrecht-Wiggans [Engelbrecht, 1980] explains that auctions model the true state of nature, which "prescribes the relevant characteristics and the number of objects being auctioned, the von Neumann and Morgenstern utility functions[2] and the number of strategic players (typically some or all of the bidders), and the behavior of any nonstrategic players (typically the auctioneer and perhaps some other players)." Engelbrecht states that players can be assumed to know the possible true states of nature and the probability distribution of these states; however, it is not known what state occurs in a given situation. Each player chooses a bidding strategy based on the observed information and the expected bids of other players.

Auctions consist of four parts—players, objects, payoff functions, and strategies. Players are defined as the number of participants each with their respective utility functions. Rational and sophisticated players are assumed. An auction may involve a single or large quantity of indivisible or divisible objects, whose true value may or may not be known to the bidder. In addition, an object's value may vary among bidders. The payoff function includes the award mechanism, reservation price, and miscellaneous costs, such as participation, preparation, and information costs. Players choose strategies to maximize their expected gain [Engelbrcht, 1980].

Auctions have been studied by two methods—theoretical and experimental. An explanation of these two approaches follows.

*Theoretical auctions*

The modeling of auctions provides a narrowly defined set of questions that enables economists to theoretically examine the implications of informational asymmetries for a pricing system. Auctions are studied because they provide an explicit model of price making. Furthermore, they have considerable empirical since huge amounts of goods are exchanged by auctions everyday. According to McAfee and McMillan [McAfee, 1987], "theoretical results in auction theory can explain the existence of certain trading institutions, and perhaps can even suggest improvements in the existing institutions: Thus auction theory has both positive

---

2 The central premise of von Neumann and Morgenstern utility functions is that people choose an alternative with highest expected utility rather than highest expected value. Note that the expected values of the outcomes of a set of alternatives do not necessarily have the same ranking as the expected utilities of the alternatives. The von Neumann-Morgenstern utility function allows the possibility to deal with the consumer's behavior under conditions of uncertainty.

"If all possible outcomes are ranked, with the least-preferred $(x_1)$ assigned a utility of 0 and the most preferred $(x_n)$ assigned a utility of 1, and if $x_i$ is the certainty equivalent of a lottery which involves winning $x_n$ with probability $r_i$ and $x_1$ with probability $(1 - r_1)$, then the von Neumann-Morgenstern utility index assigns a value of $r_i$ as the utility of $x_i$.

A von Neumann-Morgenstern utility index is *unique up to linear transformations.* In other words, any utility values can be assigned to $x_1$ and $x_n$, as long as the value assigned to $x_n$ is greater than the value assigned to $x_1$ and both values are greater than or equal to 0. The utility value assigned to $x_i$s then the expectation over the utility values assigned to $x_1$ and $x_n$, where the probabilities are the probabilities from the lottery over $x_1$ and $x_n$, for which $x_i$ is the certainty equivalent [Binger, 1988]."(see also: [Binger, 1988]; [Henderson, 1980]; [Varian, 1992]).

and normative aspects." The problem with auction theory is that complexity becomes prohibitive for anything but the simplest auctions.

*Experimental auctions*

In addition to theoretical research of auctions, economists also conduct experimental studies. These studies use human subjects to submit bids (offers) in a laboratory setting. The subjects are awarded differing amounts of currency on the basis of their efforts. Monetary rewards can be used to induce value in the form of any predesigned demand or supply function on each subject, if the following assumptions hold:

- Each experimental subject desires to maximize his earnings
- The experimental task required to earn money is simple enough so that the desired subjective transaction costs are negligible.

In fact, appropriate reward structures allow complete control over the experimental demand and supply conditions. They also allow researchers to study the effect of any given configuration on price adjustment behavior [Smith, 1974].

Experimental methods are useful in determining the equilibrium and dynamic properties of market price behavior. Vernon Smith [Smith, 1974] suggests distinguishing between three "treatment" variables in the use of experimental methods:

- The aggregation of individual values to form market values, that is, the aggregate supply and demand conditions that bind price-quantity behavior.
- The institution of contract, that is, the set of rules that specifies how the individual participants communicate, exchange information, and form binding contracts.
- The market structure, that is, the number of participants and their relative capability of influencing relative demand and supply capacity.

Smith has furthered experimental research by implementing the use of computers for selection of optimal contracts. This research is performed in a laboratory setting with individuals acting as buyers, sellers, or both. Participants enter their bid (offer) quotes into a computer terminal. A central computer then applies an optimization algorithm to determine the prices and allocations that maximize gains on the basis of the bids (offers), and budget and capacity constraints of the individual market participants. These experimental market mechanisms produce strong equilibrating tendencies, even when rental rewards to buyers and sellers are not balanced. These tendencies become stronger with increases in excess supply [Smith, 1965].

Both experimental and theoretical research focus on various types of auction mechanisms. Of course, not all types of auctions are feasible for implementation in the electric power industry. However, the description and evaluation of various contract institutions is necessary to understand if and how auctions might be used for pricing electric power.

### *2.2.2 Auction Mechanisms*

The following auctions are presented as *bid auctions* since buyers submit bids and sellers accept bids but cannot make offers. These auctions could also be *offer auctions* where the roles of buyers and sellers are reversed. One-sided auctions tend to favor the silent side—buyers (sellers) wait until offers (bids) are below (above) the competitive equilibrium (CE) price. Thus, the silent role acts as an aid to tacit collusion [Coppinger, 1980].

*The four standard auctions*

There are four standard types of auctions. These include the English auction, the Dutch auction, the first-price sealed-bid auction, and the second-price sealed-bid auction. Table 2.2 summarizes the four standard auctions.

Table 2.2 Comparison of Auction Methodologies.

| AUCTION TYPE | PROCEDURE | ACCEPTED BID PRICE |
|---|---|---|
| English (oral, open or ascending-bid) | Bids progressively raised until one bidder remains | Value of second highest bid |
| Dutch | Price decreased until first bidder accepts | Value of first bid |
| First-price sealed-bid | Bidders submit sealed bids | Price of highest bid |
| Second-price sealed-bid | Bidders submit sealed bids | Price of second highest bid |

**English auction**

The English auction (oral, open, or ascending-bid auction) is the form most commonly used for selling goods. Bids are progressively announced until no purchaser wishes to make a higher bid. Therefore, the bidding stops at a level approximately equal to the valuation of the item by the second highest bidder. The object is then purchased at that price by the bidder for whom it has the highest value. This result is Pareto-optimal[3] since the bidder with the highest valuation obtains the object [Vickrey, 1961]. An essential feature of the English auction is that, at any point in time, each bidder knows the level of the current best bid and the price at which other bidders dropped out [McAfee, 1987]. Two examples of goods commonly priced by the English auction are antiques and artwork.

**Dutch auction**

In the Dutch auction (descending-bid auction) an auctioneer calls a decreasing set of bids starting with an initial high price until one bidder accepts the current price. Thus, the first and only bid concludes the transaction. The Dutch auction is actually a "game." Each bidder must bid a value which gives his greatest expected gain [Vickrey, 1961]. A bid for an item at full value to the bidder maximizes the probability of obtaining the item, but the gain is guaranteed to be zero. The possibility of gain emerges as the price is progressively lowered. Dutch auctions require appraisal of the market situation as a whole in addition to an appraisal of what the article is worth to the bidder himself. Neglecting this general appraisal increases chances of not achieving optimal allocation. Dutch auctions are often used for the sale of cut flowers in the Netherlands and fish in Israel.

**First-price sealed-bid auction**

The first-price sealed-bid auction requires potential buyers to submit sealed bids. The highest bidder is awarded the item for the price he bid. A buyer cannot observe the bids of the other participants [McAfee, 1987]. The granting of mineral rights to government-owned land is determined by the use of first-price sealed-bid auctions.

Smith presents a slight variation of the first-price sealed-bid auction called a *discriminative sealed-bid auction* [Smith, 1974]. In this case, the seller offers a quantity $Q$ of a homogenous commodity. Buyers are then invited to tender bids at a stated price for a stated quantity. Bids are arrayed from highest to lowest and the first $Q$ bid units are accepted. Ties at the lowest bid are accepted on a random basis.

**Second-price sealed-bid auction**

In this type of auction, bidders submit sealed bids after being informed that the highest bidder wins the item but pays a price equal to the second highest bid. Second-price sealed-bid auctions have significant theoretical properties but are seldom used in practice [McAfee, 1987, Smith, 1974].

---

[3] The distribution of goods is said to be *Pareto-optimal* if every possible reallocation of goods that increases the utility of one or more consumers would result in a utility reduction for at least one other consumer [Henderson, 1980, p. 287]. For auctions this meansPareto-optimality is achieved when the winner is the bidder who values the object the most.

Second highest bid price auctions are not automatically self-policing. Assurance must be given to the top bidder by showing him what the second highest bid was. In addition, the possibility of a "schill" to jack the price up by the submission of a late bid just under the top bid must be prevented. Finally, the top bidder and the agent handling the sale must be prevented from showing the top bid to the second highest bidder so as to be able to set the price at the third highest bid [Vickrey, 1961, Engelbrecht, 1980].

Smith also presents a variant of the second-price sealed-bid auction for a homogenous commodity. It is called the *competitive sealed-bid auction*, which is the same as the discriminative sealed-bid auction except that all bids are filled at the price of the lowest-accepted bid [Smith, 1974].

Vickrey [Vickrey, 1961] suggests first-rejected bid pricing as an alternative to lowest-accepted bid pricing. In special cases, where each purchaser wants a specified quantity or none at all and the total amount to be sold is fixed in advance, Pareto optimality is achieved by establishing the first-rejected bid as the price. In spite of the lower price, the higher level of bids induced by this method results in a price that averages out at the same level as obtained with the Dutch auction, individual bid pricing, or last-accepted bid pricing.

If there is no collusion, a bidder's optimal strategy in progressive sealed-bid auctions is to make his bid equal his full value of the item or contract; bidding more or less would be disadvantageous [Vickrey, 1961].

*Variations of the four standard auctions*

McAfee and McMillan mention a number of variations on the four standard forms presented above. First, a seller may impose a reserve price and discard all bids if they are too low. Second, bidders could be allowed a limited amount of time to submit bids. Third, an auctioneer might charge an entry fee to bidders for the right to participate. Fourth, royalties may be assigned that is, a payment in addition to the bid may be added to the overall price. This extra fee depends on something correlated with the true value of the item. Fifth, an auctioneer of the English auction may set a minimum acceptable increment to the highest existing bid. Finally, a seller might offer shares in an item rather than selling the item as a unit [McAfee, 1987].

*Multiple auctions*

Vickrey discusses the institution, called a multiple auction, which occurs when there is more than one identical object to be sold but each bidder has use for only one object [Vickrey, 1961]. He presents two variations on this pricing scheme—simultaneous and successive—which are considered next.

In the *simultaneous auction*, $m$ items are put up simultaneously and each bidder can raise his bid even if this does not make his bid the highest. Once bids are no longer being raised, the highest *m* bidders are awarded the items at a uniform price equal to the *(m+1)st* highest bid.

Because of possible minor variations in the items, the *successive auction* is more often used. In this type of auction, each bidder must consider either aggressively bidding for the present item or waiting until the next item is available, hopefully at a lower price. Thus, speculation is present at all but the last auction. To avoid information complications that might be inferred from the price of the item in the first auction, the first auction is by the sealed bids of the *N* bidders. The price of the first item is the value of the highest bid of the *N-1* unsuccessful bidders. Vickrey points out that a problem associated with the *successive auction* approach is that it is "other than Pareto-optimal in the nonsymmetrical case [Vickrey, 1961]." (The nonsymmetrical case refers to bidders with asymmetrical valuations of the item to be sold.)

*Multiple sales by sealed bids*

Typically, a number of bids--starting from the highest--are accepted with each transaction price being the price of the individual accepted bid. An alternative method is to set a uniform price for all accepted bids at the level of the last accepted bid. This price level avoids price discrimination among the various buyers and also reduces the probability that a bidder's own bid will affect the price he receives. This uniform price approach causes bids closer to the full value of the bidder, which improves the chances of optimal allocation and reduces the effort and expense involved in appraising the general market [Vickrey, 1961].

For maximum advantage, the uniform price charged to successful bidders should be the first-rejected bid. This motivates the bidder to submit a bid that represents the item's full value to him. In the long run, "first-rejected bid" pricing yields just as high an average price as the greedier method of "last-accepted bid" pricing. This result assumes that each bidder is interested in one unit and there is no collusion among bidders. If a bidder desires more than one unit, it would be advantageous to bid at his full *average* value of the quantity. The reason for this is that he would prefer this quantity to be allotted at any price lower than his bid rather than be excluded altogether. In addition, a change in his bid price, within the range in which he would be successful, would not affect the contract price [Vickrey, 1961].

*Double auctions*

In this market institution, several buyers and several sellers submit bids and offers. In each trading period of specified duration, any buyer can make an oral bid for one unit of a homogenous commodity at any time; any seller can make an oral offer for a unit at any time. A binding contract is made if any buyer (seller) accepts any offer (bid). New bids or offers are not required to provide better terms than the previous ones. Finally, only one bid or offer is outstanding at a time [Smith, 1982, Smith, 1974].

A variation of the double auction is to write all bids and offers on a blackboard. Any new bid (offer) is allowed only if it is higher (lower) than the previous ones. Submitted bids (offers) cannot be withdrawn. Once a contract occurs the auction begins anew. Bids left standing at time of contract may be reentered but are not binding [Smith, 1982; Smith, 1974].

Few theoretical results on double auctions exist since modeling strategic behavior on both sides of the market is difficult. The main result relates to the efficiency of double auctions with sealed bids. Robert Wilson [Wilson, 1985] found that "if there are sufficiently many buyers and sellers, then there is no other trading mechanism that would increase some traders' expected gains from trade without lowering other traders' expected gains from trade." Thus, the double auction is Pareto optimal.

Wilson also showed that the double auction satisfies the stronger criterion of ex ante efficiency when there are equal numbers of buyers and sellers with valuations distributed uniformly. That is, the expected gains from trade are maximized [Wilson, 1986].

Although the theory of double auctions is not well-developed, experimental results for double auctions have shed considerable light on the subject. Vernon Smith presents the following propositions, based on experimental results [Smith, 1982; Smith, 1974]:

> Proposition 1. Contract prices converge quickly to "near" the theoretical (Supply = Demand) equilibrium level. The more slowly converging markets are associated with very asymmetric supply and demand (producer surplus is substantially different from consumer surplus).
> Proposition 2. Quantities exchanged per period rarely differ from the theoretical (S = D) equilibrium by more than a single unit in any trading period.
> Proposition 3. A variation on the double auction rules in which a bid (offer) is not admissible unless it provides better terms than the previous bid (offer) does not appear to provide any significant increase in the convergence rate of contract prices.
> Proposition 4. The sampling variation (among different subject groups) in market price adjustment paths is considerable, but the variation in equilibrium prices (contract prices in the final period of trading) is minor.
> Proposition 5. Contract price convergence is more likely to be from below (above) when producer's surplus is greater (less) than consumer's surplus.
> Proposition 6: Allocations and prices converge to levels near the competitive equilibrium (CE) prediction. This convergence is rapid, occurring in three to four trading periods or less when subjects are experienced with the institution (but not the particular induced values).
> Proposition 7: Convergence to CE prices and allocations (Proposition 6) occurs with as few as six to eight agents (most experiments have used eight), and as few as two sellers." Note that every agent is a price maker

(announcing bids and offers) as well as a price taker (announcing acceptances).

Proposition 8: Complete information on the theoretical supply and demand conditions of a market (i.e., agent knowledge of the induced values and costs of all agents) is neither necessary nor sufficient for the rapid convergence property in Proposition 6.

Proposition 9: Experiments with one seller and five buyers do not achieve monopoly outcomes, although some replications achieve the CE outcome. Buyers tend to withhold purchases (and repeatedly signal with high bids) giving the seller a reduced profit, especially at the higher prices. This encourages contracts near the CE price, but normally at a loss in efficiency because of the withheld demand. (In "textbook monopolies" demand is known by the single monopolistic seller, but in double auctions demand is revealed as the auction takes place.)

Proposition 10: Experiments with four buyers and four sellers in which the sellers (or the buyers) are allowed to "conspire" (i.e., engage in premarket and between market conversation about pricing strategies) do not converge to the monopoly (or monopsony) outcome; neither do they seem to converge dependably to the CE. Furthermore, the conspiring group often makes less than the CE profit.

Proposition 11: Binding price ceilings (floors) yield contract price sequences which converge to the ceiling price from below (above). If the price ceiling (floor) is nonbinding, i.e., if it is above (below) the CE price, prices converge to the CE, but along a path which is below (above) the price path in a market without a price ceiling (floor). If a binding price ceiling (floor) is removed, this causes a temporary explosive increase (decrease) in contract prices before the CE price is approached. Thus, nonbinding price ceilings (floors) do not serve as points in the sense that buyers are attracted to prices at the ceiling (floor) [Coursey, 1983].

Proposition 12: Asset markets with eight or nine agents converge slowly (eight or more two-period trading cycles) toward the CE (rational expectations) price and efficiency determined by the cumulative two-period dividend value of the asset Convergence is greatly hastened by introducing a first-period "futures" market in second-period holdings, which enables second-period dividend values to be reflected in (or discounted by) period one asset prices more quickly.

Proposition 13: Asset markets with nine or 12 agents in which the asset yields an uncertain state-contingent dividend, known in advance only by a subset of insiders (three or six), converges toward the CE (rational expectations) price and efficiency.

The experimental results summarized in the above double auction propositions support the Hayek hypothesis: "Markets economize on information in the sense that strict privacy together with the public messages of the market are sufficient to produce efficient CE outcomes [Smith, 1982]."

Empirical evidence gives strong support for static competitive price theory when markets are organized on the principle of the double auction. In addition, information requirements for achieving competitive equilibrium prices are weak. That is, no double auction trader needs to know the valuation conditions of other traders, to understand or have knowledge of market supply and demand conditions nor does he need any trading experience.

**Sealed bid-offer (double) auctions**

Smith [Smith, 1982] experiments with two different bid-offer rule sets for double auctions:

- *P(Q)*: Each buyer (seller) submits a demand (supply) schedule, that is, specifies a bid (offer) price for each unit demanded (supplied)

- *PQ*: Each buyer (seller) submits a single bid (offer) price and corresponding quantity

With both sets of rules, *PQ* and *P(Q)*, the bids are arranged from highest to lowest and the offers from lowest to highest. A selection algorithm that handles tied bids (offers) determines a single market-clearing price and corresponding quantity. All bids (offers) equal to or greater (less) than this price are accepted. The process ends with the private communication of outcomes resulting from each individual bid (offer) and a public announcement of market-clearing price and quantity. *P(Q)* provides incentive for agents to under-reveal supply and demand. Thus outcomes are not Pareto-optimal. The *PQ* institution corrects this. Experimental results lead to the following propositions [Smith, 1982]:

> Proposition 14: Let *E[x]* be the expected efficiency of institution $x$ and *p[x]* be the price deviation from CE for institution $x$. "Based on the prior empirical performance of *DA* (double auction) and theory pertaining to *P(Q)* and *PQ*, we expect the efficiency of allocations in these three institutions to be ranked $E[DA] \approx E[PQ] > E[P(Q)]$ and the deviation of prices from the CE to be ranked $p[DA] \approx p[PQ] < p[P(Q)]$. The experimental results suggest the contrary observed ordering $E[DA] > E[P(Q)] > E[PQ]$ and $p[DA] < p[P(Q)] < p[PQ]$."

**Sealed bid-offer (double) auctions: unanimity tatonnement**

Unanimity atonement variations of the above *PQ* and *P(Q)* institutions are called *P(Q)v* and *PQv*. With these two institutions the following steps are followed [Smith, 1982]:

- Market-clearing price and quantity is determined

- A *conditional* allocation of accepted bids and offers is made

- Each agent whose bid (offer) was in part accepted is enfranchised and must vote "yes" or "no" to decide if the allocation should be finalized

- If all vote "yes," the process stops and each individual executes a long-term contract for *T* times the outcome of the trial

- Otherwise the process proceeds to another bid-offer trial with a maximum of *T* trials

Experimental results lead to the following proposition:

> Proposition 15: "Measured in terms of efficiency and deviations from the CE price, *PQv* provides no improvement over *PQ*. *P(Q)v* performs better than *P(Q)* and appears to be the equal of *DA*."

**Cyclical double auctions**

Williams and Smith [Williams, 1984] examined cyclical double auctions. Two market environments were researched:

- A market in which successive market periods are temporally isolated (autarky—units of a commodity cannot be carried over from one demand period to another)
- A market in which agents (speculators) are allowed to carry commodity units from one period to another, thus temporally linking the cyclical phases of the market

Two different experimental approaches were used: cyclical demand with stable supply and cyclical demand with cyclical supply. Trading years were divided into a low-demand season and a high-demand season.

Experimental results showed that the inclusion of speculative agents tends to significantly reduce the observed magnitude of cyclical price swings relative to markets without intertemporal speculation. Speculators were also found to cause significant increases in market efficiency [Miller, 1977]. Risk takers were the most successful speculators.

Traders made most purchases in the low price periods that resulted in immediate profits for only the sellers. Traders generally made profits on the sale of inventory units in high price periods.

In markets with stable supply, speculation caused convergence toward the zero excess demand intertemporal equilibrium price. Without speculators, prices lagged in adjusting from one cyclical phase to another. In markets with shifting supply, prices did not converge within the seven-cycle duration of most experiments. The cause attributed to the risk-averse behavior of speculators that resulted in intertemporal carry-over below the socially optimal level [Williams, 1984].

*Posted pricing*

With posted pricing each seller (buyer) independently selects a price offer (bid) without the knowledge of prices selected by competitors. These prices are posted on a "blackboard." A randomly selected buyer (seller) selects a seller (buyer) and makes that seller (buyer) a quantity offer at his posted price. The seller (buyer) then responds with a quantity acceptance that forms a binding contract. If a portion of the quantity offered is not accepted, the buyer (seller) may choose a second seller (buyer) and so on until his quantity requirement is filled. When the first buyer (seller) is finished with his contracts, a second buyer (seller) is randomly chosen, and so on, until all buyers (sellers) have completed their contracts [Smith, 1974].

Posted bids result in lower contract prices than posted offers. Thus posted pricing operates to the advantage of the price initiator. Posting "take-it-or-leave-it" prices for all offered units supports collusive coordination among independent traders [Smith, 1974].

Smith [Smith, 1982] presents the following propositions, based on experimental results:

> Proposition 1: "If $G_o^t(p)$ and $G_b^t(p)$ are the proportions of contract prices at $p$ or higher in trading period $t$ under the posted-offer and posted-bid institution, respectively, then $G_o^t(p) \geq G_b^t(p)$ for all $t > 1$."
> Proposition 2: "Experiments with single seller posted-offer pricing, in both increasing and decreasing cost environments, yield convergence to the monopoly price. This convergence appears to be faster with increasing cost than with decreasing cost. The slow convergence (at least 15 periods) in three of four replications under decreasing cost appears to be attributable to the fact that buyer withholding of purchases (more likely in earlier periods at the higher posted-price offers) impacts the seller's most profitable units."
> Proposition 3: "In a market with one seller and five buyers using posted-bid pricing, prices tend to converge to the CE price, but volume and efficiency are somewhat below the CE levels."
> Proposition 4: "In decreasing cost environments in which demand is insufficient to support more than a single seller, but the market is "contested" by two sellers with identical costs, there is a strong tendency (six experimental replications, each with 15 to 25 trading periods) for posted-offer prices to decay to the CE price range."

Posted-offer (bid) pricing indicates that prices tend to converge from above (below) the CE price to a level somewhat higher (lower) than the CE price [Coursey, 1983]. In unconstrained posted-offer markets, sellers tend to post prices relatively high compared to the CE price. Coursey and Smith [Coursey, 1983] believe prices are not posted at the ceiling price when ceiling prices are in effect because "sellers do not know what constitutes a 'high' price, that is, they do not know the CE price, nor do they know whether the ceiling price is binding or not, even if we assumed that the sellers would understand what is meant by a CE price or a binding ceiling." If price controls are enforced, the impulse to quote prices at the ceiling price must be considered against the possibility of missing out on sales because of competing sellers' lower prices.

## 2.3 Evaluation and Comparison of Auction Institutions

*Independent-private-values model*

A popular approach in the theoretical study of auctions is the use of the *independent-private-values model.* This model assumes the following [Milgrom, 1982]:

- A single, indivisible object is sold to one of several bidders
- Each bidder is risk-neutral and knows the value of the object to himself
- Values are modeled as being independently drawn from some continuous distribution
- Bidders are assumed to behave competitively

The use of the independent-private-values model gives seven significant conclusions concerning the four standard auctions. First, the Dutch auction and the first-price auction are strategically equivalent; the equilibria of the two auctions coincide. In both auctions the winning bidder is the one who submits the highest bid, the price paid is equal to the highest bid, and the bidder need not know the value of the object to himself [McAfee, 1987, Milgrom, 1982]. Smith [Coppinger, 1987; Smith, 1982] experimentally shows that while Dutch and first-price auctions are theoretically isomorphic--that is, they are subject to the same analysis and predict equivalent allocations--they are not equivalent behaviorally.

Second, English and second-price auctions are equivalent. Each bidder's equilibrium strategy is to submit a bid equal to his own valuation of the object. This is called the *dominant strategy* [4]outcome; that is each bidder has a well-defined bid regardless of how his rivals bid. The payment for both auctions equals the valuation of the bidder with the second highest value. Thus, these two auctions are equivalent provided each bidder knows the true value of the object to himself [McAfee, 1987; Milgrom, 1982]

Third, the *dominant strategy equilibrium* outcome of the English and second-price auctions are Pareto-optimal; that is, the winner is the bidder who values the

[4] A *dominant strategy* is a strategy that is always best, regardless of what the other players play.

object the most. In contrast, Dutch and first-price auctions are Pareto-optimal only when bidders are symmetric (identical valuation distributions). The first-price auction does not have a dominant equilibrium; instead it meets the weaker criterion of *Nash equilibrium*[5]: each bidder chooses his bid on the basis of his best estimate of the approach used by other bidders [McAfee, 1987; Milgrom, 1982]. Coppinger et al. [Coppinger, 1980] define Nash equilibrium behavior to be the "expected gain maximization and symmetrical expectations in which each bidder assumes that the valuations of all bidders can be regarded as having been generated by a rectangular density."

Fourth, all four standard auctions result in identical expected revenue for the seller. However, there is a practical difference. In the Dutch or first-price auctions, a bidder submits a value less than his true valuation. The value submitted is dependent on the probability distribution of the other bidders and the number of competing bidders. In contrast, a bidder in an English or second-price auction can easily decide how high to bid (up to his valuation of the object) [McAfee, 1987].

The fifth conclusion is the *revenue equivalence* result: the expected revenue of the seller is the expected value of the object for the second highest evaluator [Milgrom, 1982]. In the English or second-price auctions, the price equals the value of the second highest bidder; in Dutch or first-price auctions, the price is the expectation of the second highest valuation conditional on the winning bidder's own valuation. These two prices are equal only on average [McAfee, 1987]. This *revenue equivalence* result assumes that at equilibrium the bidder who values the object most will receive it, and that any bidder who values the object at its lowest possible level has an expected payment of zero.

Sixth, for many common sample distributions of valuations, such as normal, exponential, and uniform distribution, the four standard auctions are optimal auctions if suitable reserve prices or entry fees are used [Milgrom, 1982].

Seventh, a risk-averse buyer or seller will strictly prefer the Dutch or first-price auction to the English or second-price auction [Milgrom, 1982].

Generally the more bidders there are, the higher the valuation of the second highest bidder is. However, if bidders incur costs in preparing bids and sellers incur costs in checking a bidder's credentials, the expected net price may not increase with the number of bidders [McAfee, 1987].

---

[5] "A *Nash equilibrium* of a cooperative game is a set of strategies for each player with the property that no player can unilaterally make a higher payoff by playing another strategy. Some games have more than one Nash equilibrium."

"A *strong Nash equilibrium* is a Nash equilibrium in which each player plays a dominant strategy. There can only be one strong Nash equilibrium in a particular game."

At a *weak Nash equilibrium* of a game, players play strategies that are only appropriate if their rivals' strategies do not change [Binger, 1988].

*Variations of the independent-private-values model*

In the following subsections, one of the independent-private-values model assumptions is lifted while the others remain intact.

**Asymmetric bidders**

Asymmetrical valuation distributions for the various bidders have little effect on the English model. However, asymmetrical bidders cause the first-price sealed-bid auction to yield a different price than the English auction since revenue equivalence breaks down. The first-price sealed-bid auction's expected price can be either higher or lower than the English auction's expected price. Because of asymmetric valuations the bidder with the highest valuation is no longer guaranteed to win. Thus, the first-price sealed-bid auction does not guarantee an efficient outcome and Pareto optimality is no longer achieved. The first-price auction is workable if a successful bidder can be prevented from reselling the object or transferring it [McAfee, 1987].

**Royalties/incentives**

Theorists have also considered what occurs when payment is not a function of bids alone. According to McAfee and McMillan [McAfee, 1987], it is in the seller's best interest to condition a bidder's payments upon additional information concerning the winner's valuation. For example, the successful bidder for oil rights to government land pays the amount bid plus a royalty based on the amount of oil extracted.

A seller may observe ex post some variable (˜) that is an estimate of the winning bidder's true valuation (*v*). The winning bidder is then required to pay a price (*p*) to the seller. Thus a linear payment function is represented by

$$p = b + r\tilde{v} \qquad (2.1)$$

where $b$ is the bid and $r$ is a royalty rate. "If the distribution of the observable variable ($\tilde{v}$) is exogenous, the seller's expected revenue is an increasing function of the royalty rate [McAfee, 1987]."

A seller can take three approaches in using a payment function of the form (2.1):

- Seller sets royalty (*r*) and calls for bids (*b*)
- Seller sets fixed payment (*b*) and calls for bids on the royalty rate (*r*)
- Seller calls for simultaneous bids on the fixed payment (*b*) and the royalty rate (*r*) [McAfee, 1987]

If expected revenue increases with royalty rate, why not set royalties at 100 percent? First, sellers are unable to observe bidders' valuations ex ante. Thus, valuations and bids are decoupled and the highest bidder does not necessarily value the object the most. Secondly, the distribution of $\tilde{v}$ is not exogenous—the winning bidder can, after the auction, influence the signals of his true valuation that the seller receives. For example, the amount of oil the winning bidder will actually drill is dependent on the royalty rate. This is known as a *moral hazard* since the organizer of the auction cannot control the winning bidder afterward. Thus, royalty rates must be set with consideration of the tradeoff between the effect of royalty rates on competitive bidding and the influence of resulting moral hazards [McAfee, 1987]. Moral hazard forces the optimal royalty rate below 100 percent; royalty is zero if and only if there is an infinite number of bidders (perfect competition). Finally, the optimal royalty rate $r$ increases as bidders become more risk-averse relative to the seller [McAfee, 1987].

## 2.4 Risk Aversion

Bidders are now assumed to be risk-averse and to have von Neumann-Morgenstern utility functions. For risk-averse bidders, it is not generally true that partially resolved uncertainty reduces the risk premium [Milgrom, 1982]. With risk-averse bidders, a monopolistic, risk-neutral seller can do at least as well as when bidders are risk-neutral. This is attributable to the fact that using an English auction causes buyers to continue bidding if the price is less than their value. With risk-averse bidders, the expected revenue of first-price sealed-bid auctions is larger than the English or second-price sealed-bid auctions [McAfee, 1987]. In the first-price sealed-bid auction, a risk-averse bidder will often marginally increase his bid. In doing so, he increases his probability of a win but lowers his profit if he wins. This is to the seller's advantage.

In addition, a seller should try to organize the auction so that each bidder is unaware of the total number of bidders. The reason for this is that "under constant or decreasing absolute risk aversion, in a first-price sealed-bid auction the expected selling price is strictly higher when the bidders do not know how many other bidders there are than when they do know this [McAfee, 1987]." However the first-price sealed-bid auction with risk-averse bidders is not optimal since a seller's expected revenue is not maximized [McAfee, 1987].

If bidders are risk-averse and have constant absolute risk-aversion, then [Milgrom, 1982]:

- In the second-price and English auctions, revealing public information raises the expected price

- Among all possible information reporting policies for the seller in second-price and English auctions, full reporting leads to the highest expected price

- The expected price in the English auction is at least as large as in the second-price auction

Thus, revealing information has two effects: (1) each bidder's average profit is reduced by diluting his information advantage, and (2) bidders with constant absolute risk aversion are, on average, willing to pay more.

*General symmetric model*

Milgrom and Weber [Milgrom, 1982] present a general symmetric model that relaxes the independent-private-values model. It allows interaction among different bidder's valuations. The winning bidder's payoff may depend on his personal preferences, the preferences of others, and the intrinsic qualities of the object being sold. It is assumed that $n$ bidders compete for a single object and that each bidder possesses some information about the object. Let:

- $X = (X_1, \ldots, X_n)$ be a vector of *value estimates* or *signals* observed by individual bidders

- $S = (S_1, \ldots, S_m)$ be a vector of additional real-valued variables, which influence the value of the object to the bidders

The assumptions for the general symmetric model are [Milgrom, 1982]:

1. There is a function $u$ on $R^{m+n}$, such that for all $i$, $u_i(S,X) = u(S,X_i,\{X_j\}_{j \neq i})$. Consequently, all of the bidders' valuations depend on $S$ in the same manner, and each bidder's valuation is a symmetric function of the other bidders' signals.

2. The function $u$ is non-negative, is continuous, and nondecreasing in its variables.

3. For each $i$, $E[V_i] < \infty$
   -for private values model, $m = 0$ and each $V_i = X_i$
   -for common value model, $m = 1$ and each $V_i = S_i$. (In the common value model, bidders guess the objects true value)

4. $f$ is symmetric in its last $n$ arguments
   -$f(s,x)$ denotes the joint probability density of the random elements of the model

5. The variables $S_1, \ldots, S_m, X_1, \ldots, X_n$ are *affiliated.*

Bidder's valuations are affiliated if one bidder perceives an item has high value and, in so doing, causes other bidders to also perceive that the value of the item is high.

The actual value of the object to the bidder $i$ is $V_i = u_i(S,X)$. Let $Y_1, \ldots, Y_{n-1}$ denote the largest, . . ., smallest estimates from among $X_2, \ldots, X_n$. Then, using the symmetry assumption, bidder 1's value is:

$$V_1 = u(S_1, \ldots, S_m, X_1, Y_1, \ldots, Y_{n-1}) \tag{2.2}$$

This general model has been applied to the four standard auctions. The most important result, presented by Milgrom and Weber [Milgrom, 1982], is that a seller is best off by revealing all possible information if first-price, second-price, or English auctions are used. In the first-price auction, revealing information links the price to the information. In the second-price auction, price is linked to the estimate of the second highest bidder and revealing information links the price to the information. In the English auction, the price is linked to the estimates of all the nonwinning bidders and to the seller's estimate, if he reveals it. In all three auctions, revealing information adds a linkage and raises the expected price. In summary, the absence of linkages to other bidders' estimates in the first-price auction yields the lowest expected price. The linkage to all estimates yields the highest expected price in the English auction.

*Conclusions for the four standard auctions*

Smith [Smith, 1982] presents the following experimental propositions of the four standard auctions:

> Proposition 1: "Using the subscripts e(English), d(Dutch), 1(First), and 2(Second), and letting $E_e, E_d, E_1,$ and $E_2$ be measured either by the mean efficiency or the proportion of Pareto optimal awards, then $E_e \approx E_2 > E_1 > E_d$."
>
> Proposition 2: "English and second-price auctions which are theoretically isomorphic--that is, are subject to the same analysis and predict identical allocations--appear to be equivalent behaviorally. Dutch and first-price auctions which are theoretically isomorphic are not equivalent behaviorally."
>
> Proposition 3: "Prices, allocations, and individual bids in the first-price auction require the rejection of Nash equilibrium models of bidding behavior based on the assumption that all bidders have the same concave utility function. But the experimental results for $N>3$ bidders are consistent with a Nash equilibrium model based on the assumption that bidders have power utility functions with different coefficients of constant relative risk-aversion."

Coppinger et al. [Coppinger, 1980] note that the English auction is essentially free of strategic considerations and, of the four institutions, requires the least bidder sophistication. English auction prices tend to be slightly above the optimal price. This is because bids are raised in noninfinitesimal amounts causing some overbidding of the next-to-the-highest valuation. Second-price auctions tend to be below the optimal price since not all buyers understand that bidding full value is a dominant strategy. From a seller's point of view, the first-price sealed-bid auction would be preferred since it usually provides the highest prices—well above optimal prices. Dutch auctions tend to yield prices equal to or below optimal prices.

Milgrom and Weber [Milgrom, 1982] show different results when bidders' valuations are affiliated. Using Smith's notation in proposition 1 above, the four auctions' expected revenues are ranked as follows: $E_e > E_2 > E_1 \approx E_d$.

McAfee and McMillan [McAfee, 1987] present the following conclusions on the value of information. First, sellers can generally increase expected revenue by reporting all available information on the object's true value. Second, it is more important that bidders keep information private than precise since if his information is available to another bidder, his expected surplus is zero. Finally, the more bidders there are, the more heavily the bidder discounts his private information.

*Discrimination vs. competition in sealed-bid auctions*

Smith [Smith, 1982, Smith, 1976, Smith, 1967] compares the competitive and discriminatory pricing approaches to sealed-bid auctions. As mentioned previously, successful bids are filled at their bid prices under price discrimination. Under pure competition, all successful bids are filled at the same market-clearing price [Smith, 1982; Vickrey, 1961].

The use of discriminative rules leads to submission of lower bids then when competitive rules are used. The variance of competitive bids is consistently greater than the variance of discriminative bids. This variance widens as the proportion of rejected bids increases [Smith, 1967]. With competitive rules, profit is independent of bid price; hence, there is incentive to bid higher to assure acceptance with no penalty in the form of higher purchase cost when the bid is above the lowest accepted bid [Smith, 1974].

Smith presents the following experimental results [Smith, 1982]:

Proposition 1: "When all individual values are identical and based on a single draw from a rectangular distribution (made after all bids have been entered) the following results are obtained:

(a) If $F^t C(p)$ and $F^t D(p)$ are the proportions of accepted bids specifying a price of $p$ or higher in auction period $t$ under competitive and discriminative auction rules, respectively, the $F^t C(p) \geq F^t D(p)$ for all $t$, that is, within the acceptance sets, bids in competitive auctions are at prices at least as high as those in discriminative auctions.

(b) Seller revenue in the final ("equilibrium") auction in a sequence is greater in competitive than in discriminative auctions in eight of 14 paired experiments."

Proposition 2: "When aggregate induced demand is linear and fixed, but individual private assignments are random (i.e., the assignments are without replacement) and are made prior to the submission of bids, the bids satisfy Proposition 1(a). However, if the slope of the linear induced demand is sufficiently low (i.e., steep) seller revenue is greater in discriminative than in competitive auctions. If the slope of induced demand is increased, seller revenue becomes smaller in discriminative than in competitive auctions."

*Posted-offer vs. double auction*

As discussed earlier, two important characteristics of double auctions are their "rapid" convergence to the theoretical CE and their extremely efficient allocation of resources. Coursey and Smith [Coursey, 1983] and Ketcham et al. [Ketcham, 19] experimentally compared these properties of double auctions with those of posted-offer institutions.

To facilitate comparison to the double auction, Ketcham et al. allowed sellers, as well as buyers, in the posted-offer market to see all prices offered in both the current and previous trading period. Concerning signaling and tacit collusion, the following results are presented concerning double auctions [Ketcham, 19]:

- The signaling opportunity is the same for buyers and sellers—buyers are free to signal with low bids and sellers, with high offers

- Opportunity cost of signaling is insignificant since a bid or an offer is for a single unit and can be revised if not accepted

- It is not clear if signaling is effective in yielding collusive outcomes for either buyers or sellers except under monopolies, where buyers signal vigorously and are effective in restraining monopolist power

Posted-offer pricing results are as follows [Ketcham, 19]:

- Buyers have no opportunity to signal

- Seller can signal by increasing posted price

- Since only sellers can signal, there is the possibility of tacit collusion among sellers to coordinate an increase in price

- Since a posted price cannot be revoked during a trading period (this makes signaling costly), the incentive of the individual seller to signal is reduced and each seller wants other sellers to signal
- The costliness of signaling makes it more meaningful when it does occur and increases possible effectiveness

Price signaling, used for the purpose of tacit collusion, is more common among sellers in posted-offer markets than in double auctions markets. In addition, prices tend to be higher and efficiency lower in posted-offer markets compared to double auction markets.

Coursey and Smith [Coursey, 1983] compare the effect of price controls on the two market institutions. They find that "the dynamic effect of price ceilings in posted-offer markets is qualitatively the same as in double auction markets. In both markets the effect of a ceiling is to cause convergence from below to the constrained or unconstrained competitive price. However, this downward shift in the price convergence path is much more pronounced in double auction markets." They also report that removing binding or nonbinding price ceilings in posted-offer markets produces the jump discontinuity observed in double auctions. However, with the double auction, a positive relationship can be identified between the size of the initial jump and the degree to which the original ceiling is binding. This cannot be done with posted-offer markets.

## 2.5 Auctions in Industrial Markets

### *2.5.1 Airline Industry*

**Allocation of airport slots**

An airline's demand for a takeoff slot at an originating airport is dependent on its demand for a landing slot at the destination airport. To achieve economic efficiency, a procedure is required that allocates individual slots to those flights with the greatest demand.

Rassenti et al. [Rassenti, 1982] report that Grether, Isaac, and Plott (hereafter, GIP) [Grether, 1981; Grether, 1979] have presented a procedure for achieving this goal based on the growing body of experimental evidence on competitive sealed-bid auctions and oral double auctions. The GIP approach uses an independent primary market for slots at each airport organized as a sealed-bid competitive auction occurring at timely intervals. Since this primary market allocation does not provide for slot demand interdependence, a computerized oral double auction is proposed by GIP as an "after market" to allow airlines to buy and sell slots to each other.

Rassenti et al. [Rassenti, 1982] present two difficulties with the GIP method:

- Individual airlines may experience capital losses and gains in the process of trading airport slots in the after market

- It costs resources to trade in the after market

Ideally, a primary market would optimally allocate slots, and the after market would make marginal corrections in primary market misallocations. It would also base slot allocation adjustments on new information. Rassenti et al. [Rassenti, 1982] propose designing a "combinatorial" sealed-bid auction to serve as a primary market for slot allocation in flight-compatible packages for which airlines would submit package bids.

**Auction optimization mechanism**

To decrease reliance on an after market and to achieve greater efficiency than the GIP procedure, the following model was developed for use in a computer-assisted, primary sealed-bid auction market [Rassenti, 1982]:

- "Direct maximization of system surplus in the criterion function

- Airport coordination through consideration of resource demands in logically packaged sets

- Scheduling flexibility through contingency bids on the part of airlines"

An integer programming problem was presented.

$(P)$ Maximize $\Sigma\, c_j x_j$ (2.3)

Subject to: $\Sigma\, a_{ij} x_j \leq b_i \quad \forall\, i$ (2.4)

$\Sigma\, d_{kj} x_j \leq e_k \quad \forall \Sigma\, k$ (2.5)

$x_j \subset \{0,1\};$ (2.6)

where

$i = 1, \ldots, m$ subscripts a resource (some slot at some airport)
$j = 1, \ldots, n$ subscripts a package (set of slots) valuable to some airline
$k = 1, \ldots, l$ subscripts some logical constraint imposed on a set of packages by some airline
$a_{ij} = 1$ if package $j$ includes slot $i$, 0 otherwise
$d_{kj} = 1$ if package $j$ is in logical constraint $k$, 0 otherwise
$e_k$ = some integer $\geq 1$
$c_j$ = the bid for package $j$ by some airline

Contingency bids expressed in the set of logical constraints have two format types:

- "Accept no more than *p* of the following *q* packages;"
  constraint form: $x_a + x_b \leq 1$

- "Accept package *V* only if package *W* is accepted;"
  constraint form: $x_a + x_{ab} \leq 1$
  ($x_{ab}$ is the result of package *ab* with $c_{ab} = c_a + c_b$).

Although *P* is solvable and ensures efficient primary allocations, several questions remain:

- How should bidding airlines be induced to reveal their true values?

- How should allocated slots be priced?

- How should income be divided among participating airports?

Rassenti et al. [Rassenti, 1982] suggest the following procedure:

- Determine marginal (shadow) prices for each airport slot offered
- Charge an airline, whose package j was accepted by P, a price for j equal to the sum of the marginal prices for the slots in package j. (GIP [Grether, 1979] have shown that this provides a uniform price feature that demonstrates good demand revelation behavior in single commodity experiments)
- Return to the airport whose slot i was included in package j an amount equal to the marginal price of *I*

Since *P* is an integer program, the LaGrangian multipliers may not exist. Thus, there may not be a set of prices to support the optimal division of packages into accepted and rejected categories. To solve this problem the following two pseudo-dual programs to *P* were developed to define bid rejection prices (problem $D_R$) and acceptance prices (problem $D_A$) [Rassenti, 1982]:

$$(D_R)\ \text{Minimize} \quad \Sigma\, y_r \tag{2.7}$$

$$\text{Subject to} \quad \Sigma\, w_i\, a_{ij} \leq c_j \quad \forall\ j \ldots A \tag{2.8}$$

$$y_r \geq c_r - \Sigma w_i\, a_{ir} \quad \forall\ r \ldots R \tag{2.9}$$

$$y_r \geq 0, \quad w_i \geq 0 \tag{2.10}$$

where

the optimal solution to *P* is $\{x_j^*\}$

the set of accepted packages is $A = \{j|x_j^* = 1\}$;
the set of rejected packages is $R = \{r|x_r^* = 0\}$;
the set of lower bound slot prices (prices charged) to be determined is $\{w_i^*\}$;
the amount by which a rejected bid exceeds the market price (if at all) is $y_R$
$D_A$ is the complement of $D_R$ with respect to the accept-reject dichotomy.

$$(D_A): \quad \text{Minimize} \quad \Sigma y_j \tag{2.11}$$

$$\text{Subject to} \quad \Sigma v_i a_{ir} \leq c_r \quad \forall \; r \subset R \tag{2.12}$$

$$y_j \geq \Sigma v_i a_{ij} - c_j \; \forall \; j \subset A \tag{2.13}$$

$$y_j \geq 0, \quad v_i \geq 0 \tag{2.14}$$

where the set of upper bound slot prices to be determined is $\{v_i^*\}$, and the amount by which an accepted bid is below the upper bound slot prices (if at all) is $y_j$.

The bids are categorized as follows:

- A bid greater than the sum of its component values in the set $\{v_i^*\}$ is accepted
- A bid less than the sum of its component values in the set $\{w_i^*\}$ is rejected
- "All bids in between were in a region where acceptance or rejection might be considered independent of relative marginal value and determined by the integer constraints on efficient resource utilization [Rassenti, 1982]." These bids correspond to the difficulty with integer programming problem $P$ but are a small percentage of the bids and are known to decrease as problem size increases

**Experimental results**

Comparison of the GIP procedure and the mechanism presented by Rassenti et al. (hereafter, the RSB mechanism) [Rassenti, 1982] results in the following conclusions:

- The RSB mechanism is generally more efficient
- TheRSB mechanism does not depend as strongly on the secondary market
- Market experience is significant in determining efficiency for either mechanism, however RSB requires less training to achieve high efficiency

- The high efficiency of the RSB allocation makes speculation in the after market very risky

- Combinatorial complexity lowers GIP efficiency, while the efficiency of RSB is at least maintained and may improve

This combinatorial auction approach allows consumers to define the commodity by tendering bids for alternative packages; it eliminates the need for producers to anticipate the commodity packages valued most highly in the market. In addition, Pareto optimality of packages produced and resources allocated is guaranteed if bids represent demand and if income effects[6] can be ignored. Finally, experimental results show the mechanism is operational and demand under-revelation is minimal.

### *2.5.2 Natural Gas Industry*

**Market institutions**

Rassenti et al. [Rassenti, 1982] considered a number of different market institutions for production and exchange of natural gas (NG):

1. Bargaining environment
   With this approach, producers, pipeline transporters, and buyers negotiate and strike bilateral or multilateral bargains. These bargains may be short-term or long-term contracts. This approach affords buyers the opportunity to reduce uncertainty by using a long-term contract to lock in deliveries at a fixed price. It also unbundles interstate pipeline's transportation services from their storage and brokerage functions. In addition, FERC is friendly to bargaining. However, there are some potential problems. These problems include large information requirements for participants and the need to coordinate the purchase of gas at a field with transportation.

2. Posted-offer pricing
   Pipelines might post prices for transportation with sales subject to capacity limits. Buyers and producers then negotiate contracts directly. Buyers who have acquired natural gas would be randomly queued and buyers at the head of the queue could purchase transportation service from pipelines at a posted price.
   
   Two problems with posted-offer pricing for a natural gas pipeline network are that (1) many pipeline segments are monopolies or tight oligopolies with considerable market power and (2) entry into pipeline markets takes time. The small number of buyers at most nodes

[6] "The *income effect* associated with a price change describes how a consumer's utility-maximizing choice changes as a result only of the change in the feasible set. An increase in price reduces the feasible set, while a reduction in price increases it [Binger, 1988, p. 160]."

usually allows buyers to exercise countervailing power, but posted-offer pricing takes this countervailing power away in contrast to bargaining or auction institutions.

3. Multiple decentralized markets
   Markets can be established for each of the following:

   - Gas produced at each field
   - Gas delivered at each consumption node
   - Transportation services over each pipeline segment

4. Computer-assisted markets
   McCabe et al. [McCabe, 1990] apply a uniform-price double auction called gas auction (GA), which simultaneously allocates natural gas and pipeline capacity rights among buyers, sellers and transporters of wellhead gas. In another publication, McCabe et al. [McCabe, 1989] apply GA to a more practical pipeline network than in [McCabe, 1989]. This version of GA is called gas auction net (GAN). GAN assures that buyers only pay for deliverable gas. This competitive auction economizes information and requires fewer trades than the multiple market mechanism.

**Gas auction net (GAN)**

The purpose of implementing the GAN [McCabe, 1989] mechanism was to evaluate the price and performance characteristics of a sealed-bid/offer auction mechanism for the simultaneous allocation of natural gas and pipeline capacity rights among buyers, sellers, and transporters of wellhead gas. This auction mechanism is an extension of the competitive sealed-bid auction and the double sealed-bid auction. It fills all accepted bids to buy at a price less than or equal to the lowest accepted bid price (this reduces incentive for strategic underbidding). This type of mechanism is justified on the basis that it has received wide acceptance as a price allocation mechanism in financial markets.

The idea of an auction mechanism for a gas network was a result of the deregulation of the natural gas network, which allowed the substitution of a self-regulated market mechanism in place of regulatory constraints. An auction mechanism avoids price-discrimination since prices are justified on the basis of marginal and opportunity costs. It also provides incentives for increases in production and transportation capacity where such increases have the highest value. An auction mechanism also promotes decentralization, which allows participants to concentrate on judgments and actions based on private information. In addition, the auction is applied to the available capacities above flows already precommitted by contract. Finally, this approach implements priority responsive pricing for curtailments. If pipeline capacity is restricted, deliveries to and from wholesalers who have highest consumption or production priorities are automatically curtailed.

Some of the salient features of the GAN mechanism are summarized as follows [McCabe, 1989]:

- Consumption centers are connected to gas producing fields by a capacity constrained pipeline network

- The auction market uses "smart" computer support to process location specific bid schedules, therefore participants only need to consider their own private circumstances

- Resulting prices are non discriminatory—all sellers at a location receive the same price and all buyers at a location pay the same price

- The difference in price between any two locations reflects differences in the marginal supply price of transportation and/or pipeline capacity limitations

- The GAN auction mechanism maximizes total system aggregate gains from exchange based on bids and offers

The authors evaluate the price and efficiency characteristics of GAN as a mechanism for the exchange of rights. The rights traded may be spot gas commitments or long-term firm or interruptible commitments.

The linear program (LP) below is used to implement GAN. The LP yields optimal flows and shadow (marginal) prices for all nodes in a network. Based on the submitted bids and offers, the total surplus of the gas network is maximized.

$$\text{Maximize} \quad -\sum_{i} c_i f_i \quad \text{(total surplus)} \tag{2.15}$$

$$\text{subject to} \quad \sum_{i \in E_j} f_i - \sum_{k \in S_j} f_k = 0 \quad (\forall \text{ nodes } j) \tag{2.16}$$

$$l_i \le f_i \le u_i \quad (\forall \text{ arcs } i) \tag{2.17}$$

where

$f_i$ = flow on arc $i$
$E_j$ = set of arcs which end at node $j$
$S_j$ = set of arcs which begin at node $j$
$l_i$ = least permissible flow on arc $i$
$u_i$ = greatest permissible flow on arc $i$
$c_i$ = offer price (bids are signed negative) per unit flow on arc $i$
Note, each arc of the logical auction network represents a bid.

The implementation of the above LP problem requires an artificial node that allows physical flow to be balanced with payment flow in the network. It is also used to input the bids and offers of buyers, transporters, and producers. This artificial node can be thought of as completing a circular flow system, which allows a return flow of monetary payments from buyers to producers in exchange for gas flowing on the actual system nodes from producers to buyers.

GAN appears to discipline the behavior of the buying, selling, and transporting agents. Nothing inherently monopolistic is found about pipelines except in those parts of the network served by only one pipeline.

**Real-time implementation of GAN**

GAN is modified to a continuous real-time implementation in the paper by Rassenti et al. [Rassenti, 1992], which considers cotenancy (i.e., the sharing of pipeline capacity rights among more than one agent) and competition for NG pipeline networks. The experimental characteristics are as follows:

- A location specific, two-step, nonincreasing price-quantity bid schedule for each buyer can be entered at any time during the auction period. Bid price-quantity can be raised but not lowered

- A location specific, two-step, nondecreasing price-quantity offer schedule for each producer can be entered at any time during the auction period. Offer prices can be lowered and quantities may be increased but not vice-versa

- Pipeline owners submit arc specific, two-step nondecreasing price-quantity offer schedules for each segment of transportation capacity. Offers can be improved by increasing quantity and decreasing price. Pipeline owners may only withhold peaking capacity not normal capacity. Each segment is subject to a ceiling price

- A linear program optimally matches bids and offers for delivered gas. Nondiscriminatory prices at each node in the network and gas flows on all arcs are updated with each new bid or offer. The real-time iterative process gives subjects more information about prices and yields faster convergence

- Price allocation generalization of double auction is a quasi-tatonnement (tentative contract) process since allocations and prices are binding only at the end of each auction period, i.e., bids/offers cannot be withdrawn, but they may be improved or "bested" by someone else. Bids not initially accepted may be accepted later

- Buyer (seller) can consummate a trade at a node by bidding higher (lower) than the node price. The only difference from oral double auction in stock or commodity exchange trading is that trades are not firm until auction period ends

- At the end of the trading period, allocations and prices are set and a new auction period immediately begins

The experimental network used by Rassenti et al. [Rassenti, 1992] contains single monopoly pipelines between nodes. However, the buyers and producers have alternative routes. The experiments were designed so that pipeline owners have less than one-half of production capacity in each of the producing fields. The capacity of the network is also constrained. This sparse network, loaded near capacity, favors pipelines in the competitive auction process. Therefore, this experiment represents an unfavorable scenario and a demanding test for the market exchange mechanism.

Rassenti et al. ran experiments with and without cotenancy. Cotenancy occurs whenever two or more firms undertake a joint investment venture to construct a new pipeline, or when a pipeline user obtains rights from the original owner. In the experiments, all cotenants are free to submit price schedules of the transportation capacity for which they have user rights in excess of their own spot needs. Cotenants may be producers, consumers, or pipeline owners. The authors believe that cotenancy is a "promising vehicle by which the government might institute a competitive self-regulating property right system to replace rate-of-return regulation of a monopoly owned capital facility [Rassenti, 1992]."

Smith [Smith, 1988] proposes a set of competitive cotenancy rules:

- Each of several co-owners would acquire capacity rights to the facility in proportion to his contribution to fixed costs

- These rights are freely transferable (sold, leased or rented) subject only to antitrust limitations on ownership concentration

- Each co-owner pays his agreed share of any variable costs up to his percentage of ownership whenever his rights are exercised

- The facility is managed as a cost center by a separate operating company

- Any co-owner or outsider can increase his share of capacity utilization rights by a unilateral expansion of capacity

Cotenancy is a common property right arrangement for capturing *economies of scale*[7] while avoiding unnecessary duplication costs. Electric generation and large transmission lines often have two to seven cotenants.

Experimental results [Rassenti, 1992] show that cotenancy increases total market efficiency, increases the buyers' and producers' share of surplus, and lowers the pipelines' share of surplus. The pricing results suggest that competition via cotenancy is a viable way to limit the market power of monopoly pipeline segments. Even in experiments without cotenancy, the GAN mechanism achieved efficiencies converging to 85% of total surplus. GAN also provides signals, via node prices, that indicate the profitability of expansion. However, whether appropriate social incentives are provided is an open question.

### *2.5.2 Electric Power Industry*

McCabe et al. [McCabe, 1991] discuss two concerns for restructuring a nationally networked industry:

- Reported characteristics (bids/offers) differ from true system characteristics. Lab research has shown that although submitted information may be untrue, appropriate algorithms can maximize gains relative to true characteristics

- Treatment of natural monopolies with cotenancy so that a thinly connected network can be deregulated

It has been conceded that electric generation can be partly deregulated because of IPPs and cogeneration but that system stability cannot be left to market forces. Such arguments do not consider that every market has elements in the "property" right arrangements (rights to act) that allow a market to do its job [McCabe, 1991].

Generators can be privately owned. Phase compatibility of generators can be governed by existing contracts. Generator owners may submit a supply schedule and supply price for spinning reserve to a coordination center. Buyers may submit bids for power delivered to local buses. The dispatch center uses a mixed integer nonlinear program to maximize system surplus subject to system integrity and stability [McCabe, 1991].

There are three differences with current procedures [McCabe, 1991]:

- Dispatch center treats each generator supply schedule as a marginal cost function. However, a generator owner may be willing to operate at less

[7] "A particular production process exhibits *economies of scale* if the minimum unit cost associated with using that particular production process can only be achieved if a substantial number of units of a good are produced [Binger, 1988, p.285]."

than marginal cost during low demand periods to avoid shutdown and startup costs

- Generator owners recover fixed cost from lump sum offers for spinning reserve. (This is currently recovered by regulation.) The authors propose that "system" is the marginal supply of the most expensive generation required by the optimization routine. Then "system" is adjusted for incremental transmission loss (ITL) to determine prices for each node *k* in the network. Thus, the resulting price is $(1\text{-}ITL_k)$, where $ITL_k$ is the increase in transmission loss that occurs if one additional unit of power was injected at node *k*

- The marginal value of wholesale power to local distribution companies and private commercial/industrial users is determined by bids. Buyers with cogeneration or price-sensitive demand for power reflect these alternative opportunity costs in bids and will cycle the system on or off depending on optimal prices and allocations

Transmission prices are determined by the difference between the spot output and input prices of each line. For lines loaded below capacity, the power from node *i* to node *j* implies price $P_{ij} = (ITL_i - ITL_j)$ . When the flow on a line exceeds the reliability capacity limit, $P_{ij}$ exceeds marginal value of the power loss on the line.

Implementation of auctions in the electric power industry would require generation and transmission players to submit offers that reflect system operating conditions. For example, generation owners must submit offers that reflect ramp times, spinning reserve, maintenance costs, etc. Transmission owners must submit offers that reflect current system conditions, equipment limits, security constraints, etc. In addition, demand centers must submit bids that represent their valuation of requested demand. Finally, the auction center must only assign contracts that do not violate system security and reliability. Auction markets may be applied for all demand in the power system, or they may be applied after native load customers are supplied power by their respective utilities.

A concern for any market player is that his accepted bid (offer) may be much higher (lower) than competitors. Thus, winning a contract at a bid (offer) price much higher (lower) than the other submitted bids (offers) is called the *winner's curse* since possible profits are lost. However, in a market where valuation of an item varies for different players, the *winner's curse* need not be incorporated in the information structure [Philips, 1988, p.107]. Therefore, since the value of a unit of electric power is not the same to all market players in the industry, the *winner's curse* need not be considered. Players who submit bids (offers) that are too high (low) are responsible for their actions. However, losses incurred by such high (low) bids (offers) can be minimized with the addition of price ceilings (floors). If these ceilings (floors) are nonbinding and implemented in a double auction mechanism, CE convergence is guaranteed [Smith, 1982].

With consideration of the auction mechanisms already used in industry and with regard to the experimental results of various auction mechanisms, I believe the double auction, in a sealed-bid format, is the best for pricing electric power. First, the double auction allows all participants to submit their respective values for requested or provided services. Second, the double auction has been shown to converge quickly to competitive equilibrium prices. Third, the double auction does not require complete information of the supply and demand conditions for CE convergence. Finally, sellers with "monopolistic power" do not achieve monopolistic profits.

## 2.6 Overview of Brokerage/auction Systems

### *2.6.1 Auction as a Market Institution*

An auction market can be considered a trading institution where buyers and sellers can readily meet to maximize their trade gains. McAfee and McMilan [McAfee, 1987] define auction as "a market institution with an explicit set of rules determining resource allocation and prices on the basis of bids from market participants." In standard auction institutions such as Chicago Board of Trade (CBOT) [CBOT, 1996] and NewYork Mercantile Exchange (NYMEX) [NYMEX, 1995], all trade units are standardized. The only component of a trade unit that varies is the price. This removes all the informational asymmetries prevailing in the market institution on the trading floor. The market participants efficiently decide transactions on the basis of prices only. In another words, an auction system is a very efficient way to move from a cost-based operation to price-based operation. In this context, the study of auction institutions and their applications to the electric power system in the proposed deregulated environment becomes very significant.

Post and Sheblé [Sheblé, 1993; Post, 1994] present a detailed study of auction institutions. Post [Post, 1994] has described four standard types of auction institution: (1) English; (2) Dutch; (3) first-price sealed-bid; and (4) second-price sealed-bid. These auction mechanisms employ different methodologies of trading. In the English auction, the auction bids begin with a low price. The bids are progressively announced until no purchaser wishes to make a higher bid. The auction result is pareto-optimal since the winner is the bidder who values the trade unit the most [Henderson, 1980]. In the Dutch auction, the auctioneer calls a decreasing set of bids beginning with a high price. The bidding proceeds until one bidder accepts the current price. Thus, the Dutch auction is a game that rewards the player who wishes to maximize his expected gain (not the gain itself). In the first-price sealed-bid auction, buyers submit sealed bids. The highest bidder is awarded the item for the price he bid. The second-price sealed-bid auction is based on the premise that the highest bidder wins the item but pays a price equal to the second highest bid. Buyers submit sealed bids similar to the first-price sealed-bid auction, but with the above information at hand. Both the first-price sealed-bid auction and the second-price sealed-bid auction maximize the trade gains of the market participants.

There exists numerous variations of the four standard auction institutions, including multiple auctions [Vickery, 1961], multiple sales by sealed bids [Wilson, 1985], and double auctions [Smith, 1984, 1982]. Smith [Smith, 1974] presents a slight variation of the first-price sealed-bid auction called a discriminative sealed bid auction. In this case, the sale quantity is fixed at a specified amount. Buyers are then invited to tender bids at a stated price for a stated quantity. Bids are accepted from highest to lowest until the specified amount of bid units is exhausted. Ties at the lowest bids are accepted on a random basis. Smith also presents a variant of the second-price sealed-bid auction for a homogeneous commodity. This variant is called the competitive sealed-bid auction which is the same as the discriminative sealed bid auction except that all bids are filled at the price of the lowest accepted bid [Smith, 1974].

Of the various auction institutions, double auction and sealed-bid methods appear to be operationally and structurally suitable for the deregulated electric industry. In the double auction institution, buyers and sellers submit bids and offers. After the bids have been placed, the broker determines buyers and sellers by what is called the high-low algorithm [Smith, 1967]. The highest buy bid is matched with the lowest sell bid. The procedure continues with the next highest buy bid and the next lowest sell bid, and finishes when the highest remaining buy bid is lower than the next lowest sell bid. If a proposed match violates an operational constraint, it is omitted and the next match is determined.

Although the theory of double auctions is not well developed, experimental results for double auctions have shed considerable light on the subject. Smith has presented a number of propositions based on his experimental results [Smith, 1974, 1982]. The experiments have been conducted by implementing the use of computers for selection of optimal contracts [Smith, 1965]. This research is performed in a laboratory setting with individuals acting as buyers, sellers, or both. Participants enter their bid (offer) quotes into a computer terminal. A central computer then applies an optimization algorithm to determine the prices and allocations that maximize the gains on the basis of the bids (offers), and budget and capacity constraints of the individual market participants. These experimental market mechanisms produce strong equilibrating tendencies, even when rental rewards to buyers and sellers are not balanced.

### *2.6.2 Auctions in Industrial Markets*

In the past, many other regulated industries applied some form of auction methods to move from regulated rate of return pricing to market-based pricing. Some examples are the natural gas industry and the airline industry. Post [1994] has described the market institutions and auction optimization methods used in these industries. Most of these methods are based on competitive sealed-bid auctions and double oral auctions. Smith [1974] presents a detailed description of implementational issues involved with these methods.

The case of the natural gas industry is of particular interest since the electric and gas industries have some similar operational characteristics. Natural gas flows from wells located in distant, producing fields through pipelines to users. Interstate pipelines end at state borders or at gateways to urban markets, where gas is transferred to a distribution system for delivery to consumers. Thus, field wells, interstate pipelines, and gas distribution systems are structurally and operationally similar to the concept of GENCO, TRANSCO and DISCO respectively in the electric power system. The electric utility industry seems to follow the natural gas industry in deregulation activities and is expected to continue to do so.

A series of regulatory crises has forced deregulation in the natural gas industry: well-head prices, gas contracts, and pipeline transportation. As deregulation proceeded, the FERC came up with "open access" ruling for interstate pipelines to facilitate the implementation of gas contracts. The hypothesis, "Interstate natural gas pipelines are natural monopolies and hence they cannot be competitively organized," has been invalidated by the emergence of a competitive gas market. Competitive prices of natural gas have moved together within a band related to transportation costs, so that price differences within bands are not so large that a profit can be made by arbitrage. Also, the price differences have narrowed overtime and eventually have become correlated. The initial narrowing and eventual correlation are one of the significant properties of competitive markets. The natural gas market has shown that monopoly power of the pipelines can be made nonexistent by making transmission an asset that can be traded in a market open to producers, distributors, customers, brokers, and others.

Rassenti [1982] presents a number of different market institutions for production and exchange of natural gas, include bargaining environment, posted-offer pricing, multiple decentralized markets, and computer-assisted markets. Many of these institutions are based on the premise of sealed-bid and double auction mechanisms. McCabe [1990] has developed a computer-assisted market called gas auction network (GAN). The purpose of GAN is to evaluate the price and performance characteristics of a sealed-bid auction mechanism for the simultaneous allocation of natural gas and pipeline capacity rights among buyers, sellers, and transporters of wellhead gas. The experimental results with GAN mechanism have shown that the nodal prices tend to converge towards market equilibrium.

Discussion of operational and implementation issues concerning the various institutions in the gas industry provides good insight on how auctions can be applied to the electric utility industry. However, it is important to note that the technical constraints network (active and reactive power) flows and security constraints and nature of commodity (electric energy and ancillary services) of the utility industry are different from those in natural gas industry. Unlike natural gas, electricity cannot be stored in "energy tanks." Research into storage devices for huge amounts of electricity is still in its infancy [Bentley, 1987]. Hence, generation of electricity requires immediate consumption. The instant generation and instant distribution must be coordinated properly by the transmission system. A secure transmission of electrical power requires a number of ancillary services as described in Chapter 1. Additionally, the transmission path for a given transaction can not be chosen apriori. The path of power flow is governed by Kirchoff's laws. The path followed by energy can cause problems when wheeling power across intermediate systems resulting in voltage dipping, reactive power flow, increased losses, reliability problems, etc.

In short, all the operational issues, such as transmission access [Vojdani, 1995], reliability standards [Felak, 1992], etc. must be re-examined in the deregulated environment. In general, administration of a complete auction institution to electric power industry requires a thorough analysis in terms of mathematical framework and implementation issues. References [Sheblé, 1996; O'Neal, 1994, 1993; Wu, 1995a, 1995b] describe such implementation and technical issues in great details.

### *2.6.3 Examples of Auction Systems for Electric Energy*

The main stated purpose of the new FERC proposed rulings (open access, comparability, ancillary services, etc.) is to reduce the cost of electricity. Previous attempts by electric utilities to solve such economic problems led to the formation of power pools based on cost-based operation. Energy brokerage/auction systems is a way to reduce the operational costs in price-based framework. The function of an energy auction system is to establish multilateral transactions among participating market players in such a manner as to maximize the total trade gain. Research study has shown that the multilateral power transactions cross subsidize bilateral transactions [Wu, 1995] and hence, they are required to achieve efficiency. Auction system is an efficient way to establish multilateral contracts. Some of the real-life examples of power brokerage systems have shown significant advantages of power brokerage pools, compared to the traditionally integrated pools. This subsection presents two such real-life examples.

**2.6.3.1 Florida Power Brokerage System.** The most prominent example of power brokerage system in the United States is the Florida brokerage system [Oren, 1994; RPAI, 1979]. The goal of the Florida brokerage system was to encourage short term transfer of electric power to reduce the aggregate cost of generation. The broker of the Florida system used to ask for buy and sell quotes on an hourly basis. The hourly quotes were matched to set up bilateral contracts. A research study on the Florida brokerage system [FEPCG, 1980] indicated that the net savings in the brokerage system were almost twice that of a centrally operated pool system. The total operation costs in a brokerage type pool system turned out to be less than the centrally operated pool system. This was because the utilities retained the responsibilities of local operational decisions, such as unit commitment, fuel scheduling, etc. However, the Florida brokerage system did not survive because the market was too small to provide enough trading opportunities.

**2.6.3.2 England and Wales System in the UK.** A recent example of bidding arrangement for electric power interchange is the England and Wales system in the United Kingdom. The UK electric industry is comprised of 12 regional electricity companies (RECs) and one national grid corporation (NGC). The main activities of the RECs are distribution and supply. The NGC has all transmission assets plus 2 GW of pumped hydro facility. The generation companies make offers to the NGC. Based on these offers and its own estimate of demand, the NGC produces a "unconstrained schedule" that is used to develop the pool price for every half-hour. Actual operation is also scheduled by the NGC. Differences between actual operation and planned unconstrained schedule arise due to error in demand estimate, forced outages, and transmission related constraints. The unscheduled units are called into operation to meet the shortfall of generation and are paid their offer prices. Generating units which run in the unconstrained schedule but not in the event (spinning reserves) are paid the difference between the pool purchase price and the offer price. The pool-selling price is calculated by adding the ancillary services cost to pool purchase price.

The England and Wales power grid is the largest competitive electricity market in the world. Hence, the experiences of the UK electric market should be carefully examined to learn lessons. Although the objectives of generating companies (to maximize profit without regard for the effect of their actions on system security) differ from those of the NGC (i.e. to maintain secure and economic operation of the system), the arrangements are working. This is because there is much commonalty of purpose in practice. The market has a unique feature of determining security costs after the fact. However, exceeding transmission constraints prohibits pool prices from accounting for the true operational cost. References [Russel, 1992, 1995; Hillier, 1990] describe operational and planning issues and recent experiences with independent generators of the UK and the NGC in details.

### *2.6.4 Auction Optimization Mechanisms for Electric Energy*

Various optimization schemes have been applied to energy auction mechanisms. Reference [Doty, 1982] summarizes the application of a number of optimization techniques, such as high-low matching algorithm, network flow algorithm dynamic programming, etc. to energy auction system. These mechanisms are based on different assumptions. References [Fahd, 1991, 1992] describe the implementation of energy brokerage system using linear programming. The data required for an elementary interchange brokerage system are the amount of power available for trading, and the buy and sell quotes. Economic dispatch has been considered the source for this data in [Fahd, 1991]. [Fahd, 1992] computes the data with network constraints to provide a complete analysis of the power system. Implementation of a brokerage system using augmented LaGrangian technique has also been investigated [Anwar, 1995]. [Post, 1995] shows how a sequential sealed-bid/sealed-offer auction mechanism could be applied to the pricing of electric power. Interchange of electric power by using a double auction mechanism has been discussed in [Coppinger, 1995]. [Chattopadhyay, 1995] presents an energy brokerage system with emission trading and allocation of cost savings.

## 2.7 Schweppe's Theory of Spot Pricing [Caramanis, 1982]

Schweppe introduced the concept of optimal spot pricing in a vertically integrated industry. The proposed theory took the perspective of a global controller who wished to maximize the social welfare function by adjusting the level of each generating unit and the usage level of each consumer device. The optimal spot price denoted the summation of marginal fuel cost, energy balance, quality of supply premium, and transmission network quality of supply premium. Schweppe developed a set of rates related to spot prices and discussed their applicability in view of different customer characteristics. The proposed rates included cost of rationing, cost of equipment, and value of electricity usage in addition to the variable and fuel costs. Thus, the developed foundation was a cost-based approach for a power system operation. The introduced notion of component-based costing presented a number of issues related to customer response and utility revenues. The proposed spot pricing approach highlighted a number of issues that should also be carefully examined in a deregulated power system operation.

# 3 AUCTION AS LINEAR PROGRAM

*Even benevolent dictators charge too much.*

This chapter demonstrates how to formulate an auction an optimization problem. The main methods of solution are Linear Programming, Linear Programming Assignment Problem, Linear Programming Decomposition, LaGrangian Relaxation, and price decomposition.

The intent of this work is to demonstrate that auctions are optimization techniques of a distributed nature. They are distributed since the information is not centrally maintained as with most optimization algorithms as described academically. Auction methods are not centralized, nor are they consistent with regard to sub-problem optimization techniques. The auction is the central optimization algorithm. The players bidding process and strategies are the sub-problems that are often solved heuristically in computers external to the central computer or the human ring in a commodities exchange. The resulting comparison shows that auctions have been used for optimization for decades.

## 3.1 Linear Programming

The auction can be viewed as an assignment or a transportation problem. I prefer to view auctions as assignment problems. The problem is to assign an offer from a seller to a buyer. The objective is to maximize the satisfaction of both the buyer and the seller. This chapter outlines the formation of an auction as a Linear Programming problem. Then an example auction is solved for the power system cash market (spot) problem. Next, price decomposition is compared to auctions. Then decentralization of optimal allocation of central resources is shown to result in auction-like solutions.

## 3.2 Basic Discrete Auctions

Consider a sealed bid discrete auction where the sellers and buyers do not consider anything but profit when negotiating for a product. Consider that the seller has an amount to sell:

$a_i > 0$ for $i = 1,\dots,m$ sellers.

Also each buyer has an amount to buy:

$b_j > 0$ for $j = 1,\dots,n$ buyers.

Assume that the product is homogeneous for each seller. Also, assume that the product is non-heterogeneous among sellers for electricity. Specifically, buyers can not identify the product from each seller. The quality is similar, etc. Heterogeneous markets include many products in agriculture where the buyers can identify the producer by the product (i.e. cereal). Assume that the buyer ($j^{th}$) is offering to buy a product through a sealed bid auction. Define the sealed bid:

$c_{ij} >= 0$

as the maximum amount the buyer is willing to pay to any seller per unit of product. Heterogeneous products would be indicated by unique offers to each seller. Non-heterogeneous products would be indicated by common offers to all sellers. Assume that the offer is a sealed bid. The seller does not know who is the buyer. The auctions (assignment of buyer to seller) would result in a sale of $x_{ij}$ quantity of product to be assigned from seller "j" to buyer "i" at a price $u_{ij}$ yet to be determined. The value of the exchange is the amount willing to be paid times the amount exchanged. The cost is the price times the amount exchanged. Note that both the amount to be exchanged and the price are non-negative, and hopefully positive, quantities.

The price is to be determined by the rules of the auction mechanism. The value is the upper limit of the price a buyer is willing to pay. The rules of fair auction mechanisms have been defined by many economists. The rules must be fair or equally applied to all players. Items sold must be delivered. Items bought must be paid on delivery even if not used by buyer. Each seller and buyer gets an identical price for all units sold or bought. If the supply is greater than the demand, then the price of excess products is zero. A buyer's bid must be sufficiently high for all demand to be satisfied. Otherwise the surplus between value and price must be zero. The surplus for buyer j is defined as:

$$v_j = c_{ij} - u_i \quad \Rightarrow \quad v_j \geq 0$$

Note that this is a non-negative value as economic reason would dictate.

The Linear Programming form to solve the assignment problem is:

$$Max \quad \sum_{i=1}^{m} \sum_{j=1}^{n} c_{ij} x_{ij}$$

$$\sum_{j=1}^{n} x_{ij} \leq a_i \quad \forall_i$$

$$\sum_{i=1}^{m} x_{ij} \leq b_j \quad \forall_j$$

$$x_{ij} \geq 0 \quad \forall_{i,j}$$

Note that the decision variables are non-negative as required. The first and second constraints require delivery and purchase. However, the objective may not be clear. Why do we want to maximize the bid value? Consider Adam's "invisible hand" that causes the products to go to the highest bidders until the supply is exhausted. The objective function is considered the economic potential function that achieves a maximum value just as supply satisfies demand.

As with all linear programs, there is a dual program. Use *ui* as the dual variable for each inequality constraint for the sellers in the above formulation. Let *vj* be the dual variables associated with each inequality constraint for each buyer in the above formulation. At the optimal solution, the complementary slackness conditions hold for each seller:

$$\sum x_{ij}^* - a_i \leq 0$$

$$u_i^* \geq 0 \quad \forall_i$$

$$u_i^* \left(\sum x_{ij}^* - a_i\right) = 0$$

The first inequality is the same constraint in the LP formulation. This requires the seller to sell no more than the amount available for sale. Also, the price from the second constraint is non-negative and independent of the buyer purchasing the product. The third constraint specifies that the seller will sell all products if the price is positive. It also specifies that the price is zero if all goods are not sold.

At the optimal solution, the complementary slackness conditions hold for each buyer:

$$\sum x_{ij} - b_j \leq 0$$

$$v_j^* \geq 0$$

$$v_j^* \left(\sum x_{ij}^* - b_j\right) = 0$$

The first constraint requires that the buyer does not buy more than the amount ordered. The second constraint specifies that the value is always non-negative. The third constraint specifies that if the value is not zero, then the buyer receives the bid amount. It also specifies that the buyer does not receive all of the bid amount if the value is zero.

The dual problem has to satisfy the dual constraints:

ui + vj >= cij for all i and all j (sellers and buyers)

Since the original decision variables are the dual variables for the dual problem, the following complementary slackness conditions hold at the optimum:

$$u_i^* + v_j^* - c_{ij} \geq 0$$
$$x_{ij}^* \geq 0 \qquad \forall_{i,j}$$
$$x_{ij}^* \left(u_i^* + v_j^* - c_{ij}\right) = 0$$

If the amount assigned from seller to buyer (*xij*) is positive, then the bid price of the seller and the surplus of the buyer equals the bid of the buyer. However, if the buyer's offer price is less than the seller's offer price and the buyer's surplus, then no products are assigned.

These interpretations hold for the sum of all assignments. At the optimal solution, it is required that the dual objective function value and the primal objective function value are identical:

$$\sum \sum c_{ij} x_{ij}^* = \sum \sum u_i^* x_{ij}^* + \sum \sum v_j^* x_{ij}^*$$

This demonstrates how the bid value is allocated amongst the buyers and the sellers. Note that at the optimum solution, all of the value is distributed.

A question to ask is whether the above is "fair" to all buyers and "sellers." The fairness to each seller can be shown. First, the seller cannot sell more than offered. Second, the seller gets the same price for all products. Third, if the product has any positive value, the all units of the product are assigned for the seller. Fourth, all of the units are sold unless the product value goes to zero.

$$\sum x_{ij} \neq a_i$$
$$u_i \forall x_{ij}$$
$$u_i > 0 \Rightarrow x_{ij} = a_i$$
$$x_{ij} < a_i \Rightarrow u_i = 0$$

The fairness to each buy can also be shown. First, each buyer does not buy more than the bid quantity. Second, the buyer receives the same value for all units assigned. Third, all units bid are assigned if the value is positive. Fourth, no units are assigned if the value goes to zero.

$$\sum x_{ij} \neq b_j$$
$$v_j \forall x_{ij}$$
$$v_j > 0 \Rightarrow x_{ij} = b_j$$
$$x_{ij} < b_j \Rightarrow v_j = 0$$

The solution to the above does not yield a unique value at the optimum solution. Indeed, an infinite number of solutions is possible depending on the arbitrary constant added to the lowest dual variable. The choice for this constant depends on whether the auction is established by the sellers or the buyers. It also depends on whether there are other auctions competing with this auction, thus driving the price down. It depends on whether the supply exceeds the demand or the demand exceeds the supply. It depends on the sellers being enabled to specify a reservation price, establishing a lower limit on the market price.

Since sellers have to yield a return on investments, a reservation price is allowed to specify the lowest price for seller. If the market price is above the reservation price, then all units of the product are assigned. If the market price is below the reservation price, then the units of the product are sold to the seller. Of course, the charge for use of the auction mechanism would be included in the seller selling back to itself. This reservation price is interesting because if it is too high, the units will not be assigned from seller to buyer. The market will not clear. That is, the goods will not be assigned. This is an example of price rigidity or market control. The auction mechanism then fails. The seller then has an excess supply that cannot be moved to any buyer.

The formulation of the LP problem to include reservation prices is the same as the above with the following changes to the objective function and the first constraint:

$$Max \quad \ldots + \sum \Pi_i y_i$$
$$\sum x_{ij} + y_i = a_i$$

The objective function represents all assignments, including the assignment from the seller back to the seller at the reservation price, *pi*. The first constraint requires that all of the product units offered by the seller are sold, even if only to the seller. All other equations and interpretations are the same.

At the optimum, the dual constraints are the same, with the addition of the constraint:

$$u_i^* + v_j^* \geq c_{ij} \qquad \forall_{i,j}$$
$$u_i^* \geq \Pi_i \qquad \forall_i$$

This constraint specifies that each seller will receive at least the reservation price for each unit sold. The complementary slackness condition at the optimum holds:

$$\left(u_i^* - \Pi_i\right)y_i^* = 0 \qquad \forall_i$$

If any amount is assigned back to the seller, then the market price is the same as the reservation price. However, if the market price is better than the reservation price, then all units of the product are assigned. Thus, at the optimal solution:

$$y_i^* > 0 \quad \Rightarrow \quad u_i^* = \Pi_i$$
$$u_i^* > 0 \quad \Rightarrow \quad y_i^* = 0$$

Electric power systems, similar to most other real products, require a more complex model. Consider the following extensions of the above to include transportation costs (electric power losses and transmission rates), nonlinear production costs, and nonlinear buyer values.

## 3.3 Degeneracy and Auctions as an Assignment Problem

### *3.3.1 Degeneracy*

Degeneracy in linear programming (LP) occurs when an excess number of hyperplanes pass through an extreme point. In the case of *n* decision variables, degeneracy occurs when more than *n* hyperplanes pass through an extreme point. One cause of excess hyperplanes passing through an extreme point is that of redundant constraints. When linear programs are degenerate, the pivoting process moves from one basis to another basis and both bases represent the same extreme point. If the pivoting occurs in the same sequence over and over again, the solution stays at the same extreme point and the optimal solution cannot be found. This problem is called cycling. Rules designed to prevent cycling and to guarantee finite convergence of the simplex method can be seen in [Bazaraa, 1990].

Degeneracy at the optimal solution vertex in primal problem indicates that there may be multiple optima in the dual problem. Similarly, degeneracy at the optimal solution vertex in dual problem indicates that there are possibly multiple optima in primal problem. Multiple optima occur when hyperplane of the objective function is parallel to one of the hyperplanes that bounds the feasible region. Thus, it can be implied that if an excess number of hyperplanes pass through the optimal extreme point in the primal (dual) problem, the hyperplane of the objective function is probably parallel to one of the hyperplanes that bounds the feasible region in the dual (primal) problem. All the multiple optimal solutions yield the same optimal objective function value. Multiple primal optima and multiple dual optima occur individually or simultaneously. Multiple optima are not desired in an auction because a fair and unique solution is required. The causes of multiple optima and how multiple optima can affect solution fairness and uniqueness are explained separately for primal and dual problems below.

**3.3.1.1 Multi-optima for Primal Problems.** The primal problem in the context of this work is the auction problem. Multiple primal optimal solutions result when the hyperplane of the objective function is parallel to one of the hyperplanes that bounds the feasible region in the primal problem. Multiple primal optima can be observed when at least one of the reduced-cost coefficients (RCCs) of the optimal nonbasic variables is zero. However, a zero RCC value of at least one of the optimal nonbasic variables does not always mean that multiple primal optima will occur. Pivoting must be performed to ascertain whether the new primal optimal solution after pivoting is different from the primal optimal solution before pivoting. The primal simplex method is used for pivoting. Note that pivoting does not change the dual optimal solution because the pivoting is always on the nonbasic variable that has a zero RCC.

In an auction, multiple primal optima indicate that the optimal assignment between sellers and buyers is not unique. A variety of assignments can be specified to the accepted bid of each seller and each buyer and all assignments yield the same social surplus. This is strictly undesirable in auctions because if one of the optimal solutions is chosen, it is unfair to some other unchosen bidders who can provide the same social surplus if they are selected.

**3.3.1.2 Multi-optima for Dual Problems.** Multiple dual optima occur when an excess number of hyperplanes pass through the optimal extreme point in the primal problem. Multiple dual optima can be observed when at least one of the basic variables of the primal problem is zero. However, a zero value of at least one of the optimal basic variables does not always mean that multiple dual optima will occur. Pivoting must be performed to ascertain whether the new optimal dual solution after pivoting is different from the optimal dual solution before pivoting. The dual simplex method is used for pivoting. Note that pivoting does not change the primal optimal solution because the pivoting is always on the basic variable that has zero value.

For auctions, multiple dual optima indicate that the set of shadow prices (dual prices) is not unique. A variety of assignments can be specified to the shadow prices and all assignments yield the same social surplus (objective function). This is strictly undesirable in auctions when the shadow prices are used for pricing. Recently, shadow prices have been proposed for use in pricing transmission congestion. If one set of the shadow prices is chosen to price transmission congestion, it will be advantageous to some of the bidders and disadvantageous to some other bidders. This is because one set of shadow prices can cause a bidder to pay higher transmission congestion fee than when other sets of shadow prices are used. Note that using shadow prices to price transmission services should not be accepted for implementation.

### *3.3.2 Auctions as an Assignment Problem*

An auction can be viewed as the assignment of products from sellers to buyers. This is why it is more appropriate to treat an auction as an assignment problem. The term "assignment problem" used here is in the context of assigning products from sellers to buyers. The term "assignment problem" is different from that used in the context of the minimal cost network flow problem solution in most textbooks [Bazaraa, 1990, Hillier, 1955, Thompson, 1992]. In the context of most textbooks, the assignment problem is a special class of the minimal cost network flow problem since the solution procedure is explained and not the type of application. As a solution procedure, the assignment problem has a particular structure for a special method to solve a unique topographic tableau. The minimal cost network flow problem is a special type of linear programming problem that has unique network structures that severely modify the application of the optimization rules. Such assignment problems are a special type of the transportation problem, which is a special type of the minimal cost network flow problem. The minimal cost network flow, the transportation, and the assignment problems are referred to in this section because they have similar problem structure to that of auction problems. The transportation problem and the assignment problem are separated from the minimal cost network flow problem according to their special structures so that special methods can be applied to solve the problems. The network simplex method has been applied to solve the minimal cost network flow problem and the transportation simplex method is applied to solve the transportation problem [Bazaraa, 1990, Hillier, 1995]. The Hungarian algorithm has been applied to solve the assignment problem [Bazaraa, 1990]. The network simplex method, the transportation simplex method, and the Hungarian algorithm are special versions of the simplex method. For this work, it is not appropriate to focus on these special methods to solve the auction problem. The general auction problem is formulated without regard to special equation structures and thus it can only be solved by general simplex method.

**3.3.2.1 Assignment Problems.** Two major types of products are considered, heterogeneous products and homogeneous products. Homogeneous products are indistinguishable from each other while heterogeneous products can be distinguished by their quality or characteristics. Because heterogeneous products are distinguishable, they have different value to each seller and each buyer while homogeneous products have the same value to each seller and each buyer. Products can be traded through an exchange or traded by individuals via bilateral contracts. Trading through an exchange is more convenient for traders because the exchange gathers different types of products together, which means that traders do not waste time finding the products they desire. In addition, an exchange provides insurance to protect parties from sellers or buyers who default. For example, sellers who do not supply products according to the contracts will be fined through the exchange. Then, the exchange can distribute the money to participants, or use the fine to provide products from other sellers to the buyers for compensation. Regardless of whether products are traded through the exchange, the parties to the transactions remain identifiable.

Trading heterogeneous or homogeneous products as bilateral contract is shown in Figure 3-1. Transaction, $x_{ij}$, is defined as from seller $i$ to buyer $j$. For heterogeneous products, products from different sellers have different unit value to each customer; i.e., $x_{1j}$, $x_{2j}$, $x_{3j}$, ..., $x_{mj}$ have different unit value to buyer $j$. This is the case when products of different sellers can be distinguished by buyers. For example, electricity produced from one GENCO has higher power quality than electricity from other GENCOs. Another example involves a buyer who is concerned about environment values the electricity produced from clean energy more than the electricity produced from the energy that pollutes the environment. Another property of heterogeneous products is that products may vary from seller to seller. The more general case when products from a seller are different is considered. Specifically, each buyer has different unit price from each seller. An interesting example of this case is price discrimination. One prevalent example of price discrimination in electricity is when a seller prices electricity corresponding to a guaranteed level of reliability. For homogeneous products, products from any seller have the same unit value to each customer. In other words, $x_{ij}$ is the same for all $i$ from $1$ to $m$ and all $j$ from $1$ to $n$.

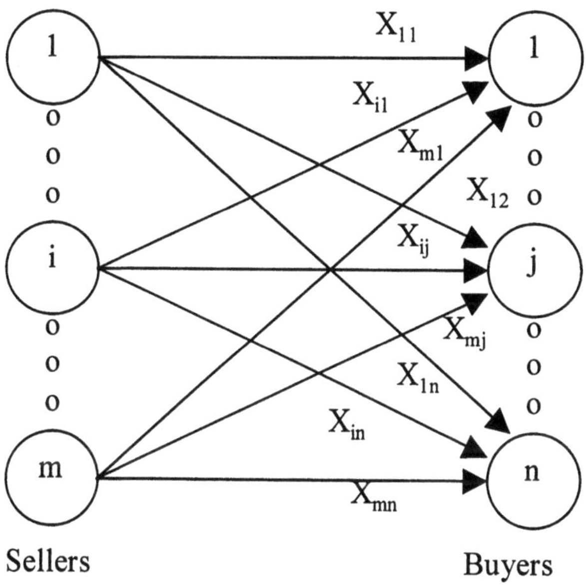


Figure 3-1. Trading Heterogeneous or Homogeneous Products Through Bilateral Contracts.

Trading homogeneous products through an exchange is shown in Figure 3-2. Sellers sell products to the exchange and buyers buy products from the exchange. Trading heterogeneous products through two exchanges is shown in Figure 3-3. In Figure 3-3, there are two classes of exchanges, Exchange a and Exchange b, which are classified according to types of products. Many classes of exchanges can be added to Figure 3-4. In addition, when classes are provided for pairs of every seller and buyer (the number of classes is $m*n$), trading through the exchange is equivalent to trading through bilateral contracts. The difference is one of convenience as explained above. However, each class of exchanges is usually provided for a group of products and so the number of classes is usually less than $m*n$. In addition, when the properties used to separate classes of exchanges are continuous quantities, they are usually discretized. For example, reliability is discretized when used to separate classes of exchanges, since reliability is measured in continuous quantities; e.g., six classes of exchanges are provided, which have reliability as 0.7, 0.75, 0.80, 0.85, 0.90, 0.95. Three classes of exchanges may be provided instead, which have reliability as 0.7, 0.8, 0.9. This results when the reliability levels 0.7, 0.75 are grouped together, and so are 0.8, 0.85, and 0.9, 0.95.

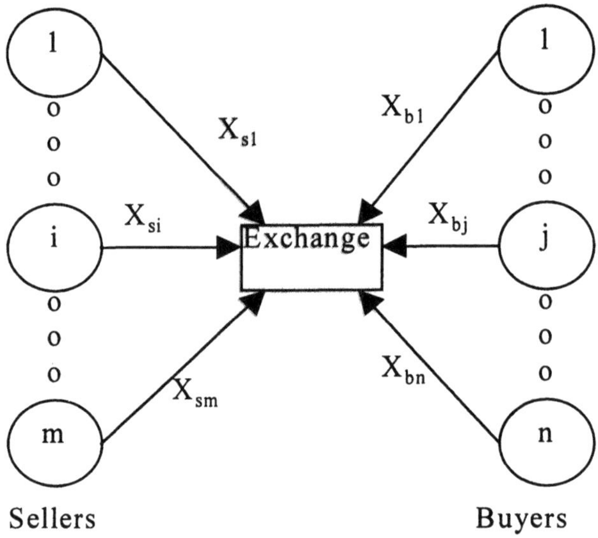


Figure 3-2. Trading Homogeneous Products Through an Exchange.

**3.3.2.2 Formulation of Assignment Problems.** This section formulates the assignment problems for heterogeneous and homogeneous products. Both primal and dual problems are shown. The formulations are for finding the partial equilibrium. For heterogeneous products, formulations for trading without or with the exchange are different except when the number of exchange classes is equal to $m*n$. For homogeneous products, formulations for trading without or with the exchange are the same. The formulations are classified in different cases. One criterion used for classification is based on the parties who specify prices. This criterion breaks the assignment problems into three cases: (a) sellers specify prices, (b) only buyers specify prices, and (c) both sellers and buyers specify prices. Sellers and buyers only know their own prices of other sellers and buyers. This is a sealed bid auction.

The effects of reservation prices are also considered. The reservation price of a seller is the lowest price at which the seller is willing to sell, and the reservation price of a buyer is the highest price at which the buyer is willing to buy. Reservation prices are considered when only sellers or buyers specify prices to ensure that parties who do not specify prices get the products at acceptable prices. Reservation prices are included by buyers when only sellers specify prices and reservation prices are included by sellers when only buyers specify prices. When both sellers and buyers specify prices, reservation prices are not needed because both parties can specify the prices according to their willingness. Not only are the reservation prices considered to ensure that parties get products at acceptable prices, but they are also useful for avoiding degeneracy problem, which will be explained in next section. Formulations for all cases are classified below. Cases 1 to 5 belong to heterogeneous products when they are traded via bilateral contracts. Cases 5 to 10 belong to homogeneous products and the formulations are applicable to when products are traded through either bilateral contracts or an exchange. Case 11 is when heterogeneous products are traded via exchanges.

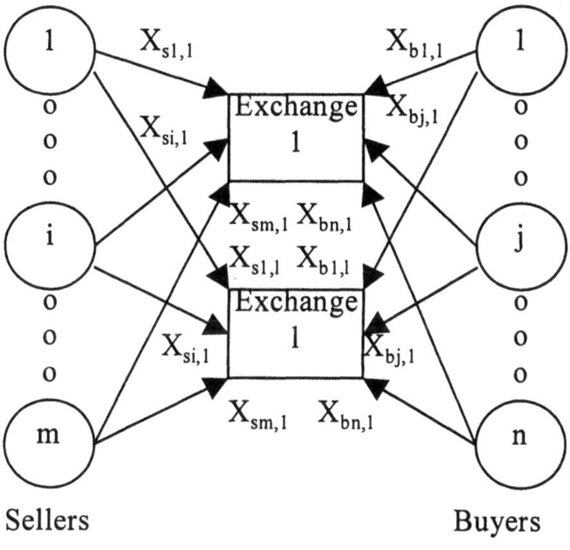


Figure 3-3. Trading Heterogeneous Products Through Exchanges.

Presently in many electric power auctions, electric power is treated as a homogeneous product and electric power is auctioned in the exchange. The ICA performs bid matching. In assignment problems, cases where only sellers specify prices are similar to single-sided auctions in which only GENCOs submit the bids. Cases where both sellers and buyers specify prices are similar to double-sided auctions in which both GENCOs and ESCOs submit the bids. The formulations of assignment problems can be applied to electric power auctions. All the formulations shown below can be applied to electric power auctions when different auction frameworks are needed. The complete formulation for each auction structure can be acquired by adding additional constraints to the formulations. Examples of additional constraints are power flow constraints and transmission line flow limits.

In the present case that electric power is treated as a homogeneous product, formulations in cases 6 to 10 can be applied to when electric power is traded via either bilateral contracts or auctions. Cases 6 and 7 are one-sided in which prices are specified by sellers and cases 8 and 9 are one-sided in which prices are specified by buyers. Case 10 is double-sided. If electric power is considered as heterogeneous products, formulations in cases 1 to 5 and in case 11 can be applied. Cases 1 to 5 are when electric power is traded via bilateral contracts and case 11 is when electric power is traded via auctions. Note that this work considers electricity as heterogeneous products in the case when electric power has different reliability level. This work does not consider other cases of heterogeneous electric power because of complexity in transmission.

The notation of symbols used in the formulation is:

| | |
|---|---|
| $c_{sij}$ | price specified by seller $i$ to buyer $j$ for a heterogeneous product |
| $c_{bij}$ | price specified by buyer $j$ to seller $i$ for a heterogeneous product |
| $c_{si}$ | price specified by seller $i$ for a homogeneous product |
| $c_{bj}$ | price specified by buyer $j$ for a homogeneous product |
| $c_{si,h}$ | price specified by seller $i$ for a heterogeneous product sold in exchange $h$ |
| $c_{bj,h}$ | price specified by buyer $j$ for a heterogeneous product bought in exchange $h$ |
| $\pi_{si}$ | reservation price specified by seller $i$ |
| $\pi_{bj}$ | reservation price specified by buyer $j$ |
| $x_{ij}$ | amount of a heterogeneous product sold from seller $i$ to buyer $j$ |
| $x_{si}$ | amount of a homogeneous product sold by seller $i$ |
| $x_{bj}$ | amount of a homogeneous product bought from buyer $j$ |
| $x_{si,h}$ | amount of a heterogeneous product sold by seller $i$ in exchange $h$ |
| $x_{bj,h}$ | amount of a heterogeneous product bought from buyer $j$ in exchange $h$ |
| $y_{si}$ | amount sold back of seller $i$ |
| $y_{bj}$ | amount bought back of buyer $j$ |
| $S_i$ | supply capacity of seller $i$ |

| | |
|---|---|
| $D_j$ | potential demand of buyer $j$ |
| $u_i$ | dual variable associated with supply constraint of seller $i$ |
| $v_j$ | dual variable associated with demand constraint of buyer $j$ |
| $w$ | dual variable associated with the constraint balancing supply and demand of a homogeneous product |
| $w_h$ | dual variable associated with the constraint balancing supply and demand of a heterogeneous product traded in exchange $h$ |
| $m$ | number of sellers |
| $n$ | number of buyers |
| $l$ | number of exchanges |

*Case 1: Heterogeneous products, trading as bilateral contracts, prices specified by sellers, without reservation prices*

Primal problem

$$\min_{x_{ij}} \sum_{i=1}^{m} \sum_{j=1}^{n} c_{sij} x_{ij}$$

s.t.

$$\sum_{j=1}^{n} x_{ij} \le S_i \qquad i=1,2,3,\ldots,m$$

$$\sum_{i=1}^{m} x_{ij} \ge D_j \qquad j=1,2,3,\ldots,n$$

$$x_{ij} \ge 0 \qquad i=1,2,3,\ldots,m \qquad j=1,2,3,\ldots,n$$

Dual problem

$$\max_{u_i, v_j} \sum_{i=1}^{m} S_i u_i + \sum_{j=1}^{n} D_j v_j$$

s.t.

$$u_i + v_j \le c_{sij} \qquad i=1,2,3,\ldots,m \qquad j=1,2,3,\ldots,n$$

$$u_i \le 0 \quad v_j \ge 0 \qquad i=1,2,3,\ldots,m \qquad j=1,2,3,\ldots,n$$

*Case 2: Heterogeneous products, trading as bilateral contracts, prices specified by sellers, with reservation prices*

Primal problem

$$\min_{x_{ij}, y_{bj}} \sum_{i=1}^{m} \sum_{j=1}^{n} c_{sij} x_{ij} + \sum_{j=1}^{n} \pi_{bj} y_{bj}$$

s.t.

$$\sum_{j=1}^{n} x_{ij} \le S_i \qquad i=1,2,3,\ldots,m$$

$$\sum_{i=1}^{m} x_{ij} + y_{bj} \ge D_j \qquad j=1,2,3,\ldots,n$$

$$x_{ij} \ge 0 \quad y_{bj} \ge 0 \qquad i=1,2,3,\ldots,m \quad j=1,2,3,\ldots,n$$

Dual problem

$$\max_{u_i, v_j} \sum_{i=1}^{m} S_i u_i + \sum_{j=1}^{n} D_j v_j$$

s.t.

$$u_i + v_j \le c_{sij} \qquad i=1,2,3,\ldots,m \quad j=1,2,3,\ldots,n$$

$$v_j \le \pi_{bj} \qquad j=1,2,3,\ldots,n$$

$$u_i \le 0 \quad v_j \ge 0 \qquad i=1,2,3,\ldots,m \quad j=1,2,3,\ldots,n$$

*Case 3: Heterogeneous products, trading as bilateral contracts, prices specified by buyers, without reservation prices*

Primal problem

$$\max_{x_{ij}} \sum_{i=1}^{m} \sum_{j=1}^{n} c_{bij} x_{ij}$$

s.t.

$$\sum_{j=1}^{n} x_{ij} \le S_i \qquad i=1,2,3,\ldots,m$$

$$\sum_{i=1}^{m} x_{ij} \le D_j \qquad j=1,2,3,\ldots,n$$

$$x_{ij} \ge 0 \qquad i=1,2,3,\ldots,m \quad j=1,2,3,\ldots,n$$

Dual problem

$$\min_{u_i, v_j} \sum_{i=1}^{m} S_i u_i + \sum_{j=1}^{n} D_j v_j$$

s.t.

$$u_i + v_j \ge c_{bij} \qquad i=1,2,3,\ldots,m \quad j=1,2,3,\ldots,n$$

$$u_i \ge 0 \quad v_j \ge 0 \qquad i=1,2,3,\ldots,m \quad j=1,2,3,\ldots,n$$

*Case 4: Heterogeneous products, trading as bilateral contracts, prices specified by buyers, with reservation prices*

Primal problem

$$\max_{x_{ij}, y_{si}} \sum_{i=1}^{m} \sum_{j=1}^{n} c_{bij} x_{ij} + \sum_{i=1}^{m} \pi_{si} y_{si}$$

s.t. $\sum_{j=1}^{n} x_{ij} + y_{si} \le S_i$   i=1,2,3,...,m

$\sum_{i=1}^{m} x_{ij} \le D_j$   j=1,2,3,...,n

$x_{ij} \ge 0 \quad y_{si} \ge 0$   i=1,2,3,...,m   j=1,2,3,...,n

Dual problem

$$\min_{u_i, v_j} \sum_{i=1}^{m} S_i u_i + \sum_{j=1}^{n} D_j v_j$$

s.t. $u_i + v_j \ge c_{bij}$   i=1,2,3,...,m   j=1,2,3,...,n

$u_i \ge \pi_{si}$   i=1,2,3,...,m

$u_i \ge 0, \quad v_j \ge 0$   i=1,2,3,...,m   j=1,2,3,...,n

*Case 5: Heterogeneous products, trading as bilateral contracts, prices specified by sellers and buyers, without reservation prices*

Primal problem

$$\max_{x_{ij}} \sum_{i=1}^{m} \sum_{j=1}^{n} (c_{bij} - c_{sij}) x_{ij}$$

s.t.

$\sum_{j=1}^{n} x_{ij} \le S_i$   i=1,2,3,...,m

$\sum_{i=1}^{m} x_{ij} \le D_j$   j=1,2,3,...,n

$x_{ij} \ge 0$   i=1,2,3,...,m   j=1,2,3,...,n

Dual problem

$$\min_{u_i, v_j} \sum_{i=1}^{m} S_i u_i + \sum_{j=1}^{n} D_j v_j$$

s.t. $u_i + v_j \ge c_{bij} - c_{sij}$   i=1,2,3,...,m   j=1,2,3,...,n

$u_i \ge 0 \quad v_j \ge 0$   i=1,2,3,...,m   j=1,2,3,...,n

*Case 6: Homogeneous products, prices specified by sellers, without reservation prices*

Primal problem

$$\min_{x_{si}} \sum_{i=1}^{m} c_{si} x_{si}$$

s.t.

$$x_{si} \le S_i \qquad i=1,2,3,\dots,m$$

$$\sum_{i=1}^{m} x_{si} \ge \sum_{j=1}^{n} D_j$$

$$x_{si} \ge 0 \qquad i=1,2,3,\dots,m$$

Dual problem

$$\max_{u_i, v} \sum_{i=1}^{m} S_i u_i + \left( \sum_{j=1}^{n} D_j \right) v$$

s.t.

$$u_i + v \le c_{si} \qquad i=1,2,3,\dots,m$$

$$u_i \le 0 \quad v \ge 0 \qquad i=1,2,3,\dots,m$$

*Case 7: Homogeneous products, prices specified by sellers, with a reservation price*

Primal problem

$$\min_{x_{si}, y_b} \sum_{i=1}^{m} c_{si} x_{si} + \pi_b y_b$$

s.t.

$$x_{si} \le S_i \qquad i=1,2,3,\dots,m$$

$$\sum_{i=1}^{m} x_{si} + y_b \ge \sum_{j=1}^{n} D_j$$

$$x_{si} \ge 0 \quad y_b \ge 0 \qquad i=1,2,3,\dots,m$$

Dual problem

$$\max_{u_i, v} \sum_{i=1}^{m} S_i u_i + \left( \sum_{j=1}^{n} D_j \right) v$$

s.t.

$$u_i + v \le c_{si} \qquad i=1,2,3,\dots,m$$

$$v \le \pi_b$$

$$u_i \le 0 \quad v \ge 0 \qquad i=1,2,3,\dots,m$$

*Case 8: Homogeneous products, prices specified by buyers, without reservation prices*

Primal problem

$$\max_{x_{bj}} \sum_{j=1}^{n} c_{bj} x_{bj}$$

s.t.

$$x_{bj} \le D_j \qquad j=1,2,3,\dots,n$$

$$\sum_{j=1}^{n} x_{bj} \le \sum_{i=1}^{m} S_i$$

$$x_{bj} \ge 0 \qquad j=1,2,3,\dots,n$$

Dual problem

$$\min_{u,v_j} \sum_{j=1}^{n} D_j v_j + \left( \sum_{i=1}^{m} S_i \right) u$$

s.t.

$$u + v_j \ge c_{bj} \qquad j=1,2,3,\dots,n$$

$$u \ge 0 \quad v_j \ge 0 \qquad j=1,2,3,\dots,n$$

*Case 9: Homogeneous products, prices specified by buyers, with a reservation price*

Primal problem

$$\max_{x_{bj}, y_s} \sum_{j=1}^{n} c_{bj} x_{bj} + \pi_s y_s$$

s.t.

$$x_{bj} \le D_j \qquad j=1,2,3,\dots,n$$

$$\sum_{j=1}^{n} x_{bj} + y_s \le \sum_{i=1}^{m} S_i$$

$$x_{bj} \ge 0 \quad y_s \ge 0 \qquad j=1,2,3,\dots,n$$

Dual problem

$$\min_{u,v_j} \sum_{j=1}^{n} D_j v_j + \left( \sum_{i=1}^{m} S_i \right) u$$

s.t.

$$u + v_j \ge c_{bj} \qquad j=1,2,3,\dots,n$$

$$u \ge \pi_s$$

$$u \ge 0 \quad v_j \ge 0 \qquad j=1,2,3,\dots,n$$

*Case 10: Homogeneous products, prices specified by sellers and buyers, without reservation prices*

Primal problem

$$\max_{x_{si},x_{bj}} \sum_{j=1}^{n} c_{bj}x_{bj} - \sum_{i=1}^{m} c_{si}x_{si}$$

s.t.

$$x_{si} \le S_i \qquad i=1,2,3,\ldots,m$$

$$x_{bj} \le D_j \qquad j=1,2,3,\ldots,n$$

$$\sum_{i=1}^{m} x_{si} - \sum_{j=1}^{n} x_{bj} = 0$$

$$x_{si} \ge 0 \quad x_{bj} \ge 0 \qquad i=1,2,3,\ldots,m \qquad j=1,2,3,\ldots,n$$

Dual problem

$$\min_{u_i,v_j,w} \sum_{i=1}^{m} S_i u_i + \sum_{j=1}^{n} D_j v_j$$

s.t.

$$-u_i - w \le c_{si} \qquad i=1,2,3,\ldots,m$$

$$v_j - w \ge c_{bj} \qquad j=1,2,3,\ldots,n$$

$$u_i \ge 0 \quad v_j \ge 0 \quad \text{w free} \quad i=1,2,3,\ldots,m \qquad j=1,2,3,\ldots,n$$

*Case 11: Heterogeneous products, trading through the exchange, prices specified by sellers and buyers, without reservation prices*

Primal problem

$$\max_{x_{si,h},x_{bj,h}} \sum_{h=1}^{\ell} \left[ \sum_{j=1}^{n} c_{bj,h}x_{bj,h} - \sum_{i=1}^{m} c_{si,h}x_{si,h} \right]$$

s.t.

$$\sum_{i=1}^{m} x_{si,h} - \sum_{j=1}^{n} x_{bj,h} = 0 \quad h=1,2,3,\ldots,\ell$$

$$\sum_{h=1}^{\ell} x_{si,h} \le S_i \qquad i=1,2,3,\ldots,m$$

$$\sum_{h=1}^{\ell} x_{bj,h} \le D_j \qquad j=1,2,3,\ldots,n$$

$$x_{si,h} \ge 0 \quad x_{bj,h} \ge 0 \qquad i=1,2,3,\ldots,m \qquad j=1,2,3,\ldots,n$$

$$h=1,2,3,\ldots,\ell$$

Dual problem

$$\min_{u_i, v_j, w_h} \sum_{i=1}^{m} S_i u_i + \sum_{j=1}^{n} D_j v_j$$

s.t.

$$-u_i - w_h \le c_{si,h} \qquad i=1,2,3,\ldots,m \qquad h=1,2,3,\ldots,\ell$$

$$v_j - w_h \ge c_{bj,h} \qquad j=1,2,3,\ldots,n$$

$$h=1,2,3,\ldots,\ell$$

$$u_i \ge 0 \quad v_j \ge 0 \quad w_h \text{ free} \qquad i=1,2,3,\ldots,m \qquad j=1,2,3,\ldots,n$$

$$h=1,2,3,\ldots,\ell$$

**3.3.2.3 Degeneracy in Assignment Problems.** In the primal problems of all cases shown above, degeneracy can occur at the optimum, which in turn causes multiple dual optima. The degeneracy problem in this context results from redundant constraints. There are three major sets of constraints, supply constraints, demand constraints, and non-negativity constraints. Supply and demand constraints are redundant when some conditions are met. The explanations are divided into two parts, one that does not consider reservation prices and one that does consider reservation prices. Cases 1, 3, 5, 6, 8, 10, and 11 do not include reservation prices and cases 2, 4, 7, and 9 do include reservation prices.

First, consider the cases without reservation prices. In cases 3 and 5, summing the supply constraints together and summing the demand constraints together yields redundant set of equations. The supply and demand constraints are equivalent to the total supply capacity set equal to the total potential demand. This illustrates that redundant constraints occur when the total capacity and the total potential demand are the same. The total supply capacity is not always equal to the total potential demand. It is important to point out that the total supply is always equal to the total demand at the equilibrium to have market cleared; however, the total supply capacity is not necessary equal to the total potential demand. Explaining this in the context of auctions, the total bid amount of sellers is not necessarily equal to that of buyers but the total amount sold by sellers is always equal to the total amount bought by buyers (assuming lossless cases). Because the objective function is a maximization one, if the total supply capacity is equal to the total potential demand, both sets of supply and demand constraints are binding at optimality. Otherwise, the set of constraints with the smaller total is binding at optimality.

$$\sum_{i=1}^{m} \sum_{j=1}^{n} x_{ij} \le \sum_{i=1}^{m} S_i \tag{3.3}$$

$$\sum_{i=1}^{m} \sum_{j=1}^{n} x_{ij} \le \sum_{j=1}^{n} D_j \tag{3.4}$$

$$\sum_{i=1}^{m} S_i = \sum_{j=1}^{n} D_j \tag{3.5}$$

$$u'_i = u_i - k; \quad v'_j = v_j + k; \quad k \in R, k \neq 0 \tag{3.6}$$

$$\begin{aligned}\sum_{i=1}^{m} S_i u'_i + \sum_{j=1}^{n} D_j v'_j &= \sum_{i=1}^{m} S_i (u_i - k) + \sum_{j=1}^{n} D_j (v_j + k) \\ &= \sum_{i=1}^{m} S_i u_i + \sum_{j=1}^{n} D_j v_j - k\sum_{i=1}^{m} S_i + k\sum_{j=1}^{n} D_j\end{aligned} \tag{3.7}$$

If condition (3.5) is satisfied, cases 1, 6, 8, 10, and 11 are also primal degenerate at the optimal solution. Although in cases 1 and 6, demands are in the forms of greater or equal constraints, they are binding if condition (3.5) is satisfied. There is only one demand constraint in case 6 and one supply constraint in case 8; however, this does not affect the application of condition (3.5). There is one additional constraint in cases 10 and 11. This constraint is to assure that the total supply is equal to the total demand; thus, it does not affect the application of condition (3.5).

This illustrates one type of degeneracy of optimal primal solution. The application can also be applied to the dual problem. It can be concluded that multiple dual optima occur when condition (3.5) is satisfied. In general, it is difficult to know all the dual optima from the degenerate primal unless pivoting is performed. Because the structure of dual problems in all cases is unique, the pattern of all the dual optimal solutions can be determined. Investigation of the dual problems in all cases without reservation prices show that the objective functions are the same. The constraints (not including constraints of dual variable signs) are all in the form that the left-hand-side is $u_i+v_j$, and the right-hand-side is the price. The constraints can be in either of the two standard forms: less than or equal to, or greater than or equal to. In cases 10 and 11, there is two sets of constraints and $u_i$ or $v_j$ are separated in each set of constraints. However, two sets of constraints can be summed and result in the specific form as mentioned. From this investigation, it can be inferred that the value of every $u_i$ can be reduced by a same constant and that the value of $v_j$ can be increased by the same constant (3.6). The constant (k) can be any positive or negative number for which the dual variable signs are not violated. This ensures value of $u_i+v_j$ does not change and so this does not affect the constraint. For the objective function, (3.7) shows that the dual objective function does not change after shifting dual variables according to (3.6), as long as condition (3.5) is satisfied.

Next, the cases with reservation prices (cases 2, 4, 7, and 9) are detailed. The reservation prices appear as limits for either $u_i$ or $v_j$ in dual problems. If any of the limit constraints is not binding, multiple dual optima can occur if condition (3.5) is satisfied. This is because if any of the limit constraints is not binding, values of the dual variables can still be shifted as long as they satisfy the limit constraints. In terms of the primal problem, if any of the limit constraints is not binding, the variables representing sold back or bought back values ($y_{bj}$, $y_{si}$, $y_b$, and $y_s$) are zero, according to complementary slackness. This will make the structure of the primal problem the same as the cases without reservation prices. On the other hand, if any of the limit constraints is binding, it will determine a unique value of the associated dual variable and in turn determine the values of all other dual variables. In addition, the binding limit constraint does not require the variables representing sold back or bought back values ($y_{bj}$, $y_{si}$, $y_b$, and $y_s$) to be zeros. This proves that the binding limit constraint prevents multiple optimal dual solutions even that condition (3.5) is satisfied. The limit constraint is binding when either any specified price by the buyer to the seller is less than the reservation price of the associated seller or any specified price by the seller to the buyer is greater than the reservation price of the associated buyer.

In summary, degeneracy in primal optimal solution occurs when the total supply capacity is equal to the total potential demand (3.5). This also causes multiple dual optima. Fortunately, the total supply capacity is not always equal to the total potential demand. Although the total supply is always equal to the demand (in lossless cases), but the total supply capacity and the total potential demand are not always equal to each other. Multiple solutions are not desired in auctions as explained in previous section. The interesting question is how to handle the problem of multiple dual optimal solutions when (3.5) is satisfied. It is not possible to eliminate just one of either the supply or demand constraints because the choice of elimination is equivalent to specifying the shifted constant. This would be unfair to the bidders. One possible solution is the ICA can ask the bidders to resubmit the bids if the ICA discovers that the total supply capacity is equal to the total potential demand, which can be easily detected.

### 3.4 More General Models

Industrial modeling of power plants normally includes more realistic functions representing the input output characteristics of the units. Generally, piece-wise linear curves are often used. The changes to incorporate such a model into the above formulation is shown below. Again, given the following notation:

m=number of sellers
n= number of buyers
si = supply for seller i
$\gamma_i + \delta_i s_i$ = supply (shadow or marginal) price for seller i for the range of [0,si]
$b_j$ = demand by buyer j
$c_{ij}$ = cost of delivery from seller i to buyer j
$x_{ij}$ = quantity to be sold from seller i to buyer j.

Note that the constants for the range change depending on the range of operation for the unit given previous contracts and bidding awards.

The objective function is the same as before, to minimize social welfare:

$$\text{Minimize} \quad \sum_{i=1}^{m} \left(\gamma_i s_i + 0.5\,\delta_i s_i^2\right) + \sum_{i=1}^{m} \sum_{j=1}^{n} c_{ij} x_{ij}$$

subject to:

$$\sum_{j=1}^{n} (x_{ij} - s_i) \le 0, \quad i = 1,2,\ldots,m$$

$$\sum_{i=1}^{m} x_{ij} \ge b_j, \quad j = 1,2,\ldots,n$$

$$x_{ij}, s_i \ge 0, \qquad i = 1,2,\ldots,m, \quad j = 1,2,\ldots,n$$

The KKT conditions required at the optimal solution are modified. The optimal shipments can not exceed supplier capacity. If the dual variable is nonzero, then the supply capacity is completely used.

$$u_i^* \le 0, \quad \sum_{j=1}^{n} x_{ij}^* - s_i^* \le 0, \; u_i^* \left( \sum_{j=1}^{n} x_{ij}^* - s_i^* \right) = 0,$$
$$\textit{for } i = 1,2,\ldots,m$$

The commodity shipped is equal to the demand if the dual variable is nonzero.

$$v_j^* \ge 0, \quad \sum_{i=1}^{m} x_{ij}^* - b_j^* \ge 0, \; v_j^* \left( \sum_{i=1}^{m} x_{ij}^* - b_j^* \right) = 0,$$
$$\textit{for } j = 1,2,\ldots,n$$

Consider the dual relationships to measure the value added when the commodity is transported from the supplier to the buyer. The value added can not exceed the transportation cost.

$$u_i^* + v_j^* \le c_{ij}, \quad x_{ij}^* \left(c_{ij} - u_i^* - v_j^*\right) = 0,$$
$$\textit{for } \; i = 1,2,\ldots,m \; \textit{ and } \; j = 1,2,\ldots,n$$

As are the KKT conditions for the dual problem:

$$-u_i^* \le \gamma_i + \delta_i s_i^*, \quad s_i^* \left(\gamma_i + \delta_i s_i^* + u_i^*\right) = 0,$$
$$\textit{for } \; i = 1,2,\ldots,m$$

The transmission network can be modeled as nonlinear shipping costs to account for the power lost and for the revenue collected for the use of each piece of equipment. The only change is to include the nonlinear shipping cost, $d_{ij}$. The objective function is only changed slightly to include the transmission costs:

$$\text{Minimize} \quad \sum_{i=1}^{m} \left(\gamma_i s_i + 0.5\, \delta_i s_i^2\right) + \sum_{i=1}^{m} \sum_{j=1}^{n} \left(c_{ij} x_{ij} + d_{ij} x_i^2\right)$$

subject to:

$$\sum_{j=1}^{n} (x_{ij} - s_i) \le 0, \quad i = 1,2,\ldots,m$$

$$\sum_{i=1}^{m} x_{ij} \ge b_j, \quad j = 1,2,\ldots,n$$

$$x_{ij}, s_i \ge 0, \qquad i = 1,2,\ldots,m, \quad j = 1,2,\ldots,n$$

The KKT conditions are easily extended from the previous case:

$$u_i^* + v_j^* \le, \quad u_{ij}^*\left(c_{ij} + 2d_{ij}x_{ij}^* - u_i^* - v_j^*\right) \; = 0,$$
$$\mathit{for}\; i = 1,2,\ldots,m \quad \mathit{and} \quad j = 1,2,\ldots,n$$

The marginal shipping cost is the first term in the second equation above:

$$c_{ij} + 2d_{ij}x_{ij}^*$$

Customers will not always consume the same amount of commodity as the price increases. A more complete model should inlcude a nonlinear market demand where the demand decreases as the price increases. Let α be the constant and β the slope of the demand price curve. Then the change to the objective function is another simple extension of the previous case:

$$\text{Minimize} \quad \sum_{i=1}^{m} \left(\gamma_i s_i + 0.5\, \delta_i s_i^2\right) + \sum_{i=1}^{m} \sum_{j=1}^{n} c_{ij} x_{ij} - \sum_{j-1}^{n} \left(\alpha_j d_j - 0.5 \beta_j d_j^2\right)$$

subject to:

$$\sum_{j=1}^{n} x_{ij} - s_i \le 0, \qquad i = 1,2,\ldots,m$$

$$\sum_{i=1}^{m} x_{ij} - d_j \ge 0, \quad j = 1,2,\ldots,n$$

$$x_{ij}, s_i, d_j \ge 0, \qquad i = 1,2,\ldots,m \;\mathit{and}\; j = 1,2,\ldots,n$$

Again, the KKT conditions at the optimal solution hold:

$$\sum_{i=1}^{m} x_{ij}^{*} - d_{j}^{*} \geq 0, \; v_{j}^{*}\left(\sum_{i=1}^{m} x_{ij}^{*} - d_{j}^{*}\right) = 0,$$
$$for \;\; j = 1,2,\ldots,n$$

$$v_{j}^{*} - \alpha_{j} + \beta_{j} d_{j}^{*} \geq 0, \quad d_{j}^{*}\left(v_{j}^{*} - \alpha_{j} + \beta_{j} d_{j}^{*}\right) = 0,$$
$$for \;\; j = 1,2,\ldots,n$$

The linearized power flow or available transfer capacity relationship could be used in these equations to solve for the market clearing prices for a region.

## 3.5 Vertically Integrated Industry

It is most instructive to compare auctions used by companies in a horizontally organized industry to a vertically integrated monopoly. Consider a company consisting of three divisions. The result of applying Linear Programming to optimize the profit of a three-division company yields the following generalized LP structure for the constraint matrix:

| | | | |
|---|---|---|---|
| $A_{11}$ | 0 | 0 | 0 |
| 0 | $A_{22}$ | 0 | 0 |
| 0 | 0 | $A_{33}$ | 0 |
| $A_{41}$ | $A_{42}$ | $A_{43}$ | $A_{44}$ |

Note that the diagonal sub-matrices consist of those constraints applicable to an individual division. The last row of matrices consist of those constraints on organizational resources needed by two or more divisions. These are the linking constraints.

These are considered multidivisional problems because the special structure can be exploited for computational efficiency. This problem is almost decomposable into separate problems. Multidivisional problems exhibit this block diagonal structure even in non-monopolistic industries. Decentralization of production decision often leads to these types of problems.

Consider two units in a power plant within a horizontally integrated company. The company is considered horizontally integrated since each division has to make its own decisions with regard to operation. However, each division uses the common fuel supply, common crew for startup, common equipment for maintenance, etc.

Consider the optimization of profit for each unit separately. Note that the operation of one unit is independent of the other unit. The objective is to maximize profit subject to operational constraints:

Maximize $\hat{p}^{(1)}x^{(1)} + \hat{p}^{(2)}x^{(2)}$

subject to

$$A^{(1)}x^{(1)} \le b^{(1)}$$
$$A^{(2)}x^{(2)} \le b^{(2)}$$
$$D^{(1)}x^{(1)} + D^{(2)}x^{(2)} \le d$$
$$x^{(1)},\ x^{(2)} \ge 0$$

The first two constraints list the restrictions on each unit. The third constraint lists the restrictions on the common resources. These are the master (Dantzig-Wolfe) or corporate constraints.

The corresponding dual problem is a minimization subject to the proper pricing of the common resources:

Minimize $u^{(1)}b^{(1)} + u^{(2)}b^{(2)} + vd$

subject to

$$u^{(1)}A^{(1)} + vD^{(1)} \ge \hat{p}^{(1)}$$
$$u^{(2)}A^{(2)} + vD^{(2)} \ge \hat{p}^{(2)}$$
$$u^{(1)}, u^{(2)}, v \ge 0$$

Both have a staircase structure. If the common resource constraints are absent, then the problem can be completely decomposed.

Consider a solution found with a resulting vector of transfer prices v*.

Transfer prices are charged by the corporation for the common resources to each division. Substitute and re-arrange:

Minimize $u^{(1)}b^{(1)} + u^{(2)}b^{(2)} +$ constant

subject to

$$u^{(1)}A^{(1)} \ge \hat{p}^{(1)} - v^* D^{(1)} = p^{\#(1)}$$
$$u^{(2)}A^{(2)} \ge \hat{p}^{(2)} - v^* D^{(2)} = p^{\#(2)}$$
$$u^{(1)}, u^{(2)} \ge 0$$

Where # denotes corrected prices. The duality of this problem is easily found.

Maximize $p^{\#(2)}x^{(1)} + p^{\#(2)}x^{(2)}$

subject to

$$A^{(1)}x^{(1)} \le b^{(1)}$$
$$A^{(2)}x^{(2)} \le b^{(2)}$$
$$x^{(1)},\ x^{(2)} \ge 0$$

Note that these are decomposable programs. Thus, they can be solved separately.

So treat division 1 as decision-making unit (dmu) 1 and division 2 as dmu 2. Solve each of the problems separately:

Maximize $p^{\#(1)}x^{(1)}$

subject to $A^{(1)}x^{(1)} \le b^{(1)}$

$x^{(1)} \ge 0$

Maximize $p^{\#(2)}x^{(2)}$

subject to $A^{(2)}x^{(2)} \le b^{(2)}$

$x^{(2)} \ge 0$

Thus, the transfer prices have enabled formal decomposition of the multidivisional problem.

An interpretation of these results is to consider the splitting of the vertically integrated company into two or more decision-making units with the sole responsibility of maximizing the dmu profit. The center requests each division to forecast the amount of central resources needed. After the request, the price for central resources is established by recovery of all central expenses. This price is sent to the divisions to enable them to maximize their profit given the new information. The divisions can then resubmit new bids based on this new information. The central allocation process is then restarted with the new division demands. The process continues until the divisions do not change their production based on the central prices.

Such a multidivisional problem is a coherent decentralization if the optimal solutions to the corrected divisional problems constitute an optimal solution to the original, global problem. Such a solution may not exist if the original problem has a degenerate solution that is not at a vertex, but on a line (plane) connecting two vertices. In such cases, the centralized coordinator has to give the two divisions additional instructions on which solutions to select. Otherwise, the centralized organization does not have to know anything about the problems in either division.

This decomposition algorithm is called the Dantzig-Wolfe mechanism. The Dantzig-Wolfe algorithm is a price-based decomposition as opposed to a resource directive that allocate scarce resources by right-hand-side allocation.

It should be noted that the advantages of large multi divisional companies may be lost without proper consideration of the use of central resources. It is also true that corporate restructuring may be indicated if such decentralization provides a better profit than a centrally administered organization. The Dantzig-Wolfe algorithm is discussed below.

### 3.6 Decentralization in a Nonlinear Model

The problem of not finding a coherent decentralization by price alone can be alleviated by requiring a unique solution at a vertex or at a singular solution point. The key is to replace the central objective function by a strictly concave, nonlinear objective function. The central objective function is still the summation of divisional objective functions. Under such circumstances, price decomposition is effected.

Assume that two of the central resources increase linearly with output:

$$\hat{p}_j^{(1)} = p_j^{(1)} - \gamma_j^{(1)} - \delta_j^{(1)} x_j^{(1)}$$
$$\hat{p}_j^{(2)} = p_j^{(2)} - \gamma_j^{(2)} - \delta_j^{(2)} x_j^{(2)} \qquad \text{for j=1, ..., n}$$

The return is now the last unit of commodity produced by the division. Both gamma and delta are known positive values. The master program can now be stated as:

$$\text{Maximize} \quad \sum_{j=1}^{n} \left\{ p_j^{(1)} x_j^{(1)} - \left[ \gamma_j^{(1)} x_j^{(1)} + 0.5\, \delta_j^{(1)} \left(x_j^{(1)}\right)^2 \right] \right\} + \sum_{j=1}^{n} \left\{ p_j^{(2)} x_j^{(2)} - \left[ \gamma_j^{(2)} x_j^{(2)} + 0.5\, \delta_j^{(2)} \left(x_j^{(2)}\right)^2 \right] \right\}$$

$$\text{subject to} \quad \begin{aligned} A^{(1)} x^{(1)} &\le b^{(1)} \\ A^{(2)} x^{(2)} &\le b^{(2)} \\ D^{(1)} x^{(1)} + D^{(2)} x^{(2)} &\le d \\ x^{(1)},\ x^{(2)} &\ge 0 \end{aligned}$$

Note that the objective function is immediately separable. There is one term for each division. Since the parameters are positive, then the objective function is strictly concave. The optimum will be unique but may not be at a vertex. It may be an interior point. The corrected prices are now stated and substituted into the above formulation to form two completely decentralized sub-problems.:

$$\hat{p}^{(1)} - v^* D^{(1)}$$
$$\hat{p}^{(2)} - v^* D^{(2)}$$

The decentralized problems are:

Unit 1:

$$\text{Maximize} \quad \left(p^{(1)} - v^* D^{(1)}\right) x^{(1)} - \sum_{j=1}^{n} \left[ \gamma_j^{(1)} x_j^{(1)} + 0.5\, \delta_j^{(1)} \left(x_j^{(1)}\right)^2 \right]$$

subject to
$$A^{(1)}x^{(1)} \le b^{(1)}$$
$$x^{(1)} \le 0$$

Unit 2:

Maximize
$$\left(p^{(2)} - v^{*}D^{(2)}\right)x^{(2)} - \sum_{j=1}^{n}\left[\gamma_j^{(2)}x_j^{(2)} + 0.5\,\delta_j^{(2)}\left(x_j^{(2)}\right)^2\right]$$

subject to
$$A^{(2)}x^{(2)} \le b^{(2)}$$
$$x^{(2)} \le 0$$

Consider a company with vertical integration. Such a company would mine coal and deliver the coal to a power plant. Alternatively, such a company may own natural gas wells connect directly to a power plant. The LP model of such a company follows:

Maximize
$$-\hat{c}^{(1)}x^{(1)} + \hat{p}^{(2)}x^{(2)}$$

subject to
$$A^{(1)}x^{(1)} \le b^{(1)}$$
$$-x^{(1)} + T^{(2)}x^{(2)} \le 0$$
$$A^{(2)}x^{(2)} \le b^{(2)}$$
$$x^{(1)},\ x^{(2)} \ge 0$$

where the "c" is the cost of the delivering division and “p” is the profit of the production division.

C = rh +c
P = p - rh - c

"T" is the amount of division one production used by division two. Note that we assume that the product of division one does not go to the outside world and that all of division two output goes to the outside world. The first two constraints are the inter-division constraints and the variables are the amount of product transferred.

The dual LP is found as before:

Minimize
$$u^{(1)}b^{(1)} + u^{(3)}b^{(2)}$$

subject to
$$u^{(1)}A^{(1)} - u^{(2)} \ge -\hat{c}^{(1)}$$
$$u^{(2)}T^{(2)} + u^{(3)}A^{(2)} \ge \hat{p}^{(2)}$$
$$u^{(1)}, u^{(2)}, u^{(3)} \ge 0$$

Assume that the solution to this dual is known. Thus, the dual variables are known and used as prices:

$$\begin{aligned} \text{Minimize} \quad & u^{(1)}b^{(1)} + u^{(3)}b^{(2)} \\ \text{subject to} \quad & u^{(1)}A^{(1)} \geq -\hat{c}^{(1)} + u*^{(2)} = p^{(1)} \\ & u^{(3)}A^{(2)} \geq \hat{p}^{(2)} - u*^{(2)}\,T^{(2)} = p^{(2)} \\ & u^{(1)}, u^{(2)} \geq 0 \end{aligned}$$

The dual LP problem follows:

$$\begin{aligned} \text{Maximize} \quad & p^{(1)}x^{(1)} + p^{(3)}x^{(2)} \\ \text{subject to} \quad & A^{(1)}x^{(1)} \leq b^{(1)} \\ & A^{(3)}x^{(2)} \leq b^{(2)} \\ & x^{(1)}, x^{(2)} \geq 0 \end{aligned}$$

The merger of many companies is often called into question when the interdivisional products are compared to external offerings. If such comparisons are made, then the corporation can afford such products from the outside market at price q. The eternal question: Should the products be obtained inter-divisionally or using the outside market? This question is often called the "make or buy" question. The vertically integrated company model is amended to include this outside possibility:

$$\begin{aligned} \text{Maximize} \quad & \hat{p}^{(1)}x^{(1)} + \hat{p}^{(2)}x^{(2)} - q^{(3)}y \\ \text{subject to} \quad & A^{(1)}x^{(1)} \leq b^{(1)} \\ & -x^{(1)} + T^{(2)}x^{(2)} - y = 0 \\ & A^{(2)}x^{(2)} \leq b^{(2)} \\ & y \leq Y \\ & x^{(1)},\ x^{(2)}y \geq 0 \end{aligned}$$

where Y is the maximum amount available from the outside market. Of course, it may be most beneficial to deliver the entire internal product to the outside market. Normally, the result is mixed with some of the internal product being used and some of the outside market being used. Reliability of production schedules may dictate such a combination if price penalties are present, as is often true with electricity.

# 4 ECONOMIC DISPATCH, UNIT COMMITMENT, AND OPTIMAL POWER FLOW AS AUCTIONS

*Once understood, it is obvious. Why have I missed the obvious?*

## 4.1 Economic Dispatch

The Economic Dispatch algorithm is the most used optimization for real-time and for planning. Economic Dispatch is investigated to determine if any comparison could be made with the auction problem by a more appropriate choice of algorithm. The algorithms which we have evaluated for the Economic Dispatch problem are presented in order of complexity [Sheblé, 1986; Wood, 1996].

The general formulation includes transmission losses modeled as a set of penalty factors.

Minimize $PI = \sum_{n=1}^{N} y_n(u_n)$

Subject to $\sum_{n=1}^{N} u_n PF_n = BLD + LOSS$

Where:

(1) Unit cost function:

$y_n = f_n e_n F_n(u_n)$

where: $y_n$ = production cost

$u_n$ = production power

$f_n$ = fuel cost

$e_n$ = efficiency

$F_n$ = heat rate curve

(2) Unit capacity limits:

$\overline{U}_n \geq u_n \geq \underline{U}_n$

(3) Network Loss Model:

$PF_n$ = penalty factor

(4) Power system model: BLD + LOSS

Where: BLD = the base load demand (system load), and LOSS = the transmission loss.
The unit limitations apply to all units n-1,...,N.

The unit models for each the input output conversion curve from generation level to cost (I/O curve) has not been standardized at this point. Instead, many different curves are in use as shown in Table 4-1. However, the piecewise linear curve model is very popular.

Table 4-1. Energy Conversion Curve Models

| Input/Output Curve Function $(F_n(u_n))$ | Incremental Heat Rate Curve Function $(f_n(u_n))$ |
|---|---|

(1) Polynomial Curve

$$F_n(u_n) = a_n u_n^{\ 4} + b_n u_n^{\ 3} + c_n u_n^{\ 2} + d_n u_n + e_n$$
$$f_n(u_n) = 4a_n u_n^{\ 3} + 3b_n u_n^{\ 2} + 2c_n u_n + d_n$$

(2) Cubic Curve/Quadratic Curve

$$F_n(u_n) = a_n u_n^{\ 3} + b_n u_n^{\ 2} + c_n u_n + d_n$$
$$f_n(u_n) = 3a_n u_n^{\ 2} + 2b_n u_n + c_n$$

(3) Quadratic Curve/Linear Curve

$$F_n(u_n) = a_n u_n^{\ 2} + b_n u_n + c_n$$
$$f_n(u_n) = 2a_n u_n + b_n$$

(4) Linear Curve/Constant

$$F_n(u_n) = a_n u_n + b_n$$
$$f_n(u_n) = a_n$$

The penalty factors are computed from on-line measurements in many energy control centers. However, this calculation has not been standardized. We have assumed that the penalty factors are stored for a variety of network conditions and operating conditions. The format we have assumed is shown in Tables 4-2.

Table 4-2. Penalty Factor Matrix

| Network Topology 1 | | All lines in service | | | |
|---|---|---|---|---|---|
| Generation Range (MW) | | 500-600 | 600-760 | 760-925 | 925-1225 |
| Interchange | -225,-125 | PFIV #1 | PFIV #2 | PFIV #3 | PFIV #4 |
| | -125,-50 | PFIV #5 | PFIV #6 | PFIV #7 | PFIV #8 |
| Range | -50,50 | PFIV #9 | PFIV #10 | PFIV #11 | PFIV #12 |
| | 50,125 | PFIV #13 | PFIV #14 | PFIV #15 | PFIV #16 |
| (MW) | 125,176 | PFIV #17 | PFIV #18 | PFIV #19 | PFIV #20 |

| Network Topology 2 | | 765 kV line 1024-E out of service | | | |
|---|---|---|---|---|---|
| Generation Range (MW) | | 500-600 | 600-760 | 760-925 | 925-1225 |
| Interchange | -225,-125 | PFIV #1 | PFIV #2 | PFIV #3 | PFIV #4 |
| | -125,-50 | PFIV #5 | PFIV #6 | PFIV #7 | PFIV #8 |
| Range | -50,50 | PFIV #9 | PFIV #10 | PFIV #11 | PFIV #12 |
| | 50,125 | PFIV #13 | PFIV #14 | PFIV #15 | PFIV #16 |
| (MW) | 125,176 | PFIV #17 | PFIV #18 | PFIV #19 | PFIV #20 |

| Network Topology 3 | | 345 kV line 1036-N out of service | | | |
|---|---|---|---|---|---|
| Generation Range (MW) | | 500-600 | 600-760 | 760-925 | 925-1225 |
| Interchange | -225,-125 | PFIV #1 | PFIV #2 | PFIV #3 | PFIV #4 |
| | -125,-50 | PFIV #5 | PFIV #6 | PFIV #7 | PFIV #8 |
| Range | -50,50 | PFIV #9 | PFIV #10 | PFIV #11 | PFIV #12 |
| | 50,125 | PFIV #13 | PFIV #14 | PFIV #15 | PFIV #16 |
| (MW) | 125,176 | PFIV #17 | PFIV #18 | PFIV #19 | PFIV #20 |

Note: The penalty factor information vector (PFIV) contains all of the penalty factors.

The basic optimization formulation is to include one of the curves from Table 4-1 into the overall set of relationships. The objective function is augmented with the constraint, the first derivatives are found and set equal to zero. Mathematically, this satisfies the necessary condition but not the sufficient condition.

The second-order derivatives would have to be taken to satisfy the sufficient condition. Most techniques rely only on the necessary condition to find the optimal solution. If the energy conversion curve has favorable characteristics (as defined below), then the sufficient condition may be satisfied simultaneously with the necessary condition. The inclusion of system losses simply requires the addition of the penalty factors. The penalty factor (PF), the incremental transmission loss factor (ITL), and the network sensitivity factor are all related. Any technique, which produces one of the three, may be used. This research assumes that the loss model parameters (PF, ITL, or NS) are available and included as part of the input data.

The type of algorithms considered are based upon directed search techniques. Directed search techniques use information available from the objective function to improve the solution either without forcing the solution to be an infeasible solution along the way or by forcing the solution to become feasible.

The first algorithm considered is the steepest descent gradient search. This algorithm starts from a known feasible solution and experimentally searches for the best direction to improve each variable (unit generation). Some of the more natural solutions to start from include initial conditions, the optimal solution for the previous stage (hour), or the optimal solution for the previous combination. Any of the above energy conversion curve models may be used as long as the first derivative exists. The strength of this approach includes simplicity, easy to model unique unit operating constraints, will converge if the set of energy conversion curves is monotonically increasing and sufficiently smooth, solution accuracy, and some sensitivity information. The weakness of this approach includes slow convergence, inaccuracies of curve representation for non polynomial models, the models have to be regenerated whenever a unit status changes (e.g., the system loss model changes, fuel cost changes, fuel type changes) for piecewise linear models.

(1) Augment objective function with constraint

$$\underset{u}{MIN}\,\pi = \sum_{n=1}^{N} F_n(u_n) - \lambda\left(\sum_{n=1}^{N} u_n - BLD\right)$$

(2) Find all partial derivatives and equate to zero

$$\frac{d\pi}{du_n} = \frac{dF}{du_n} - \lambda(1) = 0 \mid n = 1,\ldots,N$$

$$\frac{d\pi}{d\lambda} = -\left(\sum_{n=1}^{N} u_n - BLD\right) = 0$$

The next algorithm considered was the first-order gradient method. This algorithm can be obtained from the Taylor series expansion [Wood, 1996].

(1) Augment objective function with constraint

$$\underset{u_n}{MIN}\,\pi = \sum_{n=1}^{N} F_n(u_n) - \lambda\left(\sum_{n=1}^{N} u_n - BLD - LOSS\right)$$

(2) Find all partial derivatives and equate to zero

$$\frac{d\pi}{du_n} = \frac{dF_n}{du_n} - \lambda * \left(1 - \frac{dLOSS}{du_n}\right) = 0 \mid n = 1,\ldots,N$$

$$\frac{d\pi}{d\lambda} = -\left( \sum_{n=1}^{N} u_n - BLD - LOSS \right) = 0$$

Note: The loss coefficient is often rewritten into an equivalent form:

$$\frac{dLOSS}{du_n} = [1 - ITL_n] = 1 / PF_n$$

(3) Find starting point, $\underline{u}$

(4) Compute gradient, $GRAD(\underline{u}) = \sum_{n=1}^{N} \frac{df}{du_i} e_i$

*(5)* Update generation of each unit *(n=1,...,N)*

$$\hat{u}_n = u_n + s * GRAD(\underline{u}_n)$$

$$\hat{u}_n = \begin{cases} u_n \min \textit{ if } u_n < u_n \min \\ u_n = \hat{u}_n \\ u_n \max \textit{ if } u_n > u_n \max \end{cases}$$

$$\hat{P}TOT = PTOT + \hat{u}_n$$

(6) If $P\hat{T}OT > GTBD$ then go to Step 5, else go to Step 6.

(7) $\hat{s} = s / 2.$, go to Step 3.

(8) Update $\underline{u}$ and PTOT.

(9) If $ABS(GTBD - PTOT) \leq \varepsilon$, then go to Step 10.

$$\delta(P_u) = \begin{cases} P_u - P_u \min \textit{ if } \frac{df_u}{dp_u} < 0 \\ u_n = \hat{u}_n \\ P_u \max - P_u \textit{if } \frac{df_u}{dP_u} > 0 \end{cases}$$

(10) Calculate maximum step size:

$$\delta = \min\{\delta(P_u), \delta(P_1)\}$$

(11) Update solution:

$$P_u^{i+1} = P_u^i + \delta; P_1^{i+1} = P_1^i + \delta$$

(12) If $ABS(\delta^i - \delta^{i+1}) \leq O_3$, then go to Step 12.

(13) it = it + 1;
If IT > ITMAX then flag nonconvergence and return, else go to Step 2.

(14) Flag convergence and return.

This method starts from any feasible solution. The generating units, which will change the objective function the most, are selected and the unit generations changed accordingly. This method is obviously very similar to the previous method except that feasibility is never lost.

Any of the above energy conversion curve models may be used as long as the first derivative exists. The strength of this approach includes simplicity, easy to model unique unit operating constraints, will converge if the set of energy conversion curves is monotonically increasing and sufficiently smooth, solution accuracy, and some sensitivity information. The weakness of this approach includes slow convergence, inaccuracies of curve representation for non-polynomial models, the models have to be regenerated whenever a unit status changes (e.g., the system loss model changes, fuel cost changes, fuel type changes) for piecewise linear models.

(1) Find feasible starting point, $\underline{u}$

(2) Compute gradient, $GRAD(\underline{u}) = \sum_{n=1}^{N} \frac{df}{du_i} e_i$

(3) Check for convergence at a stationary point (local optimum or saddle point)
If $GRAD(\underline{u}_i)| < O_2$ for all i=1,...,N
then go to 12.

(4) Select: $P = \sum_{u}^{N} \left\{ \sum_{n=1} (\underline{u}_n) \right\}$

(5) Calculate relative costs: $c_i = \frac{df_i}{dp_i} - \frac{df_u}{dp_i}|$ For all i=1,...,N and i <> u

(6) Select: $P_1 = \sum_{n=1}^{N} (c_n) n <> u$

(7) Calculate generation change of each unit (1 and u):

(8) IT = IT + 1; if IT > ITMAX then go to Step 9, else go to Step 2.

(9) Flag nonconvergence and return.

(10) Flag convergence, save optimum solution, and return.

If the objective function is additively separable, then each function can be linearized separately by a set of points for each variable [Lasdon, 1983, Neuhauser, 1988]. If the objective functions are linear functions, then this grid linearization process can be transformed into the Dantzig-Wolfe decomposition algorithm. The Dantzig and Wolfe decomposition principle [Luenberger, 1984] is considered to be the most efficient solution algorithm for convex simplex problems of linear objective functions with linear constraints. The methods to achieve such efficiencies are listed in the text by Lasdon [1970]. The approach is to form an equivalent "master program," which has only a few rows but many columns. The technique to achieve efficiency is to generate the columns only when needed by the simplex algorithm.

The main linear programming (LP) application that has also been most successful is the application of the Dantzig-Wolfe Decomposition to the Economic Dispatch algorithm. This is due partially to the predominant use of piecewise linear curves. Another reason is due to the convexity of the energy conversion curves even though this convexity is normally forced since the raw data used to generate the curves are normally generated by a step function and even though there is considerable measurement error. The convexity is often required for the real-time control algorithms. This algorithm was tested for this research.

All of the energy conversion curve models may be used, but any LP-based technique will work best with curves that are nearly linear. Linear Programming methods are used extensively in optimization. Linear Programming based algorithms are the techniques used for most commercial optimization packages. The literature abounds with information from special coding techniques to special solution algorithms for specially structured problems (e.g., network flows). The strengths of any LP-based algorithms include simplicity, easy to add models for unique unit operating constraints, convergence is easily detected, convergence to a global optimum is guaranteed if the set of energy conversion curves is monotonically increasing (convex), solution speed, solution accuracy if the curves are nearly linear, and a wealth of sensitivity information.

The weakness of these algorithms is predominantly due to the energy conversion curve representation. These weaknesses may include slow convergence if the curves are highly nonlinear, storage requirement for the linearized models, inaccuracies of curve representation, and need for regeneration whenever a unit status changes (e.g., the system loss model changes, fuel cost changes, fuel type changes).

Original problem: $$MIN\ z = \sum_{n=1}^{N} c_n * u_n$$

Such that: $$\sum_{n=1}^{N} A_n * u_n = b_0$$

$$B_n * u_n = b_n = b_0$$

$$u_n \geq 0$$

For all n=1,...,N
N ≥ 1

Master problem: $$\underset{j}{MIN} \sum_{\varepsilon B} (f_j * 1)$$

Such that:

$$\sum_{j \varepsilon B} (1_j) = 1$$

$$\sum_{j \varepsilon B} (1_j) = 1$$

$$1_j \geq 0$$

For all j $\varepsilon$ B

And each subproblem:

$$MIN\ z_i = (c_i - p_1 * A_i) * u_i$$

such that:

$$B_i * u_i = b_i$$

$$u_i \geq 0$$

For all i=1,...N

All of these techniques, and many others, estimate the system incremental cost ($\lambda$). This system incremental cost is called the *shadow price* in any operations research text and is related to the price of the commodity under certain situations. It is the price if the market has reached ideal conditions and perfect competition is attained. Otherwise, it is only an indicator of the price. This is because the system incremental cost does not capture any of the fixed costs that are lost after taking the derivative. Thus, the system $\lambda$ is an estimate of the price under ideal conditions. However, the algorithms show a very strong similarity to the auction methods of the previous chapter. The system $\lambda$ is adjusted until feasibility occurs. Alternatively, think of feasibility as an agreement to trade contracts. An auctioneer and a supplier agree to the price when supply equals demand. Specifically, conservation of energy is achieved. If both buyer and seller offer bids, then we have an auction as originally established by the Florida Brokerage system [Wood, 1996].

## 4.2 Unit Commitment [Dekrajangpetch, 1998, 1988]

The unit commitment problem has been approached by many techniques but only acceptably solved by two techniques: dynamic programming and LaGrangian relaxation. The problem of scheduling electric power units is due to the integer nature of the problem that a unit can either be off-line or on-line. The modeling of thermal power plants, for accurate scheduling, is complicated.

The modeling used in this work is a simplified model of a thermal power unit. The unit's minimum and maximum capability will be included.

$$U_i^t P_i^{\min} \leq P_i^t \leq U_i^t P_i^{\max} \qquad \text{for } i=1\ldots N \text{ and } t=1\ldots T$$

The unit's minimum uptime and downtime will be included.

The unit's production cost will be represented by a single quadratic curve.

P_cost = F(Pi) = aiPi2 + biPi + ci

The unit's startup cost and shutdown cost will be treated as a transition cost.

T_cost = Startup_cost (i) if the unit starts this hour,

or

T_cost = Shutdown_cost (i) if the unit shuts down this hour.

The modeling of the power system used in this work is simply the conservation of energy.

$$P^t_{demand} = \sum_{i=1}^{N} P_i^t U_i^t \qquad \text{for } t = 1 \ldots T$$

Other transmission models could be included inter area and intra area flows, power flow limitations, transmission security constraints, fuel limitations, environmental limitations, spinning reserve, ready reserve, individual equipment flows, etc. Spinning reserve will be discussed in the next chapter.

The easiest manner of showing the mathematical formulation is to use a state variable, U(i,j), to show the status of a unit, i, at a given hour, j. This variable then has the following values:

U(i,j) = -1 if unit *i* is off-line during hour j,
U(i,j) = +1 if unit *i* is on-line during hour j.

The resulting objective function for this unit commitment problem is as follows:

$$\sum_{t=1}^{T} \sum_{i=1}^{N} \left[F_i\left(P_i^t\right) + \text{Startup cost}_{i,t}\right] U_i^t = F\left(P_i^t, U_i^t\right)$$

The constraints are rewritten to include the unit status:

$$P^t_{load} - \sum_{i=1}^{N} P_i^t U_i^t = 0 \qquad \text{for } t = 1 \ldots T$$

$$U_i^t P_i^{\min} \le P_i^t \le U_i^t P_i^{\max} \qquad \text{for } i = 1 \ldots N \text{ and } t = 1 \ldots T$$

The LaGrangian function from the problem statement, excluding unit limitations, is as follows:

$$L(P, U, \lambda) = F\left(P_i^t, U_i^t\right) + \sum_{t=1}^{T} \lambda^t \left(P^t_{load} - \sum_{i=1}^{N} P_i^t U_i^t\right)$$

that is minimized subject to the unit limitations. Note that the unit limitations are independent of the output or of the status of other units. The conservation of energy function is a coupling constraint. Changing one of the unit's status or output changes one or all of the other units.

The LaGrangian relaxation method temporarily ignores the coupling constraints and solves the problem as if there was not any coupling. A dual optimization procedure is used to search for the constrained optimum by maximizing the LaGrangian over the multipliers while minimizing with respect to all other variables:

$$q^*(\lambda) = \max_{\lambda^t} q(\lambda)$$

$$q(\lambda) = \min_{P_i^t, U_i^t} L(P, U, \lambda)$$

This is a two-step procedure, one step for each optimization. First, find the value of λ(t) that makes the objective value as large a number as possible. Second, search for the minimum value of L by adjusting the output and the status of each unit. Assume for the second step that the value of λ is fixed.

Many methods will update the dual variables. This work uses the simplest technique to compare the procedure with auctions. Consider the following popular formula based on steepest ascent techniques:

$$\lambda^t = \lambda^t + \left[\frac{d}{d\lambda} q(\lambda)\right]\alpha$$

The step length is heuristically changed by the following rules:

If $\frac{d}{d\lambda} q(\lambda) > 0$, then let $\alpha = 0.001$.

If $\frac{d}{d\lambda} q(\lambda) < 0$, then let $\alpha = 0.05$.

Note that the dual variables are updated individually for each hour. The normal solution procedure is to continue updating the dual variables until the duality gap is "sufficiently small":

$$\frac{J^* - q^*}{q^*} \leq \varepsilon$$

Also note that the updates are the inverse value of those values normally shown [Wood, 1996] because we do not wish to overestimate any of the dual variables since the outer loop stops whenever the duality gap fails to improve significantly.

The generators on during this iteration when the dual variable is higher than necessary will be paid a higher price than necessary. This is shown in more detail below.

The LaGrangian is rewritten to eliminate the constant term by first ignoring the hard unit limit constraints:

$$L=\sum_{t=1}^{T}\sum_{i=1}^{N}\left[F_i\left(P_i^t\right)+\text{Startup cost}_{i,t}\right]U_i^t+\sum_{t=1}^{T}\lambda^t\left(P_{load}^t-\sum_{i=1}^{N}P_i^tU_i^t\right)$$

After expanding the formulation, the second term is a fixed constant, not a function of generation, and can be dropped:

$$L=\sum_{t=1}^{T}\sum_{i=1}^{N}\left[F_i\left(P_i^t\right)+\text{Startup cost}_{i,t}\right]U_i^t+\sum_{t=1}^{T}\lambda^tP_{load}^t-\sum_{t=1}^{T}\sum_{i=1}^{N}\lambda^tP_i^tU_i^t$$

Rearrange the terms to show the sequence of optimizations, one inside the other.

$$L=\sum_{i=1}^{N}\left(\sum_{t=1}^{T}\left\{\left[F_i\left(P_i^t\right)+\text{Startup cost}_{i,t}\right]U_i^t-\lambda^tP_i^tU_i^t\right\}\right)$$

The innermost term is solved for each generating unit independently of all other units.

$$\sum_{i=1}^{T}\left\{\left[F_i\left(P_i^t\right)+\text{Startup cost}_{i,t}\right]U_i^t-\lambda^tP_i^tU_i^t\right\}$$

The minimum of the LaGrangian is found by minimizing the generation cost for each unit over the complete time horizon of the study.

$$\min q(\lambda)=\sum_{i=1}^{N}\min\sum_{t=1}^{T}\left\{\left[F_i\left(P_i^t\right)+\text{Startup cost}_{i,t}\right]U_i^t-\lambda^tP_i^tU_i^t\right\}$$

The unit limits are used to constrain the solution.

$$U_i^tP_i^{\min}\le P_i^t\le U_i^tP_i^{\max}\qquad \text{for } t=1\ldots T$$

Dynamic programming very easily solves the solution since only two states are for each stage. Now the curse of dimensionality has been inherently removed. Normally, a stage is an hour of operation, but any defined time interval can be used. The initial condition of the unit is often obtained from the real-time status of the unit. Specifically, is the unit on or off now? At each stage, the minimization is with respect to generation output only.

$$\min\left[F_i(P_i) - \lambda^t P_i^t\right]$$

Remember that $\lambda$ is a constant, so set the first derivative to zero.

$$\frac{d}{dP_i^t}\left[F_i(P_i) - \lambda^t P_i^t\right] = \frac{d}{dP_i^t} F_i\left(P_i^t\right) - \lambda^t = 0$$

The solution is trivial.

$$\frac{d}{dP_i^t i} F_i\left(P_{opt}\right) = \lambda^t$$

Based on our knowledge of optimization, there are three cases of concern:

If $P_i^{opt} \le P_i^{\min}$, then the output is constrained to the minimum:

$$\min\left[F_i(P_i) - \lambda^t P_i^t\right] = F_i\left(P_i^{\min}\right) - \lambda^t P_i^{\min}$$

If $P_i^{\min} \le P_i^{opt} \le P_i^{\max}$, then the output is set to the equal incremental cost:

$$\min\left[F_i(P_i) - \lambda^t P_i^t\right] = F_i\left(P_i^{opt}\right) - \lambda^t P_i^{opt}$$

If $P_i^{opt} \le P_i^{\max}$, then the output is constrained to the maximum:

$$\min\left[F_i(P_i) - \lambda^t P_i^t\right] = F_i\left(P_i^{\max}\right) - \lambda^t P_i^{\max}$$

The overall procedure can be shown by the data flows in Figure 4-1. Note that the process is to pick a set of dual variable values, one for each hour, request generation bids from each unit, add the bids to determine if the system constraints are satisfied, increase the value of the dual variables for those hours when the bids are not sufficient, decrease the value of the dual variables for those hours when the bids are excessive, and repeat the process until all constraints are satisfied for all hours. This is a one-sided (generation only) auction.

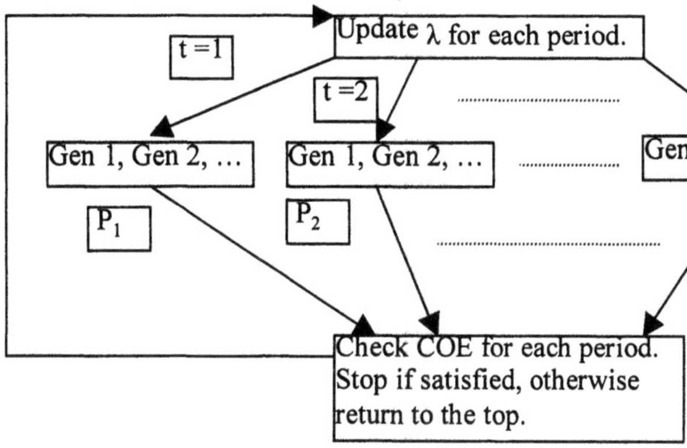


Figure 4-1. LRUC Data Flow.

This can be changed to a two-side (double) auction simply by enabling the loads (buyers) equal bidding rights. The only change to this diagram is that the term "Gen" for generator has to be replaced with "Co" or some other generic name for company.

Finally, this would be more of a standard auction if the boxes representing the companies bidding where removed from a central location and enabled to be solved anywhere. As an example, the bids could be transmitted from the generation company's computers as Figure 4-3 shows.

Then each company could decide when to bid generation to maximize profit. The bidding strategy essentially replaces the dynamic programming solution of the LaGrangian relaxation process. A more complete description, including multiple rounds, is shown in Figure 4-4. Note that a multiple round market emulates the multiple iterations of the LaGrangian relaxation solution procedure.

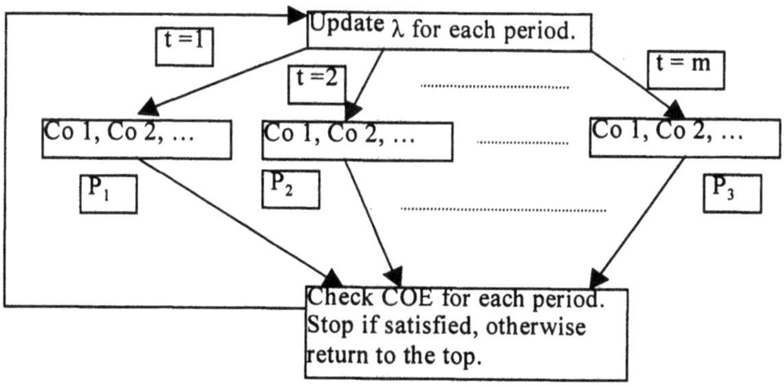


Figure 4-2. Double Auction Unit Commitment.

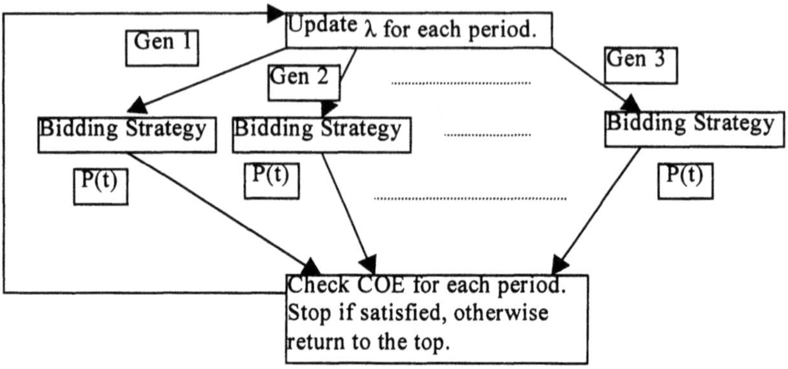


Figure 4-3. Decentralized Unit Commitment as Auction.

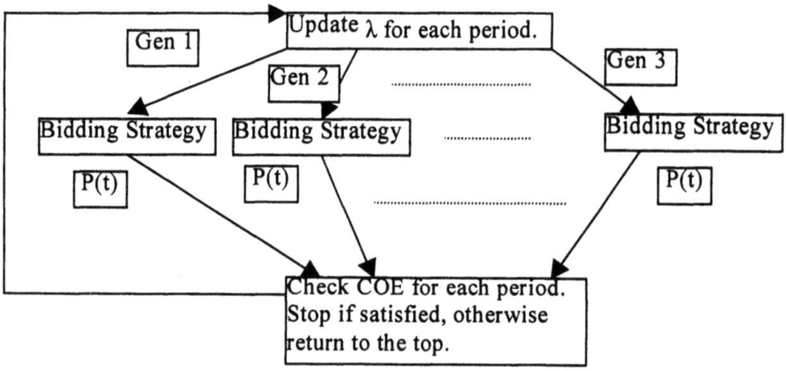


Figure 4-4. Multiple Round Double Auction.

Give model data to central authority, number of parameters (e.g., 32), Bid is not disclosed, Central authority matches bids, unit commitment, power flow, security constrained dispatch, Central authority posts matched bids. Pros, Centralized Data, One Set of Programs, Looks Like Old Environment, Cons, Audit of Data, Incentive for New Programs, Correction of Bugs, Responsibility for Equipment Constraints, One-Shot Bidding, Contingencies, TRANSCOs, etc. Matching of Sealed Bids, Adjust Matched Bids for Transmission Equipment Usage. Ancillary Services, Losses, Frequency Regulation, Voltage Support, Stranded costs.

### *4.2.1 Auction Mechanisms for Electricity*

One sided and double sided auctions have been investigated. One sided auctions allow only the GENCOs to bid. Double sided allow the GENCOs and the ESCOs to bid.

**4.2.1.1 Sequential Sealed-Bid and Sealed-Offer Auctions.** This approach of applying auctions to the pricing of electric power uses the following three-step process:

(1) A sealed-bid auction optimally allocates supply and demand independent of the transmission network.

(2) The supply allocations are adjusted for losses.

(3) A sealed-offer auction optimally selects flows on the transmission system.

*Sealed-bid auction: demand and supply.* An LP solution to a sealed-bid auction model is presented . Linear programming is used to compute the optimal exchange of a commodity. Assume there are $m$ generators and the output (MWh) each has to sell from bus $i$ is as follows:

$$p_i > 0 \quad \text{for } i = 1, 2, \ldots, m \tag{4.1}$$

Generators also specify a *reservation price*, $\pi_i$, below which they will not sell electric power. Note that buyers do not know the reservation price prior to submitting their bids.

Frequently sellers specify a reservation price. If the market price equals the reservation price, some portion of the supply may remain unsold. Thus the seller "sells the goods to himself" at the reservation price. Reservation prices are an example of *price rigidity* and can cause the market formation to fail to reach an equilibrium in which the market clears and all goods are sold [Thompson, 1992].

Also $n$ demand centers desire the following quantity of power.

$$d_j > 0 \qquad \text{for } j = 1, 2, \ldots, n \tag{4.2}$$

At the auction center, buyer $j$ submits a sealed bid of

$$c_{ij} \geq 0 \tag{4.3}$$

for one MW of generator $i$'s power. Let each generator $i$ sell

$$x_{ij} \geq 0 \tag{4.4}$$

MWs to buyer $j$.

An LP problem that maximizes total surplus uses the following form [Thompson, 1992]:

$$\text{maximize} \quad \sum_{i=1}^{m} \sum_{j=1}^{n} c_{ij} x_{ij} + \sum_{i=1}^{m} \pi_i y_i \tag{4.5}$$

subject to
$$\sum_{j=1}^{n} x_{ij} + y_i = P_i \qquad \forall i \tag{4.6}$$

$$\sum_{j}^{n} x_{ij} \geq \underline{P}_i \qquad \forall i \tag{4.7}$$

$$\sum_{i}^{m} x_{ij} \leq d_j \qquad \forall j \tag{4.8}$$

$$x_{ij} 0, \quad y_i \geq 0 \qquad \forall i, j \tag{4.9}$$

where:

$y_i$ = the unsold MWs offered by generator $i$

$\underline{P}_i$ = minimum generation of generator $i$

The objective function (4.5) maximizes total surplus in the system. Constraint (4.6) ensures that the sum of units sold and bought back by generator $i$ equals the amount, $P_i$, originally offered for sale. Equation (4.7) requires generator $i$ to produce at a level greater than or equal to its minimum generation level. Equation (4.8) limits the MW units received by node $j$ to $d_j$. Equation (4.9) requires the decision (exogenous) variables to be positive.

This linear program must be used in an iterative fashion. After the first iteration, any generator scheduled to supply at prices below its reservation price should be removed from the LP. If more than one generator is included in this condition, only the violating generator with the highest reservation price should be removed. The LP is iteratively solved until the reservation prices of the generators are not violated.

Once system demand and supply is obtained with the above LP, generation must be adjusted for losses. First, penalty factors are used to determine system losses. Second, losses are accounted for by increasing generation of the most economical generator in operation by the amount of losses.

*Sealed-offer auction: transportation*

The sealed-bid auction, used to determine the optimal supply and demand of electricity, yields a fixed inelastic demand for delivery services. Transmission owners submit sealed-offer schedules and the transmission market is cleared at a delivery price that equalizes the inelastic demand quantity and the supply of delivery services.

The LP used to maximize total surplus takes the form of the classical transshipment model with the addition of capacity constraints. This approach assumes transmission owners are able to control the power flows on transmission lines. This can be accomplished with flexible AC transmission system (FACTS) devices.

The LP takes the following form:

$$\text{Minimize} \quad \sum_{i=1}^{n} \sum_{j=1}^{n} to_{ij}\, f_{ij} \tag{4.10}$$

Subject to

$$\sum_{j=1}^{n} f_{ij} - \sum_{k=1}^{n} f_{ki} = s_i \quad \forall\, i \tag{4.11}$$

$$f_{ij} \le f_{ij}^{max} - f_{ij}^{0} \quad \forall\, i,j \tag{4.12}$$

$$f_{ij} \ge 0 \tag{4.13}$$

where: $ij$ = transmission line from node $i$ to node $j$

$f_{ij}$ = flow on line $ij$

$f_{ij}^{0}$ = initial flow on line $ij$

$f_{ij}^{max}$ = maximum flow on line $ij$

$s_i$ = MW supply at node i; $s_i < 0$ denotes demand

$to_{ij}$ = transmission offer price per MW flow on line $ij$

The objective function (4.10) minimizes the cost of transmitting electric power from the generators to the demand centers. The first constraint (4.11) requires the flow out of node $i$ to equal $s_i$. The second constraint (4.12) reflects line capacity. The third constraint (4.13) requires the decision (exogenous) variable to be positive.

*4.2.1.2 Sealed-Bid Offer/Double Auction*

This approach is similar to that used by McCabe et al. [McCabe, 1989] in the natural gas industry. Representatives of generation, transmission, and distribution submit bids/offers to a regional transmission group (RTG) that acts as the auctioneer. The RTG is responsible for identifying the equilibrium price and allocation so that total surplus is maximized.

*Players*

*Generation.* Owners of generation sites submit price/quantity offers for the electric power they are able to supply. Generation may be from independent power producers (IPPs), qualifying facilities (QFs), exempt wholesale generators (EWGs), etc.

*Transmission.* Transmission owners submit price/quantity offers for the transmission of electric power. To decide the amount of power they are capable of transmitting and, thus, the price for their offers, transmission companies use the following approach to determine the validity of their offers [Sheblé, 1994a]:

(1) Solve an AC/DC power flow for the power system (including existing transaction flows but not proposed offers).

(2) Create constraints to restrict flow on all overloaded lines.

(3) Implement step 2 by the use of distribution factors.

(4) If there is no change in system limitations (with respect to base case power flow) then quit, else continue.

(5) To determine optimal $\Delta P_i^+$ and $\Delta P_i^-$, solve the following constrained economic dispatch LP:

$$\text{Minimize} \quad \sum_{\forall i} \left(\alpha_i \, \Delta P_i^+ - \alpha_i \, \Delta P_i^-\right) \tag{4.14}$$

$$\text{subject to} \quad 0 \le \Delta P_i^+ \le \overline{P}_i - P_i^0 \quad (\forall\, i) \tag{4.15}$$

$$0 \le \Delta P_i^- \le P_i^0 - \underline{P}_i \quad (\forall\, i) \tag{4.16}$$

$$\sum_{\forall i} \left[\left(1 - \frac{\Sigma P_L}{\Sigma P_i}\right)\left(\Delta P_i^+ - \Delta P_i^-\right)\right] = D + P_L^0 + I + \sum_{\forall i} P_i^0 \tag{4.17}$$

$$\sum_{\forall i} a_{li}\left(\Delta P_i^+ - \Delta P_i^-\right) \le f_l^{max} - f_l^0 \quad \forall\, l \tag{4.18}$$

$$\sum_{\forall i} a_{li}\left(\Delta P_i^+ - \Delta P_i^-\right) \ge -f_l^{max} - f_l^0 \quad \forall\, l \tag{4.19}$$

where:

$a_{li}$ = generation shift factor (the sensitivity of the flow on line $l$ to a change in generation at bus $i$);

$$a_{li} = \frac{\Delta f_l}{\Delta P_i}$$

$D$ = load demand

$f_l^0$ = base case flow on line $l$

$f_l^{max}$ = maximum flow on line $l$

$i$ = bus index

$I$ = interchange (out of system)

$l$ = line index

$\overline{P}_i$ = maximum generation at bus $i$

$\underline{P}_i$ = minimum generation at bus $i$

$P_i^0$ = base case generation at bus $i$

$P_L$ = losses

$P_L^0$ = base case losses

$\Delta P_i^+$ = positive change in generation at bus $i$

$\Delta P_i^-$ = negative change in generation at bus $i$

The objective function (4.14) maximizes the shift in generation for the submitted offer prices (maximized since offers are negative). Equations (4.15) and (4.16) represent the generation shift limits. Equation (4.17) ensures the conservation of energy. Equations (4.18) and (4.19) represent the line flow constraints. For an explanation of these constraints and the LP formulation see [Wood, 1996].

(6) Adjust control variables [generation, tap-changing under-load transformers (TCUL), quadrature phase-shifting transformers (QPS), high voltage DC transmission lines (HVDC), flexible AC transmission (FACTS)].

(7) Return to step 1 and repeat.

Transmission and distribution owners require all line flows to be within their limits. This is done by solving the LP problem with constraints for each overloaded line. After solving it, the generation shifts should be made and a new base case load flow run. Any "new" overloaded lines should be included when re-executing the LP. Classical approaches [Wood, 1996] suggest keeping all line flow constraints from the first LP as well as the "new" overloads as a safety measure. This process, called an *iterative constraint search,* is repeated until the network has no overloads.

ESCOs and representatives of coops, municipalities, and industry submit price/quantity bids for the electric power needed to meet their demand.

The ISO implements the following procedure [Debs, 1998]:

(1) Collect bids/offers from all players.

(2) Identify maximum potential exchange between areas by the use of an optimal power flow.

(3) Obtain distribution factors for entire system by the use of a decoupled power flow.

(4) Obtain system interchange and precontract generation from all players.

(5) Run constrained economic dispatch LP (4.14) through (4.19) for the entire system.

(6) Adjust contracts according to Step 5 changes to the base case.

(7) Report results to players.

The two applications of auctions presented in this chapter approach the pricing of electric power from different angles. The sequential sealed-bid/sealed-offer approach considers the supply and demand allocations to be independent of electric power transmission; however, demand centers may submit bids that reflect expected transmission prices. The sealed-bid/offer double auction simultaneously allocates the supply, demand, and transmission of electric power. Based on similar approaches in the natural gas industry [Post, 1993], the simultaneous double auction is expected to achieve more efficient results.

*4.2.1.3 Results*

To illustrate the methods discussed above, a 12-bus, 7-generator system will be used. Table 1 lists the generation and load data of the system. Table 2 contains the transmission line data.

Seven different companies may own the seven generators of this power system. Alternatively, generation companies may own two or more generators at various geographical locations that are scheduled using economic dispatch. The necessary condition is that enough generation companies exist so that a competitive environment is realized.

Table 4-3. Generation and Load Data.

| BUS | LOAD (MW) | MAXIMUM GENERATION (MW) | MINIMUM GENERATION (MW) | PENALTY FACTOR |
|---|---|---|---|---|
| 1 | 0.0 | 0.0 | 0.0 | 1.00 |
| 2 | 350.0 | 400.0 | 100.0 | 0.97 |
| 3 | 0.0 | 400.0 | 100.0 | 0.97 |
| 4 | 0.0 | 400.0 | 100.0 | 0.96 |
| 5 | 0.0 | 600.0 | 150.0 | 0.95 |
| 6 | 250.0 | 400.0 | 100.0 | 0.96 |
| 7 | 100.0 | 0.0 | 0.0 | 1.01 |
| 8 | 200.0 | 0.0 | 0.0 | 1.01 |
| 9 | 200.0 | 0.0 | 0.0 | 1.02 |
| 10 | 0.0 | 0.0 | 0.0 | 1.00 |
| 11 | 300.0 | 200.0 | 50.0 | 0.98 |
| 12 | 100.0 | 200.0 | 50.0 | 0.99 |
| TOTAL | 1500.0 | 2600.0 | 650.0 | - |

Table 4-4. Transmission Line Data.

| LINE # | FROM BUS | TO BUS | CAPACITY (MW) | LENGTH (MILES) |
|---|---|---|---|---|
| 1 | 1 | 2 | 300.0 | 40.0 |
| 2 | 1 | 4 | 300.0 | 120.0 |
| 3 | 1 | 12 | 300.0 | 30.0 |
| 4 | 2 | 3 | 300.0 | 60.0 |
| 5 | 3 | 11 | 500.0 | 90.0 |
| 6 | 3 | 4 | 100.0 | 50.0 |
| 7 | 5 | 7 | 300.0 | 50.0 |
| 8 | 5 | 6 | 500.0 | 90.0 |
| 9 | 6 | 10 | 500.0 | 60.0 |
| 10 | 7 | 8 | 300.0 | 30.0 |
| 11 | 8 | 12 | 100.0 | 60.0 |
| 12 | 9 | 10 | 500.0 | 60.0 |
| 13 | 9 | 12 | 300.0 | 70.0 |
| 14 | 11 | 12 | 100.0 | 27.0 |
| 15 (example 2 only) | 4 | 5 | 500.0 | 40.0 |

### *Sequential sealed-bid and sealed-offer auctions*

The example system is solved by the sequential auction approach of Chapter 3. This approach forms a basis against which other auction mechanisms, such as double auctions, can be evaluated.

*Sealed-bid auction: demand and supply.* The sealed-bid auction was used to optimally allocate the supply and demand for the system. The following assumptions were made:

- Generator owners offer all available power, subject only to the reservation price.
- Each load center submits bids, in units of R/MW, for each generation bus.

Table 4-5 contains the sealed bids and generation reservation prices, $\pi_i$, submitted to the auctioneer.

Table 4-5. Sealed-Bids and Generator Reservation Prices.

| GEN. BUS | LOAD BUS 2 | LOAD BUS 6 | LOAD BUS 7 | LOAD BUS 8 | LOAD BUS 9 | LOAD BUS 11 | LOAD BUS 12 | RESERVED PRICE $\pi_i$ |
|---|---|---|---|---|---|---|---|---|
| 2 | 10.0 | 9.4 | 9.0 | 9.2 | 9.4 | 9.2 | 9.3 | 9.4 |
| 3 | 9.5 | 9.1 | 8.9 | 9.2 | 9.3 | 9.4 | 9.4 | 9.4 |
| 4 | 9.1 | 8.9 | 9.0 | 9.3 | 9.4 | 9.3 | 9.2 | 9.4 |
| 5 | 9.0 | 10.6 | 9.5 | 9.4 | 9.4 | 9.1 | 9.2 | 9.8 |
| 6 | 8.0 | 11.0 | 9.4 | 9.3 | 9.6 | 9.0 | 9.3 | 9.4 |
| 11 | 9.0 | 10.0 | 9.3 | 9.4 | 9.6 | 9.4 | 9.4 | 9.3 |
| 12 | 9.1 | 10.3 | 9.4 | 9.6 | 9.7 | 9.4 | 9.4 | 9.3 |

The bids of Table 4-5 have been arbitrarily assigned by the author. This would not be the case if auctions were used for pricing electricity. However, the intent of this example is to show how an auction might be implemented, not how bids should be submitted. Thus, the reservation prices in Table 4-5 were set. Next, the "sealed bids" were assigned so that some were acceptable and others were not. In this example, bids are submitted as a single-step process. However, bids (offers) may be piecewise linear functions that reflect different prices for specified blocks of power.

The LP based on (4.5) through (4.9) was used to determine optimal supply and demand. However, the first iteration results violate the reservation prices of the generators at bus 4 and bus 5. Since the reservation price of bus 5 generation is larger than that of bus 4, the LP was solved a second time without generation at bus 5. This results in a schedule that violates the reservation price of generation at bus 4. After withdrawing the generation at bus 4, the LP was solved a third time and, this time, no reservation prices were violated. Results for this example are presented in Table 4-6 and Table 4-7. Table 4-6 indicates that all the load was met except the demand at bus 7. Table 4-7 contains the supply allocations and shows which bids, of Table 3, were filled.

Table 4-6. Demand Allocation.

| BUS # | DESIRED DEMAND (MW) | SUPPLIED DEMAND (MW) |
|---|---|---|
| 2 | 350 | 350 |
| 6 | 250 | 250 |
| 7 | 100 | 0 |
| 8 | 200 | 200 |
| 9 | 200 | 200 |
| 11 | 300 | 300 |
| 12 | 100 | 100 |
| **TOTAL** | 1500 | 1400 |

Next, generation was adjusted for losses. Each operating generator's output was multiplied by its corresponding penalty factor (see Table 4-3). The sum of these values gave the system losses. These losses presented an additional demand which were assigned to the generator with the lowest, unmatched high-bid (with consideration of the reservation price). In this example, generation at bus 3 was assigned an additional load of 40 MW. This was supplied at a price of 9.4 R/MW. Thus the total value of sales for the system increased to R 16,106.00, and the total amount of power generated at bus 3 increased to 290 MW.

*Sealed-offer auction: transportation.* The loss adjustment discussed in the previous section results in the supply and demand allocation needed for the transportation model. These values are listed in Table 4-8. Note that the generation supply values were adjusted by penalty factors and, in the case of bus 3 generation, adjusted for the increased supply necessitated by losses. (The generators actually produced what is shown as "amount sold" in Table 4-7, except for generator 3, which produced 290 MW).

Table 4-7. Supply Allocation.

| GENERATION AT BUS | AMOUNT OFFERED (MW) | AMOUNT SOLD (MW) | PRICE (R/MW) | SOLD TO LOAD CENTER AT BUS |
|---|---|---|---|---|
| 2 | 400 | 350 | 10.0 | 2 |
| 3 | 400 | 150 | 9.4 | 11 |
| | | 100 | 9.4 | 12 |
| 4 | 400 | 0 | 0.0 | - |
| 5 | 600 | 0 | 0.0 | - |
| 6 | 400 | 250 | 11.0 | 6 |
| | | 150 | 9.6 | 9 |
| 11 | 200 | 50 | 9.6 | 9 |
| | | 150 | 9.4 | 11 |
| 12 | 200 | 200 | 9.6 | 8 |
| TOTAL VALUE OF SALES (R) | before loss adjustment: | 15,730.00 | after loss adjustment: | 16,106.00 |

Table 4-8. Supply and Demand for Transportation Problem.

| BUS | DEMAND (MW) | SUPPLY (MW) | TOTAL SUPPLY (supply- demand) |
|---|---|---|---|
| 1 | 0 | 0 | 0 |
| 2 | 350 | 339.5 | -10.5 |
| 3 | 0 | 282.5 | 242.5 |
| 4 | 0 | 0 | 0 |
| 5 | 0 | 0 | 0 |
| 6 | 250 | 384.0 | 134.0 |
| 7 | 0 | 0 | 0 |
| 8 | 200 | 0 | -200.0 |
| 9 | 200 | 0 | -200.0 |
| 10 | 0 | 0 | 0 |
| 11 | 300 | 196.0 | -104.0 |
| 12 | 100 | 198.0 | 98.0 |
| TOTAL | 1400 | 1400 | 0 |

The transportation problem represented by (4.10) through (4.13) was used to optimally allocate the transmission of electricity. The "total supply" column of Table 4-8 represents the supply, $s_i$, of (4.11). For the example power system, initial flows are zero and maximum line capacities are given in Table 4-4. The cost-per MW-mile, $c$, for each transmission line was given by transmission companies A, B, and C. The costs per MW-mile multiplied by the transmission length gives the transmission offers, *to,* shown in Table 4-9 below. Offer prices are assumed to be independent of the flow direction on a line. This need not be the case. Transmission owners may wish to encourage the sale of power in a specific direction on a line. For example, a heavily loaded line transporting power from node $i$ to node $j$ may be relieved by submission of a low offer price for transmission from node $j$ to node $i$.

**Example 1**
The transportation problem was solved, and the results are shown in Table 4-10. With the sequential auction approach, the total cost of supplying and transporting electricity was R 19,412.90.

**Example 2**
The example system was modified with the addition of line 15 between bus 4 and bus 5 (data for this line is included in Table 4-4 and Table 4-9).

Company B's addition of line 15 to the power system increases its market power. Company A must now compete with two lines, owned by company B, for the transportation of generation from buses 2, 3, and 4 to the demand centers. Company C is also affected by the addition but not as much as Company A. In this example system, Company B has an advantage because it controls nearly all of the interchange between the areas served by Company A and Company C.

Company B submitted a low offer price for line 15 in an attempt to encourage use of their other transmission lines. The transportation problem was solved, and the results for example 1 and example 2 are shown in Table 4-10.

Company B's addition of line 15 to the power system caused an increase in power routed through Company B's lines. The addition changed the flows in Company C's transmission lines and reduced the use of Company A's transmission lines. In addition, the cost of transmission was reduced from R 3,306.90 to R 3,030.42. Therefore, the total cost of supplying and transporting electricity was R 19,136.42, a savings of R 276.48 over example 1.

Example 2 assumes that the supply and demand allocations would not be affected by the addition of line 15. However, the bids for generation submitted by demand centers might reflect the expected change in transportation prices caused by the addition of line 15.

## 4.3 Auction Mechanism with Classical Optimization

The two types of auctions presently implemented can be derived from many optimization techniques (LaGrangian Relaxation Unit Commitment, Dantzig-Wolfe, and Pricing Mechanism Decomposition). The connection between hourly auctions and unit commitment is most clearly visualized by examining the data flow as shown in Figure 4-4. Note that the price (lambda) is fixed for each period, then each generator unit is requested to provide the amount of generation available for that price. This is a non discriminating auction since the same price is paid to all GENCOs.

Table 4-9. Transmission Costs.

| LINE # | COST PER MW-MILE | LENGTH (MILES) | *to* (R/MW) |
|---|---|---|---|
| 1 | 0.05 | 40.0 | 2.0 |
| 2 | 0.04 | 120.0 | 4.8 |
| 3 | 0.06 | 30.0 | 1.8 |
| 4 | 0.03 | 60.0 | 1.8 |
| 5 | 0.04 | 90.0 | 3.6 |
| 6 | 0.08 | 50.0 | 4.0 |
| 7 | 0.06 | 50.0 | 3.0 |
| 8 | 0.03 | 90.0 | 3.6 |
| 9 | 0.04 | 60.0 | 2.4 |
| 10 | 0.05 | 30.0 | 1.5 |
| 11 | 0.09 | 60.0 | 5.4 |
| 12 | 0.03 | 60.0 | 1.8 |
| 13 | 0.05 | 70.0 | 3.5 |
| 14 | 0.07 | 27.0 | 1.9 |
| 15 (ex. 2 only) | 0.02 | 40.0 | 0.8 |

Table 4-10. Transmission Results (Example 1).

| LINE # | FROM BUS | TO BUS | FLOW (MW) |
|---|---|---|---|
| 1 | 1 | 2 | -68.0 |
| 2 | 1 | 4 | 0.0 |
| 3 | 1 | 12 | 68.0 |
| 4 | 2 | 3 | -78.5 |
| 5 | 3 | 11 | 204.0 |
| 6 | 3 | 4 | 0.0 |
| 7 | 5 | 7 | 100.0 |
| 8 | 5 | 6 | -100.0 |
| 9 | 6 | 10 | 34.0 |
| 10 | 7 | 8 | 100.0 |
| 11 | 8 | 12 | -100.0 |
| 12 | 9 | 10 | -34.0 |
| 13 | 9 | 12 | -166.0 |
| 14 | 11 | 12 | 100.0 |
| **Total Cost (R):** | 3,306.90 | | |

Table 4-11. Transmission Results (Examples 1 & 2).

| LINE # | FROM BUS | TO BUS | FLOW (MW) (example 1) | FLOW (MW) (example 2) |
|---|---|---|---|---|
| 1 | 1 | 2 | -68.0 | 0.0 |
| 2 | 1 | 4 | 0.0 | 0.0 |
| 3 | 1 | 12 | 68.0 | 0.0 |
| 4 | 2 | 3 | -78.5 | -10.5 |
| 5 | 3 | 11 | 204.0 | 172.0 |
| 6 | 3 | 4 | 0.0 | 100.0 |
| 7 | 5 | 7 | 100.0 | 100.0 |
| 8 | 5 | 6 | -100.0 | 0.0 |
| 9 | 6 | 10 | 34.0 | 134.0 |
| 10 | 7 | 8 | 100.0 | 100.0 |
| 11 | 8 | 12 | -100.0 | -100.0 |
| 12 | 9 | 10 | -34.0 | -134.0 |
| 13 | 9 | 12 | -166.0 | -66.0 |
| 14 | 11 | 12 | 100.0 | 68.0 |
| 15 | 4 | 5 | -- | 100.0 |
| | | **Total Cost (R):** | 3,306.90 | 3,030.42 |

Figure 4-6 shows an hourly auction where the competing generating units submit bids consisting of price and quantity. If a non discriminating auction price were used, then the results would be the same as for the unit commitment in Figure 4-1. A discriminating auction would require each unit to receive the bid as the price amount.

The objective of both the unit commitment and the hourly auction is to find the optimal solution of operation subject to all operational constraints.

If ESCOs are enabled to bid, then the extensions to the unit commitment algorithm is to model purchases as negative generation. The same is true for the hourly auction. As shown in Figure 4-6, the objective of both methods is to find the optimal solution of the supply demand curves for each hour. The difference between lambda update techniques can be analyzed with respect to English and Dutch auctions.

Thus, the first place to identify bidding strategies is to find the best method for updating the LaGrangian multipliers for the unit commitment problem. However, each bidder would want to be more conservative since an overbid may destroy all profits if price discovery occurred before the bid could be corrected. Indeed, many auctions would not allow an overbid to be corrected.

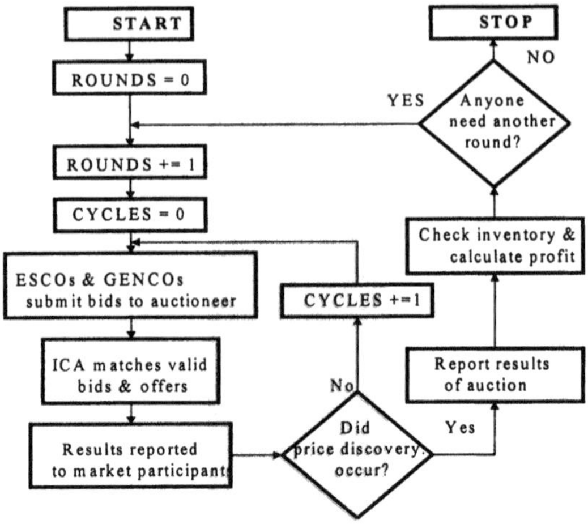


Figure 4-6. Block Diagram Of The Auction Process.

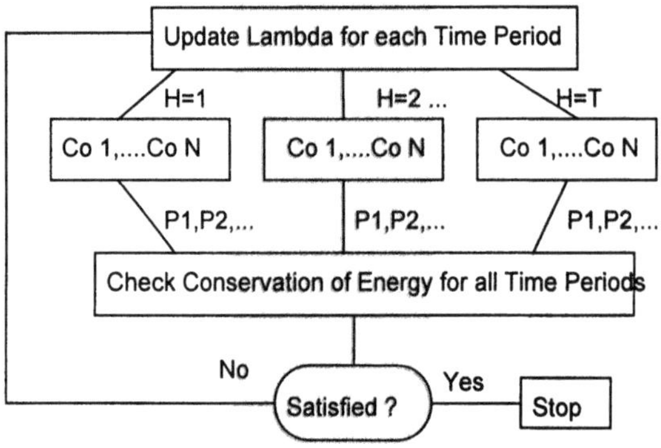


Figure 4-7. LaGrangian Relaxation Unit Commitment.

Instead of detailing the various methods to update the multipliers, which are so well documented, it is more interesting to determine what the price strategy would be for each hour if the market is dominated by one of the companies. Indeed, this has been the case in many actual markets, where one company has achieved dominance. The essential step is to identify the demand elasticity from previous market bids made by purchasing companies.

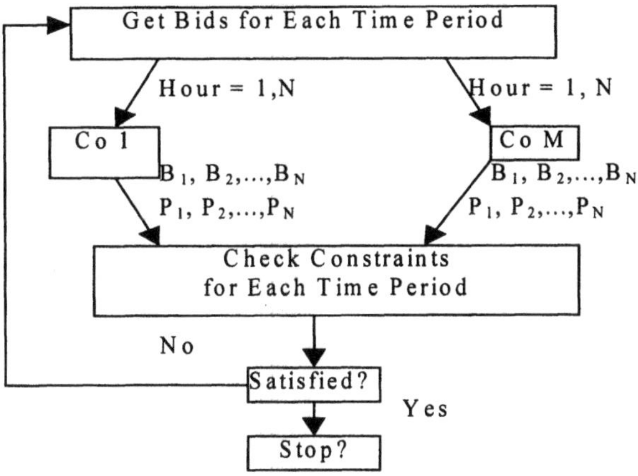


Figure 4-8. Hourly Auction.

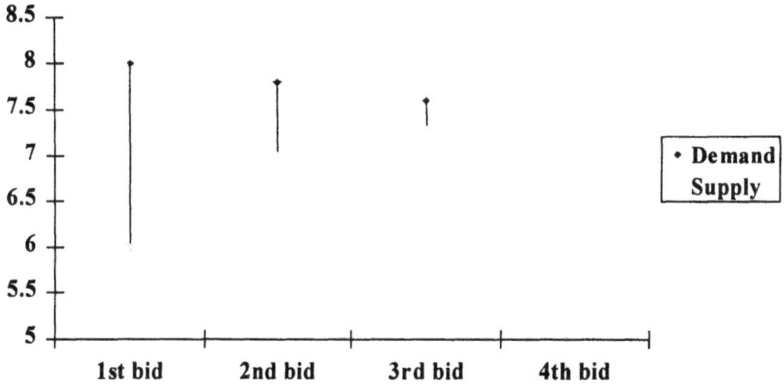


Figure 4-9. Auction Market Objective.

## 4.4 Auction Formulation

Two major types of auctions are single-sided auctions and double-sided auctions. Either generation companies (GENCOs) or energy services companies (ESCOs) submit bids to the independent contract administrator (ICA) in single-sided auctions. GENCOs and ESCOs submit bids to the ICA in double-sided auctions. The ICA performs auctions with the formulation below. This formulation is for double-sided auctions.

The supportive services included in this formulation are transmission losses, spinning reserves, and ready reserves. Transmission losses, spinning reserves, and ready reserves are bundled, i.e., traded in the same market as one contract with energy. Regulation, automatic generation control (AGC), and load following can be incorporated into the formulation just as the spinning and ready reserves if they are considered bundled. Regulation, supplemental control (AGC), and load following are for similar purposes. They differ in time frame. The time frames of regulation, AGC, and load following are 6 seconds, 30 minutes, and 60 minutes respectively. The frequency of the auction in the formulation below is every 30 minutes. There are various possible formulations. Alternative formulations are discussed below.

All of the quantities in the formulation are in per unit, except the bus angles, which are in radian and the bids' prices which are in \$/MWh. The symbols used for the formulation are shown below. Note that GENCO $i$ is at bus $i$ and ESCO $j$ is at bus $m+j$.

| | |
|---|---|
| $c_{si}$ | price of $i$th seller's bid for active power |
| $c_{bj}$ | price of $j$th buyer's bid for active power |
| $cs_{si}$ | price of $i$th seller's bid for spinning reserve |
| $cs_{bj}$ | price of $j$th buyer's bid for spinning reserve |
| $cr_{si}$ | price of $i$th seller's bid for ready reserve |
| $cr_{bj}$ | price of $j$th buyer's bid for ready reserve |
| $\Delta P_{si}$ | accepted amount of active power of $i$th seller |
| $\Delta P_{bj}$ | accepted amount of active power of $j$th buyer |
| $\Delta Pm_{si}$ | amount of active power of all other contracts of $i$th seller |
| $\Delta Pm_{bj}$ | amount of active power of all other contracts of $j$th buyer |
| $\Delta Q_{si}$ | change in reactive power of $i$th seller |
| $\Delta Q_{bj}$ | change in reactive power of $j$th buyer |
| $S_{si}$ | accepted amount of spinning reserve for $i$th seller |
| $S_{bj}$ | accepted amount of spinning reserve for $j$th buyer |
| $S_{si}^{max}$ | maximum limit of spinning reserve for $i$th seller |
| $S_{bj}^{max}$ | maximum limit of spinning reserve for $j$th buyer |
| $ks_s$ | required ratio of spinning reserve for sellers |
| $ks_b$ | required ratio of spinning reserve for buyers |
| $R_{si}$ | accepted amount of ready reserve for $i$th seller |
| $R_{bj}$ | accepted amount of ready reserve for $j$th buyer |
| $R_{si}^{max}$ | maximum limit of ready reserve for $i$th seller |
| $R_{bj}^{max}$ | maximum limit of ready reserve for $j$th buyer |
| $RR^{max}$ | maximum ramp-rate limit |
| $kr_s$ | required ratio of ready reserve for sellers |
| $kr_b$ | required ratio of ready reserve for buyers |
| $n$ | number of buyers |
| $m$ | number of sellers |
| $\Delta\delta_i$ | change in $i$th bus angle |
| $\Delta\delta^{max}$ | maximum limit of change in bus angle |
| $B'$ | matrix of negative bus susceptance matrix |
| $B_{si}$ | amount of active power submitted by $i$th seller |
| $B_{bj}$ | amount of active power submitted by $j$th buyer |

$P_{ij}^{0}$ original active line flow between buses $i$ and $j$
$\Delta P_{ij}$ change in active line flow between buses $i$ and $j$
$P_{ij}^{max}$ active flow limit of line between buses $i$ and $j$
$trc_i$ transmission cost coefficient for bus $i$
$K_p$ coefficient for active power flow constraints
$K_q$ coefficient for reactive power flow constraints
$PF_i$ power factor of $i$th bus
$PF^{min}$ minimum power factor allowed for each bus

$$Maximize \sum_{j=1}^{n} c_{bj}\Delta P_{bj} - \sum_{i=1}^{m} c_{si}\Delta P_{si} + \sum_{j=1}^{n} cs_{bj}S_{bj} - \sum_{i=1}^{m} cs_{si}S_{si} + \sum_{j=1}^{n} cr_{bj}R_{bj} - \sum_{i=1}^{m} cr_{si}R_{si} - \sum_{i=1}^{m+n} trc_i\Delta\delta_i \quad (4.20)$$

The objective function is composed of surplus in trading power, surplus in trading spinning reserve, surplus in trading ready reserve, and the negative of the transmission cost. The *surplus* is defined as the difference between the buyers' revenue and the sellers' cost. Transmission costs *(trc)* are converted from transmission costs of lines into transmission costs at buses. The negative of the transmission cost indicates that the transmission cost is minimized. The objective function (4-20) is maximized subject to the following constraints:

*A. Active Power Constraints with Losses.* The active power flow constraints and conservation of active power equation are manipulated to be constraint (4-21). Note that *Kp* is calculated from the real power loss coefficients and $B'$ matrix.

$$\sum_{i=1}^{m} Kp_i\Delta P_{si} + \sum_{j=1}^{n} Kp_{m+j}\Delta P_{bj} = 0 \quad (4.21)$$

An iterative procedure is normally used.

*B. Reactive Power Constraints with Losses.* The reactive power flow constraints and conservation of reactive power equation are manipulated to be constraint (4-23). Note that *Kq* is calculated from the reactive power loss coefficients and $B'$ matrix. An iterative procedure is normally used.

$$\sum_{i=1}^{m} Kq_i\Delta Q_{si} + \sum_{j=1}^{n} Kq_{m+j}\Delta Q_{bj} = 0 \quad (4.22)$$

*C. Transmission Line Flow Limits.* An iterative procedure is used to limit the complex flow by limiting the real and reactive flows. Equation (4-23) shows the flow limit. The change in flow, $\Delta P_{ij}$ is approximated as follows.

$$\begin{aligned} &\left|P_{ij}^{0} + \Delta P_{ij}\right| \le P_{ij}^{\max} \\ &\Delta P_{ij} = -B'_{ij}(\Delta\delta_i - \Delta\delta_j) \end{aligned} \tag{4.23}$$

Similar constraints are used for reactive flows.

*D. Sellers and Buyers' Bid Amount.* The bid offered can be used partially or completely.

$$\begin{aligned} &0 \le \Delta P_{si} \le B_{si} \\ &0 \le \Delta P_{bj} \le B_{bj} \end{aligned} \tag{4.24}$$

*E. Spinning Reserve Obligations.* Spinning reserve obligations of sellers and buyers are shown and conservation of spinning reserve follows.

$$\begin{aligned} &S_{si} = \min(ks_s \Delta P_{si}, S_{si}^{\max}) \\ &S_{bj} = \min(ks_b \Delta P_{bj}, S_{bj}^{\max}) \\ &\sum_{i=1}^{m} S_{si} - \sum_{j=1}^{n} S_{bj} = 0 \end{aligned} \tag{4.25}$$

*F. Ready Reserve Obligations.* Ready reserve obligations of sellers and buyers are shown and conservation of ready reserve follows.

$$\begin{aligned} &R_{si} = \min(kr_s \Delta P_{si}, R_{si}^{\max}) \\ &R_{bj} = \min(kr_b \Delta P_{bj}, R_{bj}^{\max}) \\ &\sum_{i=1}^{m} R_{si} - \sum_{j=1}^{n} R_{bj} = 0 \end{aligned} \tag{4.26}$$

*G. Ramp Rate.* Equation (4.27) represents ramp-rate constraints for sellers and buyers. Up-ramp-rate limit is assumed to be equal to down-ramp-rate limit. Note that *t-1* indicates the former period before auction.

$$\begin{aligned} &\left|(\Delta P_{si} + \Delta Pm_{si}) - (\Delta P_{si}^{t-1} + \Delta Pm_{si}^{t-1})\right| \le 0.5RR^{\max} \\ &\left|(\Delta P_{bj} + \Delta Pm_{bj}) - (\Delta P_{bj}^{t-1} + \Delta Pm_{bj}^{t-1})\right| \le 0.5RR^{\max} \end{aligned} \tag{4.27}$$

*H. Bus Angle Limit.* Simply limits the angle separation between any two buses.

$$|\Delta\delta_i| \leq \Delta\delta^{\max} \tag{4.28}$$

Note that stability margins may be represented with more sophisticated models.

*I. Relation between Active and Reactive Power.* Required power factor correction may be included.

$$PF_i \geq PF^{\min} \tag{4.29}$$

Note that all variables are nonnegative except all $\Delta\delta_i$ are free variables. Constraints (4.23) to (4.29) are for all $i$ from $1$ to $m$ and all $j$ from $1$ to $n$.

### *4.4.1 Handling Constraints*

Some constraints are modified so that they can be solved with LP. Some constraints that require modification are as follows.

*A. Absolute Constraints.* Many constraints are in absolute form. The general form of an absolute constraint is shown in (4-30) and it is modified to two constraints. Note that $F(P)$ indicates a function of $P$ and $U$ indicates a constant.

$$\begin{aligned} &|F(P)| \leq U \\ &-U \leq F(P) \leq U \end{aligned} \tag{4.30}$$

*B. Reserve Constraints.* Reserve constraints can be generalized as (4-31). $S$ is forced to be either $k\Delta P$ or $S^{max}$. This concept can be explained according to Figure 4-10. $S$ is equal to a constant ratio, $k$, of $\Delta P$ until $S$ is equal to $S^{max}$. In other words, $\Delta P$ is equal to $S^{max}/k$. After this, $S$ is equal to $S^{max}$ no matter how much $\Delta P$ is. Such modifications can be implemented by LP.

$$\begin{aligned} &S = \min(k\Delta P, S^{\max}) \\ &S = k\Delta P \\ &S \leq S^{\max} \end{aligned} \tag{4.31}$$

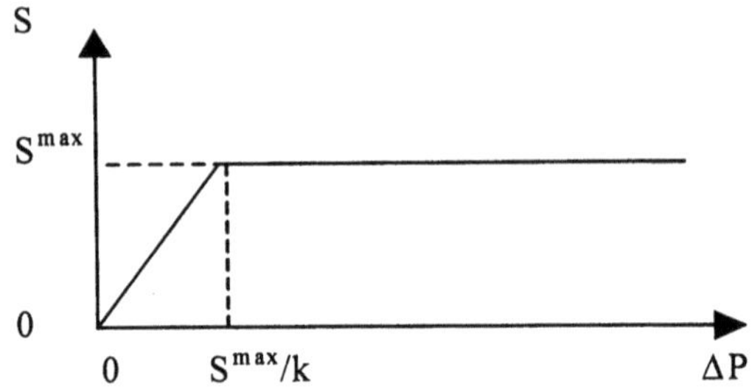


Figure 4.10. Reserve Constraint.

*c. Non-Negativity Constraints.* Free variables, $\Delta\delta_i$ are changed.

$$\Delta\delta_i = \Delta\delta_i^+ - \Delta\delta_i^- \; ; \quad \Delta\delta_i^+, \Delta\delta_i^- \geq 0 \tag{4.32}$$

*4.4.2 Structure Options*

The ICA must know all transactions each player has to operate the system economically and securely. This makes the net power at each bus a known quantity in the formulation. Thus, the ICA should supervise the implementation of every contract between parties by incorporating all of them as preconditions into the auction formulation. This will be effective only if the ICA knows all transactions and is enabled to enforce all transactions with sufficient penalties.

**4.4.2.1 Single-sided auctions vs. double-sided auctions.** Both sellers and buyers can specify the prices and amounts of bids as they desire for double-sided auctions. They can adjust the bids to maximize their own profits. For one-sided auctions, such as seller bidding only, only sellers submit price and quantity bids to the auction and buyers specify only the amount of power desired. Sellers might have to sell cheap power because all sellers submitted low price bids depends on the pricing method. This problem can be lessened if a seller can specify a reservation price. However, this is inflexible to buyers. This is true for all one-sided auctions, since one of the two sides does not submit a price. These arguments indicate the inefficiency of one-sided auctions.

**4.4.2.2 Bundled vs. unbundled ancillary services.** If an ancillary service is unbundled, the supplier and the buyer have to be identified for each ancillary service contract. This will allow the ICA to identify who is entitled to the ancillary service. If the ancillary service is bundled, the service is included with the energy contract that clearly lists the sellers and buyers. This enables the ICA to monitor the use of all ancillary services. For transmission loss service, it is hard to know the power flow amongst specific lines without knowledge of all contracts. Additionally, it is hard to monitor and accumulate all flows on an individual contract basis, especially with inadvertent flows. Thus, it is difficult to bid the losses separately.

**4.4.2.3 Spinning reserves and ready reserves.** In the formulation, bids' prices of power, spinning reserve, ready reserve are different. This is for showing the formulation generally. If bids' prices are different, players might bid in a way that they do not have to sell or buy the reserves. Should reserve transactions that have been used be the same price as the transactions that have not been used? If they are not the same, there will be more markets. The question is whether it is worthy to have more markets? Constraints for spinning reserve and ready reserve neglect losses due to these reserves. Additionally, the transmission capacity for reserves is not included. This author selects this approach because losses are small and only seldom will reserves be used.

**4.4.2.4 Frequency of auctions.** Auctions can be conducted every 15, 30, or 60 minutes. Frequent auctions cause more transaction costs while they give more flexibility in scheduling for each party.

**4.4.2.5 Ramp rate.** Ramp rate constraints are endemic to specific equipment. Some feel that the owners within the bids offered should implicitly handle equipment constraints. Thus, the ICA would not supervise the equipment use by every party.

### *4.4.3 Linear Programming or Mixed-Integer Programming*

If the bids are required to submit in blocks and one block of bid is big, auctions have to be formulated in mixed-integer programming. This will cause a lot more difficult to solve. In fact, the power the buyers receive may be not in block due to sales to cover losses. Thus, submitting bids in blocks when the block size is big may not facilitate loss charges. However, if the size of bid is of same size as solution tolerance, then the auction can still be formulated in LP and solved very efficiently.

### *4.4.4 Transmission Loss Allocation*

After the solution is found, we know the optimal transactions. The amount and cost of the loss depends on each transaction. One frequent question is how to allocate the cost of the transmission loss? One possible way is to use the information from the simplex method. While the simplex method implements the auction, the variables moving into and out of the basis at each iteration can be used to assign loss charges. This enables the identification of each seller and buyer of each transaction, as well as the cost of the transaction loss. The cost of the loss may be paid by either the seller or the buyer or by both the seller and the buyer in any proportion.

## 4.5 Utilizing Interior-Point Linear Programming Algorithm as Auction Method

### *4.5.1 Introduction*

The purpose of this work is to illustrate application of IPLP to auction methods. An extended IPLP algorithm is developed and used in the following. This extended

IPLP algorithm can find the exact optimal solution (i.e., exact optimal vertex) and can recover the optimal basis. The optimal basis can be recovered in the straightforward way. The great benefit is that sensitivity analysis can be performed after the optimal basis is found. This extended algorithm is expanded from the affine-scaling primal algorithm. The concept used in this extended algorithm to find the optimal vertex and optimal basis is simple. Note that the exact optimal vertex solution is desired for auctions because the solution must be fair and uniquely identify assignments between buyers and sellers. Thus, fairness is measured by the exactness of the solution.

### *4.5.2 Extended IPLP Algorithm*

The extended algorithm is expanded from the affine-scaling primal algorithm. The explanation of the algorithm is divided into two parts. The first part gives the basic concept of the affine-scaling primal algorithm. The explanation of the basic algorithm is based on Arbel [1993]. The basic algorithm is described along with the expanded algorithm in the second part to be the complete algorithm of IPLP used in the following auction. The expanded algorithm in the second part comes from the addition of a section to the basic algorithm so that IPLP can find an exact solution and can recover an optimal basis.

#### *A. Basic Concept*

Two major components of the affine-scaling primal algorithm are centering and projective gradient direction. Movement is made through projective gradient direction for maximizing the objective function or opposite to projective gradient direction for minimizing objective function. The projective gradient direction is used instead of gradient direction for the purpose of maintaining feasibility. Centering is performed to achieve the potential to improve objective function in each iteration.

LP using big-M method is used so that the unity vector, [1 1 ... 1], can be used as the starting solution [Arbel, 1993]. This LP has one additional variable. If this additional variable is driven to zero at the end of the algorithm, the primal problem is feasible. Otherwise, the primal problem is infeasible.

Three quantities, duality gap, primal feasibility, and dual feasibility are used as the stopping criteria for terminating the basic algorithm. These quantities are defined in (4-14) ([Arbel, 1993]). Note that $x'$ is $(n+1)$-component column vector of variables. $c'^T$ is $(n+1)$-component row vector of cost coefficients. $c'^T$ signifies the transpose of vector $c'$. $A'$ is $m \times (n+1)$ matrix of technological coefficients. $b$ is $m$-component column right hand side vector. If all three criteria are met simultaneously, the basic algorithm is terminated. In other words, the solution found is very close to the optimal solution. This type of solution is called ε-optimal solution. Although dual feasibility is desired in terminating the algorithm, sometimes it cannot be achieved. In the implemented algorithm, the dual feasibility-stopping criterion is neglected and the algorithm still works quite well with test problems.

$$duality\ gap = \frac{norm(c'^T x' - b^T y)}{1 + norm(c'^T x')}$$
$$primal\ feasibility = \frac{norm(b - A'x')}{1 + norm(b)} \qquad (4.33)$$
$$dual\ feasibility = \frac{c' - A'^T y - z}{1 + norm(c')}$$

*B. Extended Algorithm*

The extended algorithm for IPLP, affine-scaling primal algorithm, is described in Table 4-12. The left column is step number. The right column is the algorithm. Note that *(iter)* indicates the iteration number that the variable's value belongs to. Step 0 to step 4 of the algorithm is based on Arbel [1993], and step 5 to step 7 of the algorithm is developed by the authors. Two symbols, $\alpha$ and $\rho$ as used in step 4, explain interior point algorithms.

At step 4, $\alpha$ is the maximum allowable step size that maintains feasibility. Changing by step size $\alpha$ will make at least one variable hit the boundary of the feasible region. Thus, a factor $\rho$ is used to make the new solution remain inside the feasible region. The algorithm uses $\rho$ as 0.95 for the results presented.

$$\alpha = \min\left\{\frac{-x_i'(iter)}{dx_i'(iter)} : \forall dx_i'(iter) < 0,\ 1 \le i \le n+1\right\}. \qquad (4.34)$$

The extended algorithm is expanded from the basic algorithm so that the algorithm can find the exact optimal vertex. Not only can this extended algorithm find the exact solution, but also sensitivity analysis can be performed after the optimal basis is found. The algorithm is composed of two parts: the basic IPLP algorithm and the movement from the interior solution to the vertex solution.

The basic IPLP algorithm is terminated when the stopping criteria, duality gap and primal feasibility, are small enough, usually in the range of 1e-6 to 1e-8 [Arbel, 1993]. The value of the stopping criteria is denoted as $\varepsilon_1$. If $\varepsilon_1$ is set too small, numerical instability might occur. The authors handle this problem by adding an additional stopping criterion. A quantity, $z$ (defined at (7)) is calculated at every iteration and $z$ is the estimate of the reduced cost coefficient vector. When the current solution is very close to the optimal vertex, the components of $z$, which belong to the basic variables of the optimal vertex, are very close to zero. Thus, the additional stopping criterion is to measure $z$ and check to see if all the components of $z$ that belong to the basic variables are less than a value denoted as $\varepsilon_2$. In addition, the algorithm has to check that the number of components of $z$ which are less than $\varepsilon_2$ is equal to the number of constraints because the number of basic variables is equal to the number of constraints. Otherwise, degeneracy is indicated. The value of $\varepsilon_2$ is set as 1e-6 for the extended algorithm. By adding this criteria, we can use bigger value of $\varepsilon_1$ to avoid the numerical instability problem. Thus, the value of $\varepsilon_1$ is set to 1e-4, which is bigger than the suggested value, 1e-6 to 1e-8 in [Arbel, 1993].

Table 4-12. Extended Algorithm.

| | |
|---|---|
| 0 | Initialize iteration counter, *iter*=0.<br>Initialize $\varepsilon_1$ and $\varepsilon_2$ as 1e-4 and 1e-6 respectively.<br>Initialize the starting solution vector:<br><br>$x'(iter)=[1\ 1\ \ldots\ 1]^T$ |
| 1 | Increment the iteration counter, *iter*=*iter*+1.<br>Define the scaling matrix *D(iter)*:<br><br>$D(iter)=diag(\ [x'_1(iter)\ \ x'_2(iter)\ \ \ldots\ \ x'_{n+1}(iter)]\ ),$<br><br>where *diag(x')* means diagonal matrix of vector $x'$ and $x_i(iter)$ is the *i*th component of the present $x'(iter)$. |
| 2 | Calculate the dual estimate, *y(iter)*, where *y(iter)* is an *m*-component column vector:<br><br>$[A'D^2(iter)A'^T][y(iter)]=[A'D^2(iter)c']$. |
| 3 | Find the estimate of the reduced cost vector, *z(iter)* and then use it to find the primal step direction vector, *dx'(iter)*, where *z(iter)* and *dx'(iter)* are $n+1$ component column vectors:<br>$z(iter)=c'-A'^T\ y(iter)$,<br>$dx'(iter)=-D^2(iter)z(iter)$. |
| 4 | Update the solution vector:<br><br>$x'(iter+1)=x'(iter)+\rho\alpha\ dx'(iter)$. |
| 5 | Test with two criteria.<br>First criterion: duality gap and primal feasibility are less than $\varepsilon_1$.<br>Second criterion: the number of components of $z$ that are less than $\varepsilon_2$ are equal to the number of constraints.<br>If both criteria are satisfied, go to step 6. Otherwise, go to step 1. |
| 6 | The expected optimal basic variables are the variables having $z$ less than $\varepsilon_2$. Test the KKT conditions. If the KKT conditions are satisfied, the optimal basic variables have been found, go to step 7. If not, go to step 5 and reduce $\varepsilon_1$ and $\varepsilon_2$ to be one-tenth of the value previously used in step 5. |
| 7 | Find the optimal solution. $B$ is the optimal basis and $x_B$ contains values of basic variables.<br><br>$x_B=B^{-1}b$ |

The parameters $\varepsilon_1$ and $\varepsilon_2$ are important criteria for terminating the basic IPLP algorithm in step 5. The values of these two parameters are needed to be set properly so that the algorithm can find the solution. The procedure for changing these two parameters are explained in step 6.

At step 6, $\varepsilon_1$ and $\varepsilon_2$ are repeatedly reduced to be one-tenth if the KKT conditions are still not satisfied. Actually this situation is unlikely because the original values of $\varepsilon_1$ and $\varepsilon_2$ (1e-4 and 1e-6 respectively) are reasonable for most applications. These values have been tested with many examples and the test shows that these values can be used as criteria for terminating the algorithm's interior point process (step 0 to step 4) to find the exact solution at step 7.

At step 5, the number of components of $z$ that are less than $\varepsilon_2$ might be greater than the number of constraints although these components of $z$ belong to the optimal basic variables for IPLP (i.e., variables which are greater than zero). This might occur for 2 causes. First, free variables exist in the problem and each free variable, $x_i$, is implemented as $x_i^+-x_i^-$, $x_i^+$ and $x_i^-$ are non-negative. The problem occurs because both of $x_i^+$ and $x_i^-$ are greater than zero for the interior-point solution while either each of $x_i^+$ and $x_i^-$ is greater than zero and the other must be zero for the simplex solution. Second, the optimal solution is degenerate and IPLP reaches the optimal solution that is not on a vertex. The number of basic variables must be equal to the number of constraints to use (10).

These two causes are handled as follows: For each of free variable, $x_i$, only either each of $x_i^+$ or $x_i^-$ with greater value is selected. For degenerate optimal solutions, if $p$ variables are selected at step 5, and there are $m$ constraints, $p>m$, there are more than one possible optimal vertex. Some of these are feasible and some are infeasible. Trial and error is used to find one feasible optimal vertex. Trial and error might consume time when the number of possible optimal vertices is large. However, this drawback does not happen when this algorithm of IPLP is applied to auction methods because we will use the following procedure to handle degeneracy instead of trial and error. The main concept is that the degenerate solutions cannot be used for auction methods. If one of degenerate solutions is used, it will be *unfair* to the bidders who are not selected. Indeed, it would be hard to select a set of bidders without *arbitrary allocation.* Thus, when the number of components of $z$ that are less than $\varepsilon_2$ is greater than the number of constraints at step 5, the program will report that degeneracy occurs. The basic variables having $z$ less than $\varepsilon_2$ are reported, which will tell the possible bidders to be selected. The independent contract administrator can use some criteria to select the bidders. An example criterion is to select the bidders who submitted bids first.

The movement from the interior solution to the vertex solution can be performed just after the termination of the basic IPLP. Step 6 moves the solution point from the interior solution to the vertex solution. The estimated optimal basic variables are those variables satisfying the requirement that values of the reduced cost coefficients are zero. These estimated optimal basic variables must be verified with the Karush-Kuhn-Tucker (KKT) conditions for optimality. The KKT conditions are elegantly developed in Bazaraa [1996]. If the KKT conditions are satisfied, the estimated optimal basic variables are correct and the optimal solution can be calculated as explained in step 7.

At step 7, values of non-basic variables are zero. The value of objective function can be calculated from substituting the values of all the decision variables into the objective function. To avoid finding $B^{-1}$, LU or QR factorization can be used to solve (10).

The extended algorithm also can check for infeasibility, unbounded, and degeneracy of primal problem. The solution is infeasible if the selected variables at step 5 contain the artificial variable added for the extended problem. The solution is unbounded if for any iteration, all of the components of $dx'(iter)$ that are found at step 3 are all greater than or equal to zero. The solution is degenerate if the reduced cost coefficient of any of the non-basic variables while testing with the KKT conditions at step 6 is calculated to be zero. Actually, degeneracy can be detected at step 5. If the number of components of $z$ that are less than $\varepsilon_2$ is greater than the number of constraints, excluding the effect of free variables, indicates that degeneracy has occurred.

*4.5.3 Formulation*

The formulation of the auction methods used is fully developed in [Dekrajangpetch, 1997]. The transmission line flow constraints are added and always treated as hard limits. The framework of the energy market used can be seen in [Sheblé, 1994b]. The auction is double-sided. Additionally, we assume that real power and reactive power are bid in different markets. In other words, the real power and reactive power are unbundled. To operate the market securely, there must be the linking between the real and reactive markets. The linking in an unbundled set of markets is to use and estimated real power limit in the real power market and use another estimated reactive power limit in the reactive power market. Actually such series equipment should be represented with only an MVA limit. The unbundled market requires that the real and reactive power limits are estimated from the MVA limit. We will not discuss this estimation detail. We are convinced that this linking issue demonstrates the need for a market where real and reactive power are bundled. The details of such an argument is not in this section. This section demonstrates only the auction of real power market. Because of the unbundling of the real and reactive power market, the fast-decoupled power flow can be solved independently. Although the real power and reactive power markets are unbundled, the effect of the reactive power market is reflected in the shadow prices of the real power market based on the estimates of the real power flow limits. The interactions of shadow prices of both markets occur through the linking between the real and reactive power limits.

The other way to implement auctions is to bundle real and reactive power markets. The formulation for this type of auctions is illustrated in the next chapter. The comparison of unbundling and bundling real and reactive power markets are also discussed in the next chapter. For this bundled market, the interactions of shadow prices of both markets occur directly.

In this work, generation companies (GENCOs) and energy service companies (ESCOs) submit bids for auctions. For the test cases included, the interactions of transmission companies are ignored, but could easily included by incorporating the transmission costs in the objective function. The constraints are taken from the OPF. The formulation is described below:

$$
\begin{aligned}
&\min_{\Delta P_{bj},\Delta P_{si},\Delta\delta_i,\Delta\delta_j} \quad \sum_{i=1}^{m} c_{si}\Delta P_{si} - \sum_{j=1}^{n} c_{bj}\Delta P_{bj} \\
&\textit{subject to} \\
&\underline{\Delta P} - B'\underline{\Delta\delta} = \underline{0} \\
&\sum_{i=1}^{m}\Delta P_{si} - \sum_{j=1}^{n}\Delta P_{bj} - \underline{lsc} * \underline{\Delta\delta}^{T} = 0 \\
&P_{ij}^{0} - B'_{ij}(\Delta\delta_i - \Delta\delta_j) \le P_{ij}^{\max} \\
&-P_{ij}^{0} + B'_{ij}(\Delta\delta_i - \Delta\delta_j) \le P_{ij}^{\max} \\
&0 \le \Delta P_{si} \le B_{si} \\
&0 \le \Delta P_{bj} \le B_{bj}
\end{aligned}
\qquad (4.35)
$$

The symbols are clarified as follows:

| | |
|---|---|
| $c_{bj}$ | price of $j$th buyer's bid |
| $c_{si}$ | price of $i$th seller's bid |
| $\Delta P_{bj}$ | accepted amount of power of $j$th buyer |
| $\Delta P_{si}$ | accepted amount of power of $i$th seller |
| $n$ | number of buyers |
| $m$ | number of sellers |
| $\Delta\delta$ | change in bus angle |
| $B'$ | matrix containing the negative of susceptance of the bus admittance matrix (Y matrix) |
| $B_{si}$ | amount of power submitted by $i$th seller |
| $B_{bj}$ | amount of power submitted by $j$th buyer |
| $lsc$ | loss coefficient vector |
| $P_{ij}^{0}$ | original flow of line between buses $i$ and $j$ |
| $P_{ij}^{max}$ | flow limit of line between buses $i$ and $j$ |

Note the underscored variables indicate vectors. All the quantities in the formulation are in per unit, except that the bus angles are in radian and the bids' prices are in \$/MWh. The fast-decoupled power flow is used and thus the $B'$ is a constant matrix. The formulation above includes only real power relationships because the auction has been unbundled. All the variables in the formulation are non-negative except that $\Delta\delta_i$ and $\Delta\delta_j$ are free variables; thus, they are implemented as explained in section II. The objective function is the negative of trading surplus. The trading surplus is the difference between the revenue from ESCOs' bids and the cost from GENCOs' bids. The first constraint is the active power flow relationship. The second constraint restricts the slack injection. The next two constraints restrict the flow on network branches. The last two constraints restrict the bids that sellers and buyers submit.

### *4.5.4 Results*

The results are divided into two main parts. The first part illustrates the accepted bids from performing an auction. The second part illustrates the sensitivity analysis of bids' prices and lines' flow limits. A six-bus system (based on [Wood, 1996]) is used to demonstrate the implementation of network constraints. It is shown in Figure 4-11. There are three GENCOs and three ESCOs. GENCOs 1, 2, and 3 are at buses 1, 2, and 3 respectively and ESCOs 1, 2, and 3 are at buses 4, 5, and 6 respectively.

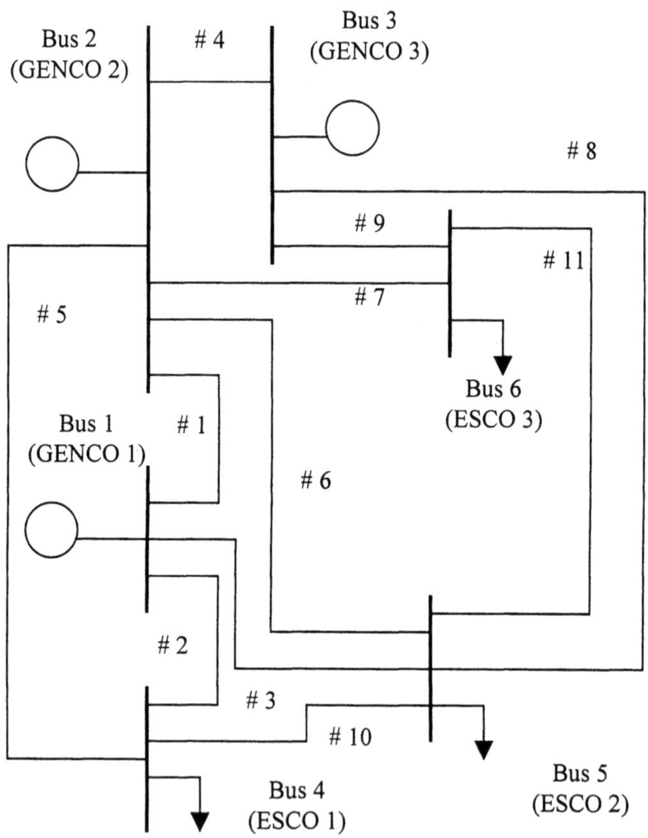


Figure 4-11. Six-bus System.

Table 5 shows line data for line resistance, line reactance, and half of line charging admittance. $P_{ij}^{max}$ is the flow limit of line. The MW-base and voltage-base are 100MW and 230 kV. Table 6 shows bus data. $V^0$ is the original bus voltage (i.e., before trading). $\delta^0$ is the original bus angle. $P^0$ is the original bus real power generation (for bus 1 to 3) and load (for bus 4 to 6). $Q^0$ is the original bus reactive power generation (for bus 1 to 3) and load (for bus 4 to 6). Note that this work only developed the real market. Thus, the reactive market has to be solved separately.

The interaction of the real and reactive markets is hard to separate as shown in [Dekrajangpetch, 1998]. This work assumed that the reactive power market was partially solved before the real market was executed. Technically, the reactive market would have to be re-executed to properly reflect the interaction of the shadow prices through the change in real power limits based on the actual MVA limits. The real market is thus reduced to a DC Power Flow model using the B' matrix.

*A. Accepted Bids*

Table 4-15 shows the submitted and accepted bids by GENCOs and ESCOs. GENCO 3 can sell all power offered for sale and ESCOs 1 and 2 can buy all power they bid on. GENCOS 1 and 2 and ESCO 3 can sell and buy only part of the desired power. The reason is not only because GENCOs 1 and 2 offer the higher bids and ESCO 3 offers the lowest bids. It also depends on the network constraints. Sometimes even the GENCO that bids the lowest or the ESCO that offers the highest bid will have their contracts limited due to the network constraints.

Table 4-13. Line Data.

| *Line No.* | *From Bus* | *To Bus* | *r (pu)* | *X (pu)* | *$Y_p/2$ (pu)* | *$P_{ij}^{max}$ (MW)* |
|---|---|---|---|---|---|---|
| 1 | 1 | 2 | 0.100 | 0.200 | 0.020 | 30.00 |
| 2 | 1 | 4 | 0.050 | 0.200 | 0.020 | 60.00 |
| 3 | 1 | 5 | 0.080 | 0.300 | 0.030 | 53.00 |
| 4 | 2 | 3 | 0.050 | 0.250 | 0.030 | 30.00 |
| 5 | 2 | 4 | 0.050 | 0.100 | 0.010 | 76.00 |
| 6 | 2 | 5 | 0.100 | 0.300 | 0.020 | 35.00 |
| 7 | 2 | 6 | 0.070 | 0.200 | 0.025 | 60.00 |
| 8 | 3 | 5 | 0.120 | 0.260 | 0.025 | 30.00 |
| 9 | 3 | 6 | 0.020 | 0.100 | 0.010 | 60.00 |
| 10 | 4 | 5 | 0.200 | 0.400 | 0.040 | 15.00 |
| 11 | 5 | 6 | 0.100 | 0.300 | 0.030 | 12.00 |

Table 4-14. Bus Data.

| *Bus* | $V^0$ *(pu)* | $\delta^0$ *(deg.)* | $P^0$ *(MW)* | $Q^0$ *(MW)* |
|---|---|---|---|---|
| 1 | 1.0500 | 0.00 | 112.62 | 34.79 |
| 2 | 1.0500 | -2.53 | 140.00 | 75.07 |
| 3 | 1.0700 | -5.15 | 60.00 | 112.34 |
| 4 | 0.9754 | -4.68 | 100.00 | 70.00 |
| 5 | 0.9677 | -6.58 | 100.00 | 70.00 |
| 6 | 0.9930 | -7.27 | 100.00 | 70.00 |

The optimal surplus is $129.13. The flows of transmission line number 5 (between buses 2 and 4) and number 9 (between buses 3 and 6) are at the limits. The shadow prices of lines numbers 5 and 9 are 4.39 $/MW and 0.09 $/MW respectively. The shadow price in this context indicates the incremental improvement in the objective function of increasing flow limit of the associated transmission line. Note that the flows and the shadow prices of lines numbers 5 and 9 that will be discussed below are based on flows in direction from bus 2 to bus 4 (for line number 5) and in direction from bus 3 to bus 6 (for line number 9

An upper-bound linear programming (UBLP), simplex program (written by the authors) is used to implement the auctions to compare to IPLP. The CPU time UBLP used to find the solution is 9.43 seconds while the IPLP uses only 0.80 seconds. The IPLP uses 8.63 fewer seconds than the UBLP even for this small problem. This shows IPLP to be very efficient.

Table 4-15. Base Case: Gencos' Bids.

| Bids | GENCO 1 | GENCO 2 | GENCO 3 |
|---|---|---|---|
| Price | 9.70 | 8.80 | 7.00 |
| Submitted Amount | 20.00 | 25.00 | 20.00 |
| Accepted Amount | 5.30 | 20.71 | 20.00 |

*B. Sensitivity Analysis*

*1) Case A: Increase in ESCO 3's Bid Price*

From Table 4-15 (base case), only part of the desired power of ESCO 3 is accepted. Case A here increases the bid price of ESCO 3 to 12.00. The new accepted bids are shown in Table 4-16. Comparing to the base case (Table 4-15), the bids of GENCO 1 and ESCO 1 are accepted less and the bids of GENCO 2 and ESCO 3 are accepted more. The bids of GENCO 3 and ESCO 2 do not change.

The optimal surplus is $154.89. The flows of transmission line number 5 (between buses 2 and 4) and number 9 (between buses 3 and 6) are at the limits. The shadow prices of lines numbers 5 and 9 are 5.75 $/MW and 6.37 $/MW respectively. These shadow prices are higher than those of the base case, especially the shadow price of line number 9. This indicates that it is worthier to increase the flow limits of lines numbers 5 and 9. In other words, the value of the objective function can be improved with the increase of the flow limits of lines numbers 5 and 9 with these new bids' prices. Note that the shadow price of line number 9 is higher than that of line number 5 in this case while the shadow price of line number 9 is lower than that of line number 5 in the base case. This indicates that increasing the flow limit of line number 9 can give higher benefit than increasing the flow limit of line number 5 in this case.

Table 4-16. Base Case: ESCOs' Bids.

| Bids | ESCO 1 | ESCO 2 | ESCO 3 |
|---|---|---|---|
| Price | 12.00 | 10.50 | 9.50 |
| Submitted Amount | 25.00 | 10.00 | 20.00 |
| Accepted Amount | 25.00 | 10.00 | 10.29 |

*2) Case B: Increase in Line Number 5's Flow Limit*

The flows of lines numbers 5 and 9 are at the limits for the base case. Case B here increases the flow limit of line number 5 from 76 MW to 80 MW. The new accepted bids are shown in Table 4-17. Comparing to the base case (Table 4-16), the bids of GENCO 1 and ESCO 3 are accepted less and the bid of GENCO 2 is accepted more. The bids of GENCO 3 and ESCOs 1 and 2 do not change. The optimal surplus for case B is $135.06, which is bigger than the optimal surplus of the base case, $129.13. This shows the benefit of increase in flow limit of line number 5.

There are not any lines operating at their limits. The spare capacities of lines numbers 5 and 9 are 2.64 MW and 0.62 MW respectively. This shows that it is not beneficial to further increase flow limit of line number 5 with these bids.

This result is useful for transmission network expansion. The result indicates the benefit from expanding the flow limit of line number 5. The same procedure of case B can be performed on line number 9. Simulation of the procedure in case B on various bidding scenarios can be used to make decisions of transmission network expansion.

Table 4-17. Case B's Bids.

| Bids | GENCO 1 | GENCO 2 | GENCO 3 |
|---|---|---|---|
| Price | 9.70 | 8.80 | 7.00 |
| Submitted Amount | 20.00 | 25.00 | 20.00 |
| Accepted Amount | 0.00 | 25.00 | 20.00 |
| Bids | ESCO 1 | ESCO 2 | ESCO 3 |
| Price | 12.00 | 10.50 | 9.50 |
| Submitted Amount | 25.00 | 10.00 | 20.00 |
| Accepted Amount | 25.00 | 10.00 | 9.48 |

### *4.5.5 Conclusions*

This work applies IPLP to auction methods. An extended algorithm of IPLP is developed and used. This extended IPLP algorithm can find the exact optimal solution (i.e., exact optimal vertex) and can recover the optimal basis. The concept used in this extended algorithm is simple. The result shows that IPLP is very efficient. The results of sensitivity analysis are very useful for adjusting the bid's price to get more amount of bid to be accepted. In addition, the results of sensitivity analysis are useful for transmission network expansion.

The comparison with UBLP has been extended to investigate the linking of buyers and sellers. If the market rules require the identification of buyers and seller uniquely, then the IPLP method has to be extended to include matching of buyers and sellers. Note that the UBLP method inherently matches buyers and sellers at each iteration. This is an interesting result of requiring the injection at the slack bus to be zero. Thus, the authors suggest that the slack bus be a tie bus without any injection for generation or demand. Specifically, the incremental losses are made up by the buyer and the seller directly. This resolves the problem of loss allocation without an arbitrary loss allocation algorithm to be executed after the fact. That is after the real power market has been resolved.

## 4.6 Market Power

### *4.6.1 Market Reach*

The goal of a firm is to maximize its profit subject to constraints. There are two major types of constraints—production constraints and non-production constraints. Production constraints include input supply limit, input capability, etc. For electricity, production constraints are comprised of fuel supply limit, generator's capability (e.g., minimum up- and down-time.), crew constraints, and other constraints depending on generating unit types. Non-production constraints could be in various forms depending on products and market rules. Transmission network constraints are the important non-production constraints that play a major role for electricity. Transmission network constraints are comprised of operational and security constraints. Security constraints are composed of reliability, voltage stability, and transient stability constraints. Thermal limits for transmission lines, voltage limits for electrical buses, power flow constraints, are examples of transmission network constraints.

Market reach is the study of how a firm can get additional high-profit customers. The firm may earn more profit from these customers either in short-term or in long-term. The firm may consider losing profit in short-term to gain profit in long-term, or the firm may consider losing some current customers to gain additional high-profit customers. This is acceptable as long as the firm can increase its total profit. In the electricity market, firms include generation companies (GENCOs) and energy services companies (ESCOs). ESCOs emerge in the deregulated market structure as the sole source of electricity to the consumer. In addition, ESCOs can sell or buy electricity to or from GENCOs. In the traditional regulated paradigm, GENCOs supply electricity to consumers following regulations specified by the regulatory commission. In the deregulated competitive market, GENCOs and ESCOs can freely select the customers they desire. The term that will be used commonly to represent both GENCO and ESCO hereafter will be "the firm".

There are factors that will limit the capability of freely selecting customers. Electricity prices in different areas are different because different inputs and production processes are used by local firms and different factors are encountered on local firms. This causes difficulty for the firm with higher cost to compete with the local firm or other firms with lower cost. This problem is more pronounced when transmission cost (i.e., transportation cost in other products) is taken into account. Transmission cost could limit the competition capability of the more distanced firm to the local firm or even a nearby firm. Even the firm with lower cost might have difficulty competing with the local firm or other firms with higher cost when the effect of the transmission cost are considered. Moreover, transmission network constraints could segment market in the way that firms in one area cannot compete with other firms in other areas. Another important factor that affects competition is market power.

Market power is the ability of a firm or a group of firms to control market in favor of the firms. The firms having market power have advantages over other firms in market reach capability. The firms having market power can utilize this advantage to enhance their ability to get customers. It is evident that the study of market power is useful for the study of market reach. Thus, the study will begin with market power study and then follow with a market reach study.

The theme of this work is to determine how generation companies (GENCOs) or energy services companies (ESCOs) can acquire additional customers or additional transactions, given the auction results. In addition, this work is to determine how GENCOs or ESCOs can utilize their market power (if they have any) to enhance their chances of success.

*4.6.1.1 Measuring Market Power*

Deregulation of electricity market breaks up the vertically integrated monopoly and this in turn reduces the firm's market power. However, market power still exists in horizontal market. Market concentration can lead to market power. The general case for market concentration is when there are a few large firms and these firms have large market shares in the market. In addition, a congested transmission network can segment market and can lead to market power even when the market is not concentrated. This is because the firms in some strategic locations can manipulate transactions to beneficially create market power for themselves. The firms, who provide critical ancillary services, e.g., reactive power for maintaining system voltages also have market power.

This section focuses on market concentration. Other causes of market power will be discussed more thoroughly as part of this work. The number of GENCOs and market share of GENCOs determine market concentration in electricity market. The indexes used to measure market concentration are described as follows.

**The Four-firm Concentration Ratio (I4).** This index is a linear summation of the market share of the four largest GENCOs in the market. The equation for the four-firm concentration ratio is shown.

$$I_4 = \sum_{1=1}^{4} s_i$$

Market share is denoted by si and the formula is shown.

$$s_i = 100 * q_i / Q$$

The output power of GENCO i and the total output of all GENCOs are denoted by qi and Q, respectively. Note that the market share and the four-firm concentration ratio are in percent unit. Because the summation is linear, this index cannot differentiate between each GENCO's market share provided that the summation is the same. For example, suppose the market is such that the four largest firms each has 15% of the market share, then the market has a four-firm concentration ratio of 60%, which is the same a market having four largest firms with market shares 57%, 1%, 1%, 1%. This gives rise to problems in measuring market power because the latter market actually has more market power than the former market. A detailed discussion can be found in Shy [1996].

A more generalized form of the four-firm concentration ratio is *m*-firm concentration ratio, where *m* is the number of the largest GENCOs in the market used to calculate the index. The number of the largest GENCOs needed to make the index indicate the right concentration level of the market is questionable. In addition, the number of largest GENCOs needed depends upon the market and varies according to time even within the same market. For example, if there are five large GENCOs having similar sizes in the market now, the four-firm concentration ratio cannot be used and it may be necessary to use five GENCOs to calculate the index instead. Talking this further, if another firm with similar size to those five largest firms enters the market, the five-firm concentration ratio can no longer be used and six GENCOs may be needed to calculate the index.

**The Herfindahl-Hirshman index (HHI).**

This index is the convex summation of all firm market shares. The equation for HHI is shown in (10). Because the summation is convex, HHI can distinguish between the effect of unequal market share among GENCOs even though the linear summation is the same. Thus, HHI can solve the problem with the four-firm concentration ratio. This is why HHI is used more widely than the four-firm concentration ratio in measuring market concentration. Note that one problem of HHI is that the data of all GENCOs is not known. This problem is also found in *m*-

firm concentration ratio. This will definitely be the case when new generating units are installed after the deregulation. A detailed discussion can be found in [Shy, 1996].

$$HHI = \sum_{i=1}^{N} (s_i)^2$$

The other related index is the Lerner index [1934], which is the index for measuring the market power of a monopoly by determining the percentage that the price deviates from the marginal cost. The Lerner index is symbolized by $\phi$ and the formula is shown below. Price in the market is denoted by *P* and marginal cost is denoted by *MC*. The Lerner index can be generalized to measure market power of an oligopoly and of GENCOs in a market with multi-GENCOs and multi-ESCOs. In this case, the index, $\phi^{mod}$ is the ratio of the difference of the price in the market and the competitive price to the price in the market and the index is shown below. Price in the market is denoted by *P* and the competitive price is denoted by

$$\phi=(P\text{-}MC)/P$$
$$\phi^{mod}=(P\text{-}P^*)/P$$

Note that the *m*-firm concentration ratio, the HHI, and the Lerner index are special cases of the gradient index and can be acquired by simple transformations on the gradient index. The gradient index is calculated for measuring the rate of potential improvement in the welfare performance of a market. Its value is sensitive to the behavior of the GENCOs in the market. The detail of gradient index is described in Dansby [1979].

There are plenty of controversial issues for calculating market share. One of them is that the GENCOs' outputs change over time. For example, the outputs might be high during peak periods and low during off-peak periods. In addition, the GENCOs' outputs change with the condition of the network. The outputs at the same period of each day may vary widely from each other. This is because the network condition might favor a GENCO in one day but might favor another GENCO a different day. This variance indicates that market power is dynamic. A GENCO that has market power one period might not have market power the next period. This indicates that the concentration measure must be calculated dynamically.

**4.6.1.2 Measuring Market-reach Ability.** Market-reach ability suggests to and from which ESCOs a GENCO can sell power, and with which of these ESCOs the GENCO can make higher profit. This concept is related to the market power and acts like a complement to market power. GENCOs that have high market power have high market-reach ability. GENCOs with market power have an advantage in reaching the customers they desire. By knowing their market-reach ability, the GENCOs can decide what to do to get higher profit. Market-reach ability can be determined by investigating past results, by calculating the transferred marginal benefits (TMB), or by performing sensitivity analysis to find the potential market-reach ability. The meanings of these terms are explained separately in each method below.

***Past Result Investigation***

By investigating the auction results, GENCOs can determine to which ESCOs they can sell power. For example, one bidding scenario in the past is that GENCO 1 increases the bid price and the bid amount is accepted more. Resulting from GENCO 1's increased bid price, GENCO 2 sells less power; ESCO 1 buys more power; and other GENCOs' and ESCOs' accepted bids stay at the same amount. This result implies that GENCO 1 can sell more power to ESCO 1 and GENCO 2 can sell less power to ESCO 1 because of increased bid price of GENCO 1.

The investigation and analysis must be performed very carefully because there are many factors that can affect results. Let us modify the above example so that this time not only GENCO 1 increases the price, other GENCOs might also change the bid prices and amount, and so do the ESCOs. This time the result is affected from the changes of every bidder, not only by GENCO 1. This prevents us from drawing our former conclusions. In addition, it is difficult to conclude which ESCOs each GENCO will reach when the results of more than bidder change. This is because it is difficult to distinguish the path of each of the power injections on the network. The past result investigation is useful for giving us an idea about the market-reach ability; however, it does not always yield any specific conclusions because of the disadvantages described.

***Transferred Marginal Benefit (TMB) Calculation***

This approach is to calculate the marginal benefit of each ESCO with respect to the location of a GENCO. In other words, the value of the marginal benefit of each ESCO is transferred to the value corresponding to the location of a GENCO. The concept is to calculate the amount of power that a GENCO has to supply to each ESCO for given amounts of loads at each ESCO, assuming that the remaining GENCOs and ESCOs are inactive. The TMB at the GENCO's location is higher than the marginal benefit at the ESCO's location because of the transmission cost and the transmission loss. The transmission cost and the transmission losses must be determined and then the TMB can be calculated as follows.

$$MB' = MB + (\Delta TRC + \Delta LC) / \Delta P$$

The TMB can be used to infer the market-reach ability in the sense that it tells the GENCO the lowest price to bid. If the TMB of an ESCO is lower than the present average cost of the GENCO, this indicates that the GENCO cannot reach this ESCO. Otherwise, the GENCO does have a chance to reach this ESCO. This TMB is important information for bid pricing in the next period. The bid price for the next period can be equal to the TMB or a little bit lower. However, this bid price might not be accepted because other GENCOs might bid lower and get their bids accepted instead. To develop the suitable bid price, the TMBs of the same ESCO at other GENCOs' locations are needed. The same method is used to find TMBs at other GENCOs' locations.

The additional great benefit from determining TMBs of an ESCO at all GENCOs' locations is able to determine dynamic market power of all GENCOs to the interesting ESCO. TMBs of all GENCOs are plotted together and the plot, together with the cost curves of all GENCOs complete the analysis. The analysis can tell which GENCO has advantage for the amount of changed load of the ESCO under consideration. Note that the TMBs of all GENCOs may not be needed because not all GENCOs can affect the bid acceptance. The GENCOs unable to affect acceptance could be the GENCOs that are located very far from the interesting ESCO or those that have more expensive costs. Trimming these GENCOs from the analysis will help speed up the analysis without loss of accuracy.

This approach can be performed for any individual ESCO that is of interest to GENCOs. The resulted TMBs are calculated under the assumption that other GENCOs and other ESCOs will not change their power. If they change their power, the result will be affected. Although other GENCOs might change their power, the resulted TMBs' curves can give the big picture of how GENCOs can affect each other and which GENCO has a greater advantage. The effect of changing other ESCOs should be included in the analysis. It can be done by estimating the reaction function of the ESCOs. However, this may result in a complex problem and could make the calculation slower. This indicates the major drawback of this method when a specific result is needed. I suggest neglecting the effect other ESCOs changes and combining the result with the sensitivity analysis. The details will be explained below.

***Sensitivity Analysis***

Many parameters can be used with sensitivity analysis to determine how to raise the market-reach ability. They will be explained later in this work. This section explains how to determine the potential market-reach ability, which is defined as the ability to reach customers when the submitted bid price is at the average cost, given that other bids do not change. This should be an indication of the potential market reach, which is the best that this GENCO can achieve all other things remaining constant. Average cost is utilized here instead of marginal cost to ensure that GENCOs can recover their fixed costs in bidding. The potential market-reach ability can tell a GENCO about its ability to best reach ESCOs. In other words, it will tell a GENCO which ESCOs would buy power from the GENCO if the GENCO bids its average price. Note that the potential ability in this context is based on the submitted bid price. There are other factors that the GENCO might be able to change to achieve a better result. However, I choose to focus on the price because it is the main parameter used for bidding. In addition, the potential ability in this context assumes that other bidders do nothing to affect the bidding results. Other bidders might enhance the GENCO's chances of bid acceptance. However, this is not considered for calculating the potential market reach because of the uncertainty involved.

The procedure for determining the potential market reach is based on the sensitivity analysis on the cost coefficient. Note that the submitted amount must exceed the new accepted amount. If the submitted amount is accepted entirely, it indicates that better result might be achievable if an additional submitted amount is provided. If the new accepted amount is equal to the submitted amount, the sensitivity analysis on the right-hand-side must be performed. *The complete algorithm for determining the potential market reach will be added in future work.* Note that the potential market-reach ability is dynamic because it depends on the based condition.

After the potential market reach is found, the market-reach ability can be studied further. The effect of other bidder's changes can be directly incorporated into the sensitivity analysis. After the incorporation, the new result tells the potential market-reach ability including the effect of other bidder's changes. *An illustrated example will be shown in future work.* Sensitivity analysis can give a more specific result for the market-reach ability analysis. However, it does not give us a handle on the big picture about the market-reach ability. Past result investigation and the TMB calculation are superior to sensitivity analysis in the context of giving the big picture but they are inferior in the context of providing the specific information. The sensitivity analysis results, combined with the past result investigation and the TMB calculation, can enhance the ability to determine the market-reach ability such as the accuracy. As suggested above, the TMB of an ESCO is calculated without incorporating the effect of ESCOs' changes. This is reasonable because the TMB will be used as information to the sensitivity analysis. The specific effect of changes of other ESCOs can be incorporated directly in the sensitivity analysis.

**4.6.1.3 Enhancement of Market-reach Ability** After the market-reach ability is known, the next step is to enhance the market-reach ability. In the short-run, this can be done by adjusting the bidding parameters. Some states in the United States have not deregulated the electric power industry. For the long-run, GENCOs in these states can install the equipment e.g., a phase-shifter transformer and/or build the new transmission lines that will increase their market reach ability.

This section uses sensitivity analysis to illustrate how to change parameters to enhance the market-reach ability. In addition, the procedure of how to take other bidders' reaction functions into consideration is also illustrated. *The detail will be added in future work.* The examples of changing parameters are described below.

*A.* *Change in submitted bid amount*
*B.* *Change in submitted bid price*
*C.* *Change in line limit*
*D.* *Add or remove line (maintenance)*
*E.* *Sensitivity to other players' bidding strategies*
*F.* *Tap changing under load*
*G.* *Phase shifter control of flow*
*H.* *Series compensation of line flows*
*I.* *Shunt compensation of line flows*

**4.6.1.4 Market Power Integration to Market-reach Ability Enhancement** As mentioned earlier in this work, some GENCOs will have market power even after deregulation. This section discusses using market power to enhance the market-reach ability. The basic form of market power studied in this section assumes a GENCO owns the generating units in multi-locations. This work studies how this multi-location GENCO can modify the bid in each location to favor the bidding. In other words, this GENCO can find a combination of the bids in all its locations to enhance its net profit. The GENCO might lose money in bidding in some locations but gain money in other locations. This is acceptable as long as the net gain exceeds the net loss. In addition, this GENCO can use the advantage to take market power away from other bidders.

This is conducted to enhance the market-reach ability of GENCOs by incorporating the market power they have. However, the ISO can conduct the same study to investigate market implementation to reduce or eliminate the bidding advantage of using market power.

### *4.6.2 Application of Auction Methods to Power System Expansion*

Auctions can be conducted for durations of any length, e.g., every 15, 30, or 60 minutes. Auction results in different periods are different because of different bids and power system conditions. Different generator and/or transmission network conditions are coupled to different contingencies. For example, one transmission line out of service and one generator out of service are considered different contingencies. If auctions are conducted hourly, there will be about 720 different auction results in one month, which are coupled to different scenarios. Auction results in a large number of periods can be aggregated by an on-peak/off-peak or weekend/weekday model. In addition, auction results can be aggregated through some other rational combination from hourly to monthly model.

There exist many choices for power system expansion. In other words, additional generators and/or lines can be added to the system in many different ways. The auction results in a large number of periods can indicate candidate generators and/or transmission lines for power system expansion. For example, if a transmission line is often congested in many auction results, the location of this transmission line can be a candidate for building a new transmission line. The dual value of the congested line indicates the benefit of increasing line capacity [Dekrajangpetch, 1998c] and it can be used for selecting the candidates for new lines. Note that the case of additional lines is different from the case of increasing transmission line capacity because additional lines change the effective impedance between buses while increasing transmission line capacity does not.

After all the new generator and/or line candidates have been identified, each of these candidates is introduced to the power system network and new auction results are calculated for each case. The new auction results are compared to the existing auction results and a justification can be made that a particular line is the optimal decision for expansion. Decision analysis method is used to justify which additional line is the optimal choice. It can be seen that a great amount of calculation is needed for calculating the new auction results for each of the new line candidates. Thus, the method of calculating auction results must be very fast and the interior-point linear programming method is very suitable with this need.

This work is intended to impact power system planning. The planning engineer can use this information combined with other power system planning methods for optimal decisions in planning. Methods of power system planning are well illustrated in Wang [1994].

## 4.7 Supporting Services

As stated in the introduction, the term supportive services are used instead of ancillary services. This is because these services are necessary for the proper, timely, reliable operation of the composite power system. It is impossible to decouple the real and reactive generation limits of the generator or transmission line as shown in many textbooks [Wood, 1996]. Indeed, the transmission limits are a function of weather and other factors. It is also impossible to schedule the energy over the period without regard to the underlying change of demand from one period to the next. The demand is constantly changing at least as a ramp between period with random fluctuations around this ramp. Thus, it is very hard to separate the reactive limits of generation from the real, as it is with all transmission and distribution equipment.

### *4.7.1 Ancillary Services*

This work focuses on four ancillary services: spinning reserve, ready reserve, transmission losses, and load following. The purpose and time frame of these services are summarized in Table 4-18. The proposed model considers transmission losses and ready reserves as bundled ancillary services. The spinning reserve and load following are treated as unbundled ancillary services with certain obligations on part of each market agent. The sellers of load following contracts are obligated to participate in area regulation contract defined later in this report. The buyers and sellers may be obligated by the ICA to buy and sell certain minimum amounts of spinning reserves in proportion to their accepted buy and sell bids respectively. This is required because an aggregate system spinning reserve cannot take care of transient and voltage instability problems. Some units need no spinning reserves whereas others may do. The ICA is to determine how much of reserves is needed and where they are needed.

The unbundled ancillary services are treated as separate commodities. In this case, there exists a separate brokerage market for unbundled ancillary services. However, the energy market and unbundled ancillary services market are coupled by their operational characteristics. Hence, the same broker governs both markets and solves the co-ordination problems between the coupled markets. The bundled ancillary services can not be traded. However, the broker procures the bundled ancillary services by single bid auction for a reliable system operation. The broker must compute the cost of such services. The cost of bundled services are jointly incurred.

Table 4-18. Time Frame and Purpose of Ancillary Services.

| Ancillary service | Time frame | Purpose |
|---|---|---|
| Spinning reserve | Instantaneous | Transient instability<br>Voltage instability |
| Load following | 5 - 10 minutes | Tie flow control<br>Frequency control |
| Ready reserve | 15 minutes | Post-contingency management |
| Transmission losses | Instantaneous | Embedded with power flow |

### *4.7.2 Assumptions*

This work aims to recommend pricing mechanism for reserve margins, transmission losses, and load following in electric power transaction. To simplify the discussion, the problem is restricted with the assumptions below; however, the theory developed is flexible enough to allow these assumptions to be relaxed if desired.

Assumption 1: The auctioneer will establish interchange schedules (multilateral transactions) between the participating agents on hourly basis. Bilateral transactions may be established outside the auction but are still subjected to scrutiny with respect to operating constraints. Furthermore, it is assumed that TRANSCOs are not economic agents, i.e. they are exogenous to the model. This assumption can be relaxed by including Transco bids and evaluating them on the $/MW-Mile basis. Post [70] has shown an auction simulation with the transmission bidding. However, the transmission bid evaluation should include the cost of control (such as phase shifters) to direct the power flow in accordance with the established contract. Since, transmission is still considered to be monopoly, this work has not included Transco bidding.

Assumption 2: The agents (ESCOs and GENCOs) are obligated to buy or sell the binding bids declared by the auctioneer. Hence, all local operational constraints (such as ramp rate constraints, emission constraints, fuel constraints, minimum down times, minimum up times, start-up procedure curves, etc.) must be considered by the agents while generating bids for the next trading session or round.

Assumption 3: The auctioneer is the sole authority to verify that the network remains in operation with the new bids in place. The control center operators are responsible to implement the contracts given by the auctioneer.

Assumption 4: GENCOs and ESCOs both are obligated to bid for electric energy, spinning reserves, and load following contracts. Additionally, GENCOs are obligated to bid for ready reserves.

Assumption 5: GENCOs are constrained to provide a certain minimum amounts of spinning reserves based on their accepted bids.

Assumption 6: The amount of spinning reserve margins needed is decided by market mechanisms. The amount of ready reserves needed is a fixed ratio of total energy bids accepted.

Assumption 7: The cost of reactive power, energy imbalances, and redispatch are computed as after the fact analysis and are added to industry cost. The allocation of industry cost among different market agents is an another issue and is assumed to be beyond the topic of this research.

### *4.7.3 Development of Auction Model*

The representation of electric power network for energy brokerage system consists of models for the GENCOs, ESCOs, TRANSCOs and the ICA. The model for transmission network includes the individual transmission lines, transformers and the current distribution of load and generations. The selected models for GENCO and ESCO include the specification of various attributes of a bid offer as shown in Figure 4-19.

**Bus Number**

Block description for sell/buy bids:

Transaction order (priority of blocks)

Block size (MW)

Block price for power/spinning reserve

Price for ready reserves (seller)

Minimum biding quantity (seller/buyer)

Maximum biding quantity (seller/buyer)

Figure 4-19. GENCO and ESCO Bid Data.

The proposed formulation is based on the idea of linearizing the power flow equations with respect to the current power flow state and solving the problem in successive incremental steps. The broker maximizes the consumer surplus in the energy market and spinning reserve market, as well as, minimizes the cost of required ready reserve for the system. This is accomplished by including the price information of energy bids and reserve margin bids in the objective function. This approach is equivalent to maximizing the amount of transactions indicating that all the potential trade gains are realized.

### *4.7.4 Structure Options*

The alternative structures for auctions have been outlined. The ICA must know all transactions each player has to operate the system economically and securely. This requires that the net power at each bus is a known quantity in the formulation. Thus, the ICA should supervise the implementation of every contract between parties by incorporating all of them as preconditions for contract acceptance. Power system operation will be effective only if the ICA knows all transactions and is enabled to enforce all transactions with sufficient financial penalties. The ICA must be authorized to terminate contracts or player actions contrary to the security and the safety of power system operation.

***4.7.4.1 Singled-Sided Auctions vs. Double-Sided Auctions*** Both sellers and buyers can specify the prices and amounts of bids as they desire for double-sided auctions. They can adjust the bids to maximize their own profits. For one-sided auctions, such as seller bidding only, only sellers submit price and quantity bids to the auction and buyers specify only the amount of commodity desired. One of the concerns is that sellers might have to sell cheap power because all sellers submitted low price bids. The bidding price does depend on the pricing method. This problem can be lessened if a seller can specify a reservation price. However, this is an inflexible auction for buyers. This is true for all one-sided auctions, since one of the two sides does not submit a price. These arguments indicate the inefficiency of one-sided auctions. Such inefficiencies and the desire to implement competitive markets support the conclusion to use double sided auctions.

***4.7.4.2 Bundled vs. Unbundled Supportive Services*** If supportive services are unbundled, the supplier and the buyer have to be identified for each supportive service contract. This will allow the ICA to identify who is entitled to the supportive service. If the supportive service is bundled, the service is included with the energy contract that clearly lists the sellers and buyers. This enables the ICA to monitor the use of all supportive services. For transmission loss service, it is hard to know the power flow amongst specific lines without knowledge of all contracts. Additionally, it is hard to monitor and accumulate all flows on an individual contract basis, especially with inadvertent flows. Thus, it is difficult to bid the losses separately.

***4.7.4.3 Spinning Reserves and Ready Reserves*** The formulation to include bids' prices of power, spinning reserve, ready reserve are possible. If bids' prices are different, players might bid in a way that they do not have to sell or buy the supportive services. Should reserve transactions that have been used be the same price as the transactions that have not been used? If they are not the same, there will be more markets. The question is whether it is worthy to have more markets? The above equations, which show inclusion of spinning reserve and ready reserve, neglect losses due to the flow resulting from the use of such supportive services. Additionally, the transmission capacity for reserves is not reserved. This author selects this approach because losses are small and only seldom will reserves be used. However, when reliability constraints are added, this author believes that such transportation reserves will be required.

***4.7.4.4 Frequency of Auctions*** Auctions can be conducted every 15, 30 or 60 minutes. Frequent auctions cause more transaction costs while they give more flexibility in scheduling for each party. Some stock market auctions occur once each day. Others occur every minute of each day. The selection is based on the need to refine the contracts as other markets and supply availability change.

#### *4.7.5 Ramp Rate*

Ramp rate constraints are endemic to specific equipment. Some feel that the owners within the bids offered should implicitly handle equipment constraints. Thus, the ICA would not supervise the equipment use by every party. This author believes that the owners should be responsible for the use of their equipment not the ICA.

#### *4.7.6 Reactive Power*

There are two possible approaches to implement reactive power. The first approach is direct implementation in the formulation. This approach has difficulty that the nonlinear equations have to be linearized. The second approach allows players to bid for reactive power. The objective function will have additional terms of reactive power bids for this approach. This approach can avoid linearizing constraints but there is one more market for the auctions that causes more complexity. This complexity is not warranted based on the models developed.

## 4.8 Linear Programming or Mixed-Integer Programming

If the bids are required to submit in blocks and one block of bid is big, auctions have to be formulated in mixed-integer programming. This will cause a lot more difficult to solve. In fact, the power the buyers receive may be not in block due to sales to cover losses. Thus, submitting bids in blocks when the block size is big may not facilitate loss charges. However, if the size of bid is of same size as solution tolerance, then the auction can still be formulated in LP and solved very efficiently.

## 4.9 Transmission Loss Allocation

After the solution is found, we know the optimal transactions. The amount and cost of the loss depends on each transaction. One frequent question is how to allocate the cost of the transmission loss? One possible way is to use the information from the simplex method. While the simplex method implements the auction, the variables moving into and out of the basis at each iteration can be used to assign loss charges. This enables the identification of each seller and buyer of each transaction, as well as the cost of the transaction loss. The cost of the loss may be paid by either the seller or the buyer or by both the seller and the buyer in any proportion.

# 5 GENERAL OPERATIONS PLANNING

*It will only take a few weeks to code the simulator...*
*Allocation, subject to reality, is not the "fun" part.*

## 5.1 General Corporate Model

The general corporate model used for GENCOs and for ESCOs is shown in Figure 5.1. The concepts of engineering microeconomics is to model the corporation as a mathematical model of input and output for a generic system. The input is always income from sales of products or services along with the raw material used to make the products. The output is the products consumed by the customers. The assets to be managed in this process include the manufacturing equipment, the maintenance equipment, labor, financial contracts, client contracts, and administrative expenses such as sales and management. This is a rather restrictive model but sufficient for the purposes of this work.

The general intent is to manage the assets of the corporation. The first task is to model the various markets. The second task is to model the corporation's capability to compete. The third task is to use scarce resources to best advantage. The fourth is to model the competition. The fifth task is to model the primary market of competition. This is done by assuming that all other markets are in equilibrium and that these markets are independent. The sixth task is to simulate the interactions of the various markets to remove the constraint that the markets are independent. Once a general market solution is complete, then the cash flow can be managed to maximize profits in a deterministic sense. Next, the optimal selection of transactions can be made to optimize profits given the interdependency of contracts. The continued maximization of profit is accomplished in real time by the procurement of "insurance contracts" either through financial markets or through commodity markets. The valuation of such contracts requires dynamic auction market simulation either by modern control theory or by classical optimization techniques.

## 5.2 Forecasting Demand, Markets, and Competitors

### *5.2.1 Basic Concepts*

The forecasting of demand and markets can be accomplished in two popular but alternative fashions: forward curves through modern control theory or classical optimization and time series analysis. The application of forward curves is to model the econometric dependencies between the observed effects and the dependent economic factors. The forward curves can be microeconomic models of each industry or of the customer based on principles of profit taking behavior. The time series analysis used by this author includes classical Box-Jenkins ARIMA models, modern control theory models, Fast Fourier Transform techniques, and artificial neural networks. These techniques are not mutually exclusive. It is possible to treat an ARMA process as a State Space model as an example.

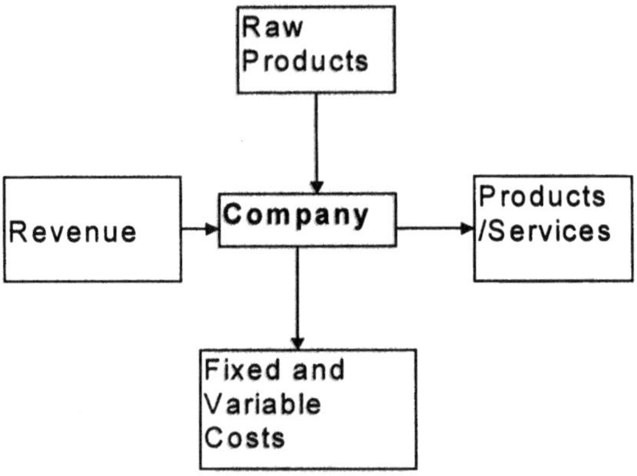


Figure 5-1. Generic Corporate Model.

The key concept is to understand the market for the products. The general model for any consumer, either direct or through a market, is that there is a preference for each product given the limited resources (cash) available. Such preferences can be visualized on preference graphs as shown in Figure 5.2. The two products are real power (P) and reactive power (Q). The budget constraint is shown as a straight line. The consumer preference to either product is the indifference curve. This indifference curve is based on the need to provide real power to the ultimate consumer with the restriction that reactive power is necessary to meet this need. All of the supportive services should be considered as products desired for procurement by ESCOs or for sale by GENCOs.

The basic aspect of pricing theory is the concept of matching supply characteristics with demand characteristics. Such matching is normally shown as the crossing of the supply curve with the demand curve as shown in Figure 5-3. The analysis of markets assume perfect competition. The supply and demand curves are the marginal cost and marginal revenue curves under such conditions. Imperfect markets use alternative models such as average cost and average revenue curves. The difference is in the handling of the fixed costs of production and of consumption. Additionally, the introduction of new technology or of dynamic markets is not included within such steady state solutions. Many excellent economic and financial texts extensively discuss these concepts. The concept of pricing necessary for this work is that the price is determined by the crossover (solution) of these two curves. The area above the supply curve but below the solution price is the producer surplus (PS). This area is labeled PS as defined by the triangle formed by PAB. The area above the price but below the demand curve is the consumer surplus (CS). This area is labeled CS as defined by the triangle formed by PAC.

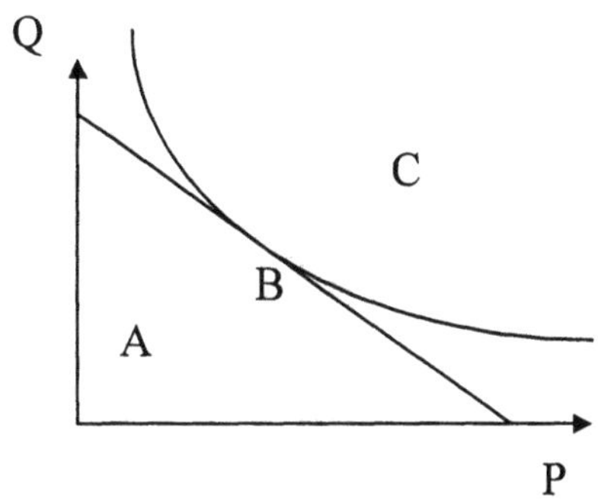


Figure 5-2. Two Product Preference Curve.

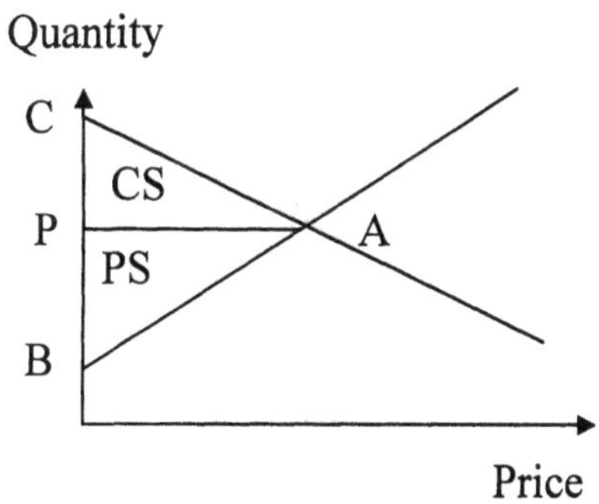


Figure 5-3. Basic Supply Demand Solution Concepts.

Under imperfect pricing conditions, the price may be determined by the supplier. If there is a single supplier setting price, then the supplier has a monopoly. If there is a single price setting supplier (price leader) and one or more price followers this is called an oligopoly. In either case, the price controller is free to set price so as to reap the largest profit possible. The price would be set ideally by the following relationship:

$$Max \quad s = P_1 q_1 - \int_0^{q_1} C(q)dq$$

The surplus that is collected due to such a set price is defined as the monopoly rent as shown in Figure 5-4. Such graphs are useful for interpreting the behavior of markets and of consumers. However, such simple graphs show only steady state equilibrium behavior. The concept used in this work is that such prices provide information for future corporate behavior. Thus, only if such information is available can a corporation respond to consumer and raw suppliers needs. The solution price can change if either the supply curve or the demand curve changes.

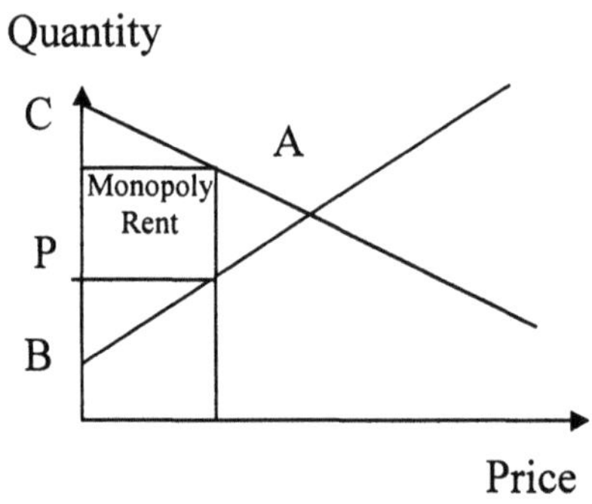


Figure 5-4. Monopoly Rent Due to Monopoly Pricing.

The industrial basic structure may require the acceptance of a fixed set of suppliers with a commodity in a monopolistic marketplace. Society allows such action when the benefits outweigh the detriments. Electric power was such an example of a monopolistic marketplace. Prevention of abusive price setting was accomplished by regulation. Such regulated corporations are called monopolies. The objective of the regulation is to prevent the assessment by the monopolistic company set a price such that the monopoly rent is unfair to the consumer.

The demand curve changes whenever the consumption value of the product(s) changes. Such consumption value can change as the result of research and development (R&D) as shown in Figure 5-5. The area between the two demand curves is the added net benefit due to R&D. This is the area formed by ABCD. The need to expend R&D is based on the need to change the product quality and use to achieve such benefits.

Improvements in product manufacturing changes the supply demand solution as shown in Figure 5-6. The area between the two supply curves is the added net benefit for improved production efficiencies. This is the area formed by ABCD. This assumes that all consumer perceived product characteristics are not changed.

Note that the main beneficiary of such changes is the consumer in both cases. This is seen by the amount of area changed. Such equilibrium economics do not engender the use of scarce cash flow for R&D to improve production or the product(s). Unfortunately, the benefits for the corporation are in the dynamic behavior of the markets as efficiencies are found and products improved. Possibly the supplier can capture a larger market share or, temporarily, use market power to capture monopolistic rents (extra profits). Then such benefits can justify R&D expenditures. Alternatively, such expenditures are necessary to remain competitive. Sometimes, pricing has to be below the price solution to remain in the market or to retain market share as new corporation enters the marketplace.

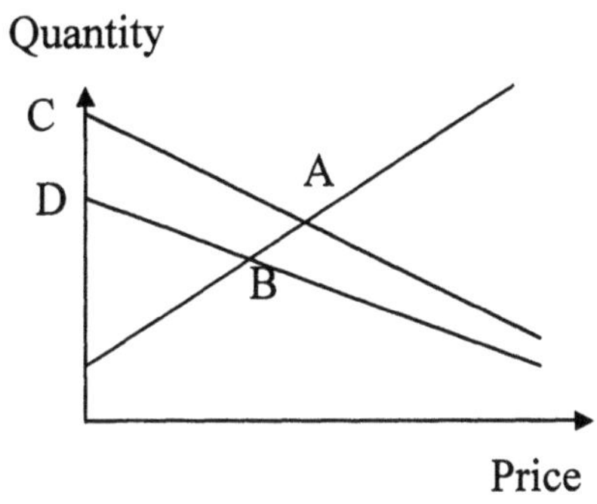


Figure 5-5. Effect of Product R&D.

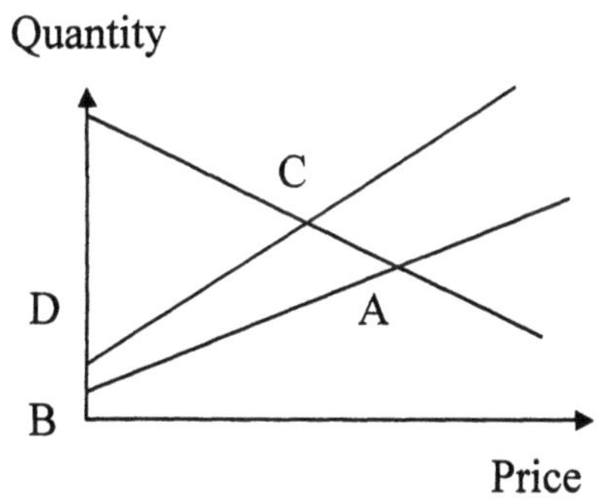


Figure 5-6. Effect of Improved Supply Curve Efficiencies.

The final concept in this section is the value of forecasts. A two period model showing the value lost to consumers resulting from inaccurate forecasts is the Hyami-Peterson model. The value of better forecasts shifts to many GENCOs involved in the supply of electricity as they can better plan their own activities. Better forecasts also provide benefit to consumers. Consider the consumption of real power (P) over the course of two periods. P1 is the consumption over period 1 and P2 is the consumption over period 2. The total consumption is limited due unless there is a change in the budget allocated. Then any forecast error leads the consumer to less value since the poor information leads to a consumption pattern that is less than optimal under conditions of perfect information.

The best solution of the above concepts would include dynamic solutions of the price setting process as a function of time. Such solutions are available from dynamic market simulations as outlined later in this chapter.

The dynamic simulation of competitive markets is best understood from the interaction of GENCOs and ESCOs as shown in Figure 5-7. The period between auctions are not defined nor is the duration of an auction. Assume for the following discussion that the auction is for the daily exchange of energy on an hourly basis. Also, assume that the auctions occur once per day. Additionally, the bids are publicly posted for the 23 hour period before the auction occurs. Let's assume that the auction takes one hour to be calculated and the awards verified and posted. (This is similar to the German stock exchange.) The factor changing between each auction would be the unit availability, the transmission system capability, and the factors included in other markets and other exogenous factors.

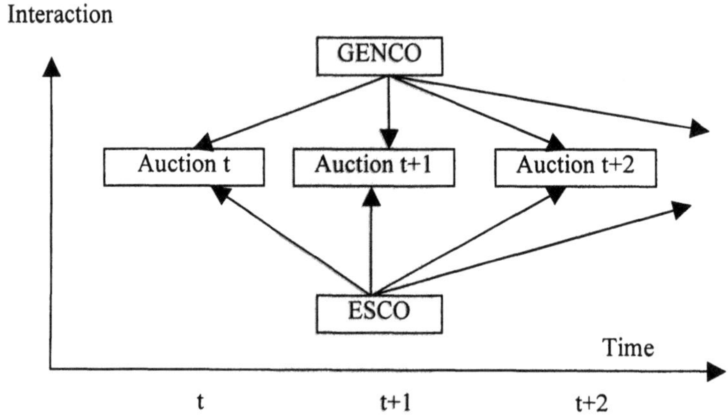


Figure 5-7. Auction Market Interactions.

The successful GENCO or ESCO will have models of these markets. Such general solutions are considered later in this chapter. The market models include forecasting consumer demand at the retail, commercial, and industrial levels. My research has used ARIMA models for load forecasting. Thus, the next section will detail how such consumer demand has been modeled for this work.

The market models will include detailed corporate models of each competitor. Many questions can be answered with such models. What can the competition do when units are outaged? What can the competition do when fuel costs change? Naturally, the successful company will have a detailed model of it's own operation. There are two methods to model companies. One method is to use *forward curves*. Another method is to use modern control theory parameter estimation. This work uses parameter estimation as shown in the next section. Additionally, the successful company will know where it is with respect to the competition. The next section below defines the production frontier as a means of comparing companies. Such studies make a significant impact on strategic planning.

Once the models for the market simulation are verified. Then the auction market simulations can be performed. The auction market simulator developed as part of this research is described in complete details. Either classical optimization theory or control theory models can be used to determine the bidding implemented by all parties. Additional methods used within my research includes: genetic algorithms, genetic programming, fuzzy logic, and artificial neural networks.

Once the results of the market simulator is analyzed, the corporate position on transaction contracts selection and cash flow can be adjusted. The research models reported within this work uses general results of Thompson [1992] converted to the electric power domain. Finally, the transaction portfolio risk has to be ascertained. The models developed for this research also relies heavily on Thompson [1992] converted to the electric power domain.

Once the risk is know, then the strategies to limit the risk can be built. This work uses the general paradigm of decision analysis to control risk by best selection of contracts. After each selection is made the risk for the new conditions has to be reassessed. The decision analysis section outlines one manner of including all of the deterministic models into a risk management package.

The final section defines how the markets interact with real-time control. Automatic generation control (AGC) is expanded into automatic network control (ANC). ANC controls all connections such that the detail of each contract is implemented faithfully by all participants. Control action by ANC could be based on supplying shortages or overages from the spot market or by simply ESCOnnecting players without sufficient contract coverage.

### *5.2.2 Modern Control Theory Models*

The modern control method for modeling markets, demand and competitors is based on finding a rational model relating the states of the model to past states, external factors, and time. As most competitive companies wish to maintain secrets (proprietary information) only certain behaviors can be observed. A most generic approach is shown below [Franklin, 1981]:

$$\begin{aligned}
\dot{x}_1 &= f_1(x_1,\ldots,x_n,u_1,\ldots,u_m,t),\\
\dot{x}_2 &= f_2(x_1,\ldots,x_n,u_1,\ldots,u_m,t),\\
&\vdots\\
\dot{x}_n &= f_n(x_1,\ldots,x_n,u_1,\ldots,u_m,t),\\
y_1 &= h_1(x_1,\ldots,x_n,u_1,\ldots,u_m,t),\\
&\vdots\\
y_p &= h_p(x_1,\ldots,x_n,u_1,\ldots,u_m,t),
\end{aligned}$$

The notation is the same as that of modern control theory. The state(s) are $x$. The input variables are $u$. The observed output is $y$. Note for simplicity, that the time variant model may be reduced to a time invariant model if the time interval over which the model is to be valid is sufficiently small that the corporate decision methods would not change. The plant model in matrix notation follows.

$$\dot{x} = f(x,u,t),$$
$$x(t_o) = x_o,$$
$$y = h(x,u,t).$$

Given the time scale under operating conditions, then the approximations of f and h do not change significantly from the initial relationships (stationary). More simply, lets drop the time dependency. Appropriate models are limited to finite ranges of input and output. Thus, it is advantageous to consider that we are working with small signals where x and u are close to their reference values.

This is the gradient of the scalar f. If f is a vector, then the Jacobian matrix is used to define all the partial derivatives.

$$f_x' = \begin{bmatrix} \frac{\delta f_1}{\delta x} \\ \frac{\delta f_2}{\delta x} \\ \vdots \\ \frac{\delta f_n}{\delta x} \end{bmatrix} = \begin{bmatrix} \frac{\delta f_i}{\delta x_j} \end{bmatrix} \text{ in row } i \text{ and column } j.$$

If the values of x and u are an equilibrium point, the following approximation can be used. Use more convenient matrix notation to show the simplicity of the calculations.

$$F = f_x'(x_o, u_o), \quad G = f_u'(x_o, u_o)$$
$$H = h_x'(x_o, u_o), \quad J = h_u'(x_o, u_o)$$

The standard state space form follows.

$$\dot{x} = Fx + Gu, \quad y = Hx + Ju$$

Most bidding models reduce to a single input and output for simulator use.

$$x_{k+1} = \Phi x_k + \Gamma u_k$$
$$y_k = Hx_k + Ju_k$$

A convenient approach is to use a discrete state space approach.

$$x_{k+1} = \Phi(\theta_1)x_k + \Gamma(\theta_1)u_k$$
$$y_k = H(\theta_1)x_k$$

The canonical form to implement a transfer function follows.

$$\Phi = \begin{bmatrix} a_1 & 1 & 0 \\ a_2 & 0 & 1 \\ a_3 & 0 & 0 \end{bmatrix}, \quad \Gamma = \begin{bmatrix} b_1 \\ b_2 \\ b_3 \end{bmatrix}, \quad H = \begin{bmatrix} 1 & 0 & 0 \end{bmatrix}$$

An alternative representation, equivalent to the auto regressive moving average (ARMA) technique used in the Box-Jenkins forecast approach, follows.

$$\Phi = \begin{bmatrix} a_1 & a_2 & a_3 & b_1 & b_2 & b_3 \\ 1 & 0 & 0 & 0 & 0 & 0 \\ 0 & 1 & 0 & 0 & 0 & 0 \\ 0 & 0 & 0 & 0 & 0 & 0 \\ 0 & 0 & 0 & 1 & 0 & 0 \\ 0 & 0 & 0 & 0 & 1 & 0 \end{bmatrix}, \quad \Gamma = \begin{bmatrix} 0 \\ 0 \\ 0 \\ 1 \\ 0 \\ 0 \end{bmatrix},$$

$$H = \begin{bmatrix} a_1 & a_2 & a_3 & b_1 & b_2 & b_3 \end{bmatrix}$$

The state is exactly given by the past inputs and outputs.

$$x_k = \begin{bmatrix} y_{k-1} & y_{k-2} & y_{k-3} & u_{k-1} & u_{k-2} & u_{k-3} \end{bmatrix}^T$$

The output equation clearly shows how the states are related.

$$\begin{aligned} y(k) &= Hx(k) \\ &= a_1 y(k-1) + a_2 y(k-2) + a_3 y(k-3) + b_1 u(k-1) + b_2 u(k-2) \\ &\quad + b_3 u(k-3) \end{aligned}$$

After selecting the model form, it is next necessary to select the techniques to estimate the parameters (θ) to best fit the observed data.

Start from the nonlinear continuous time description with parameter description (θ).

$$\dot{x} = f(x, u; \theta)$$

Define the error as the difference between the observed output and the model output.

$$\dot{x}_a - f(x_a, u_a; \theta) = e(t; \theta)$$

Define a suitable function to minimize over all of the observed time periods.

$$J(\theta) = \int_0^T e^T(t,\theta)e(t;\theta)dt$$

Since the state is the recent values of input and output in the ARMA model, it is reasonable to assume that the total state and all derivatives are available. The equation error is a linear discrete model.

$$x_a(k+1) - \Phi x_a(k) - \Gamma u_a(k) = e(k;\theta)$$

Expand this to include the ARMA parameters as shown in the following.

$$\begin{bmatrix} x_1(k+1) \\ x_2(k+1) \\ x_3(k+1) \\ x_4(k+1) \\ x_5(k+1) \\ x_6(k+1) \end{bmatrix} - \begin{bmatrix} a_1 & a_2 & a_3 & b_1 & b_2 & b_3 \\ 1 & 0 & 0 & 0 & 0 & 0 \\ 0 & 1 & 0 & 0 & 0 & 0 \\ 0 & 0 & 0 & 0 & 0 & 0 \\ 0 & 0 & 0 & 1 & 0 & 0 \\ 0 & 0 & 0 & 0 & 1 & 0 \end{bmatrix} \begin{bmatrix} x_1 \\ x_2 \\ x_3 \\ x_4 \\ x_5 \\ x_6 \end{bmatrix} - \begin{bmatrix} 0 \\ 0 \\ 0 \\ 1 \\ 0 \\ 0 \end{bmatrix} u = \begin{bmatrix} e_1 \\ e_2 \\ e_3 \\ e_4 \\ e_5 \\ e_6 \end{bmatrix}$$

The elements of equation error are all zero except the first as given in the following.

$$x_1(k+1) - a_1x_1(k) - a_2x_2(k) - a_3x_3(k) - b_1x_4(k) - b_2x_5(k) - b_3x_6(k) = e_1(k;\theta)$$

Alternatively, the output can be used instead.

$$y_a(k) - a_1y_a(k-1) - a_2y_a(k-2) - a_3y_a(k-3) - b_1u_a(k-1) - b_2u_a(k-2) - b_3u_a(k-3)$$
$$= e_1(k;\theta)$$

The performance objective is simplified for the discrete case.

$$J(\theta) = \sum_{k=0}^{N} e_1^2(k;\theta)$$

Another approach is to base the performance objective on the output error. A third approach would be to use the output prediction error. The techniques used to estimate the parameters are typically the least squares estimate, the best linear unbiased estimate, or the maximum likelihood estimate.

The resulting relationship is a model of the competing firm, the market within which to bid, or the consumer. All three are needed for market simulation (partial equilibrium solution).

### 5.3 Multiple Product Production Frontier

The production frontier is the amount of product that can be produced with the existing facilities, inventories, fuel and labor contracts, and corporate decision methods. Allow that the competitive firm has enabled each plant to make it's own decisions to maximize profit. Define DMU as decision making unit. The GENCO DMU within a company can be each power plant or a collection of power plants. Similar combinations of demand can be implemented for ESCOs. Then the various competitors can be compared by data envelopment analysis (DEA). DEA uses the LP technique to determine which DMUs are at the production frontier. The production frontier is an envelope of the input-output data points. It should be noted that not all companies will occupy the same location on the frontier. Some will find a niche in one product (e.g. load following) while others may find a unique offering in another (e.g. voltage support). It is a long term business strategy as to which products each company is to outdistance the competition.

Consider the following formulation:

| | |
|---|---|
| $i = 1,\ldots,m$ | index of the inputs of the DMU's |
| $j = 1,\ldots,n$ | index of the DMU's |
| $k = 1,\ldots,r$ | index of the outputs |
| $x_j = (x_{1j},\ldots,x_{mj})$ | (column) vector of inputs of DMUj |
| $y_j = (y_{1j},\ldots,y_{rj})$ | (column) vector of outputs of DMUj |
| $\lambda = (\lambda_1,\ldots,\lambda_n)$ | (row) vector of weights |
| $\theta$ | a scalar "shrinking factor" |

The objective is to form a weighted combination of the inputs and of the outputs. The weighted combination of input vectors:

$$x_1 \lambda_1 + \ldots + x_n \lambda_n$$

The weighted combination of output vectors:

$$y_1 \lambda_1 + \ldots + y_n \lambda_n$$

The constraints on weights:

$\lambda_j \geq 0 \quad \text{for } j = 1,\ldots,n$

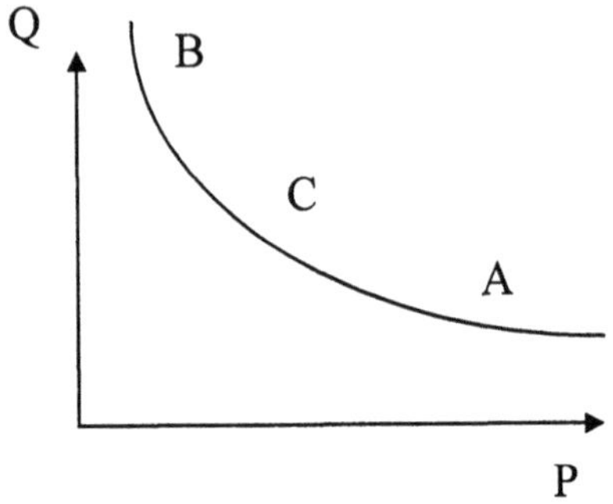


Figure 5-8. Two Product Production Frontier.

Assume each DMU exhibits constant returns to scale (at margin). Compare each DMU with composite firm, if $y^\lambda \geq y^0$ and $x^\lambda < x^0$ then a DMU is dominated or inefficient because composite firm is doing better. Better in terms of less input or more output. If no such weight vector exists, then DMU is scale efficient or undominated and is located on the production frontier. Solution requires another variable, a shrinking factor $\theta$:

$$y^\lambda \geq y^0 \qquad x^\lambda \leq \theta\, x^0 \qquad 0 \leq \theta \leq 1$$

Thus a DMU is inefficient if $\theta < 1$, efficient if $\theta >= 1$ or scale efficient. Assume that none of the real DMUs dominates any other DMU. Now problem is one of finding the smallest value of $\theta$:

Minimize $\theta$

Subject to
$$y_1 \lambda_1 + \ldots + y_n \lambda_n \geq y^0$$
$$x^0 \theta - (x_1 \lambda_1 + \ldots + x_n \lambda_n) \geq 0$$
$$\lambda_1, \ldots, \lambda_n \geq 0$$
$$\theta \quad \text{unconstrained}$$

Note that $\theta$ is nonnegative even though it is not constrained due to the second constraint. Note $\theta <= 1$ is included implicitly. Thus DMU is scale efficient if $\theta = 1$, otherwise inefficient.

Consider the dual of above:

Maximize $uy^0$

Subject to
$$vx^0 = 1$$
$$-vx_j + uy_j \leq 0 \qquad \text{for } j = 1, \ldots, n$$
$$u, v \geq 0$$

Note an equality constraint follows as the dual of an unconstrained variable.

Assume that $vx_j > 0$ for $j = 1,\ldots,n$, then:

Maximize $$\frac{uy^0}{vx^0}$$

Subject to $$\frac{uy_j}{vx_j} \le 1 \qquad \text{for } j = 1, \ldots, n$$

$$u \ge 0, \quad v > 0$$

Note that this is a fractional programming problem. The first constraint of the original dual is a normalization factor. Pick one of infinitely equivalent solutions. Dual variables are now virtual multipliers. Fractional program requires calculation of ratio which is called relative efficiency measure. Subject to constrains that the similar ratio for each DMU is less than or equal to 1.

The optimal shrinking factor θ is equal to the optimal efficiency measure when $v^*xo = 1$ is required. Duality requires that the optimal values are equal: $\theta^* = u^*y^0$. Since $v^*x^0 = 1$, we also obtain $\theta^* = u^*y^0/v^*x^0$.

## 5.4 Electric Energy Industry Modeling

Consider the coal, oil, natural gas, and electricity industries. Hydro and nuclear should also be included. Given the following multiple company, multiple market model of three fossil fuels with linear relationships (Figure 5-8). Consider a directed arc network solution that shows transportation of fuel. Add a node for conversion of fuel to electricity. The output is a node that defines the demand for the product.

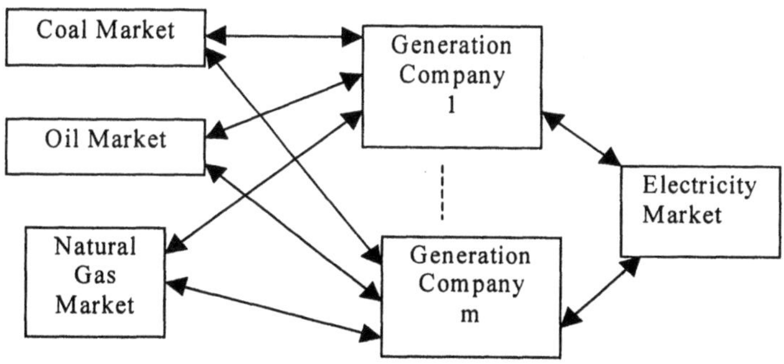


Figure 5-9. Multiple Market Model.

Use the following notation:

| | |
|---|---|
| $i, j = 1,\ldots, n$ | various stages of production and refinement of energy sources, represented as nodes in network |
| $(i,j)$ | productive activity converting energy at note $i$ into energy at node $j$. |
| $c_{ij}$, $c$ | nonenergy input costs incurred in operating activity $(i, j)$, calculated per unit of energy at node $i$; the corresponding (row) vector is written $c$ (it has as many entries as there are activities). |
| $e_{ij}$ | thermal efficiency of activity $(i, j)$, calculated as energy content of one unit of output per energy content of one unit of input. |
| $x_{ij}$, $x$ | amount of energy from node $I$ sent to productive activity $(i, j)$; the corresponding (column) vector is written $x$. |
| $b = (b_1, \ldots, b_n)$ | vector (column) of energy supplies (negative entries) and energy demands (positive entries), given and known. |

Form the node arc incidence matrix as has been done for power flow equations and other industrial interaction [Thompson, 1992, Hillier, 1996, Stagg, 191968]. The conservation of flow for each node is then written as follows:

$$\boldsymbol{Mx = b}$$

The right hand side (b) is either a minus quantity for a supply, a positive quantity for a demand, and a zero for an intermediary point when a positive sign is used for flow into a node. The network problem follows:

Minimize $\quad cx$

Subject to $\quad Mx \geq b$

$$x \geq 0$$

The solution minimizes the total nonenergy input costs. Note that the total flow into the network does not equal the total flow out due to the inefficiencies of production.

Consider the following spatial model for the fuel flow to the electric utility and the subsequent generation of electricity. Assume that each utility will select a combination of activities for generation that maximizes profits. Consider that the price of electricity at consumer point is fixed, then the objective is to minimize costs. The industrial process is to harvest raw fuel, process raw fuel for transport or for production, transport fuel to generation plant, and convert fuel to electricity. For simplicity, assume that the demand for electricity is fixed and known. However, the demand will change due to the elasticity of the product shown by the consumers through the ESCOs. Consider that neighboring companies may ship electricity for economy or for operational security. Let:

$h, k = 1,\ldots, m$ regions.
$s = 1,\ldots, r$ fossil fuel types.

$x_{hs}$ = amount (kWh) of electricity for region $h$ produced by burning fuel $s$.
$X_{hs}$ = maximum value of $x_{hs}$.
$d_h$ = demand for electricity in region $h$.
$f_{hs}$ = supply of fuel $s$ in region $h$.
$w_{hk}$ = kWh of electricity sent from region $h$ to region $k$ (for $h \neq k$).
$t_{hks}$ = number of BTU's of fuel $s$ sent from region $h$ to region $k$.
$c_{hks}$ = per unit shipping costs of sending fuel $s$ from region $h$ to region $k$.
$a_{hs}$ = amount of fuel $s$ needed to produce one unit of electricity in region $h$.
$b_{hs}$ = unit operating cost for producing electricity at region $h$ from fuel $s$.
$g_{hk}$ = unit cost of power loss when sending a kWh from region $h$ to region $k$.

The LP for minimizing the total costs follows:

$$\min_{x_{hkf},t_{hkf}} \underbrace{\sum_{h=1}^{H}\sum_{k=1}^{m}\sum_{f=1}^{r} b_{kf}\,p_{hkf}}_{\text{production cost}} + \underbrace{\sum_{h=1}^{H}\sum_{k=1}^{m}\sum_{f=1}^{r} g_{hk}\,w_{hk}\,p_{hkf}}_{\text{loss cost}} + \underbrace{\sum_{h=1}^{H}\sum_{k=1}^{m}\sum_{f=1}^{r} c_{hkf}\,t_{hkf}}_{\text{fuel cost}}$$

Subject to

$$\sum_{k=1}^{m}\sum_{f=1}^{r}(p_{hkf} - w_{hk}\,p_{hkf}) \geq d_h \quad \text{for h=1,2,3,...,H(conservation of energy)}$$

$$\left.\begin{array}{l} \sum_{k=1}^{m} a_{kf}\,p_{hkf} = \sum_{k=1}^{m} t_{hkf} \\ \sum_{k=1}^{m} t_{hkf} \end{array}\right\} \begin{array}{l} h = 1,2,3,\ldots,H \\ f = 1,2,3,\ldots,r \end{array}$$

$$p_{hkf} \leq P_{hkf}$$

$$p_{hkf}, t_{hks} \geq 0$$

Assume that the maximum supply of fuel is the same for every period and that $a_{kf}$, $b_{kf}$ are constant for every period. This provides a complete industrial model for subsequent analysis.

## 5.5 Inventory Contract Management

The critical decision is the timely procurement of resources. Resources for GENCOs are fuel supplies. Resources for ESCOs are electric energy. Consider the following [Hillier, 1996; Lieberman, 1991; Ford, 1991; Thompson, 1992] model for contract management. Let:

T = total number of time periods.
t = time period, where t = 1, 2,..., T.
$x_t^+$ = the quantity bought in period t.

$x_t^-$ = the quantity sold in period t.
$\alpha$ = unit transactions cost which has to be paid each time a transaction is made.
$p_t$ = the market price at time t; a seller gets $p_t - \alpha$; a buyer pays $p_t + \alpha$.
$d_t$ = the demand of the firm for the commodity at time t.
$I_0$ = the initial stock of the commodity.
$I_t$ = the stock of inventory held at time t.
$I_T$ = the required final inventory of the commodity.
$\bar{I}$ = the fixed warehouse capacity.
h = the unit holding cost for inventory.

Then we can formulate the following LP model based on the general temporal concept shown in Figure 5-10:

Maximize
$$\sum_{t=1}^{T} \left[(p_t - \alpha)x^- - (p_t + \alpha)x_t^+ = h I_t\right]$$

Subject to
$$x_t^- - x_t^+ + I_t - I_{t=1} = -d_t$$
$$I_t \le \bar{I}$$
$$x_t^-, x_t^+, I_t \ge 0$$

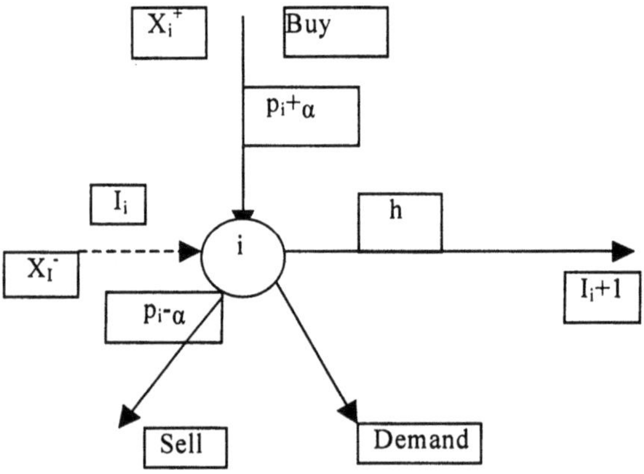


Figure 5-10. Contract Inventory Model for One Time Period.

The firm makes a decision as to how much is bought sold to the consumer or resold to the market at each time period. There are three motives for inventory transactions, speculation, and precautionary.

The result of such studies is the generation of strategies (rules) for operation. Consider the motive of pure speculation, then the normal results imply:

1. Never buy and sell at the same time
   (because this causes a per unit loss of $2\alpha$).

2. When selling, sell to empty the warehouse, or sell down to meet the final inventory demand.

3. When buying, buy to fill the warehouse, or buy to fulfill a final inventory demand.

Inventories dampen price swings. Similar price stability can also be obtained by hedging with Future contracts as discussed below. Such an alternative approach may be more profitable.

## 5.6 Cash Management

The cash management problem is one of paying all bills within contract time allowed, collecting all payments from customers with contract time allowed, and investing the surplus cash into the best short term market without having to borrow excessive cash to pay expenses. Payments from customers are received according to the contract terms. Payments to suppliers are made according to contract terms. Payments for mortgages are made according to contract terms. Investments into certificates of deposits, treasury notes, and other short term financial papers. Cash borrowed to provide payments is often obtained at a higher interest rate than mortgages. This rate is often 2-3% above prime the last month. The mortgage rate is often 2-3% above some expected average of prime over the last year.

Timely payments by customers can include the optimal location of lock-boxes at which customers can make deposits or pay bills. An optimal portfolio of short-term investments, cash on-hand and bank loans is another critical issue. This section focuses on the optimal maturity structure of a portfolio containing both short-term claims and short term debt.

Assume that any purchase of a claim remains perfectly illiquid until it is redeemed. Specifically, it can not be converted before that time into cash. The claim plus interest is paid at maturity. Any contracted debt can not be shifted or sold but remains on the books until maturity. Interest for the entire borrowing period is deducted up front so that the net cash proceeds at the time of the borrowing equal the principal minus the total interest that will be charged. Us the notation:

| | |
|---|---|
| $s, t = 1,\ldots, n$ | indices of time periods, $n$ is the last period (the planning horizon) |
| $x_{st}$ with $s < t$ | purchase of claim in time period $s$, maturing in time period $t$ |
| $r_{st}$ with $s < t$ | rate of interest obtained from investment $x_{st}$, $s < t$ |
| $x_{st}$ with $s > t$ | issue of debt (borrowing) in time period $t$ falling due in period $s$ |
| $r_{st}$ with $s > t$ | rate of interest charged on debt issued in period $t$ and falling due in period $s$ |
| $a_t$ | cash receipts in period $t$. |
| $b_t$ | cash outlays in period $t$. |
| $P, L$ | $P$ is the total buildup of cash available in period $n$ |

The temporal maturity management problem maximizes the total buildup of cash at the planning horizon. A budget constraint in each time period states the total sum of cash funds coming available must equal the sum of the cash needs. Funds coming include cash receipts, maturing claims redeemed, and debt issued. Cash needs include cash outlays, repayment of maturing debt, and new purchases of investments. The primal problem is:

Maximize $P - L$

Subject to
$$\sum_{s=1}^{t-1} (1+r_{st})x_{st} + \sum_{s=t+1}^{n} (1-r_{st})x_{st} - \sum_{s=1}^{t-1} x_{ts} - \sum_{s=t+1}^{n} x_{st} = -a_t + b_t, \quad t = 1,\ldots, n\text{-}1$$

$$\sum_{s=1}^{n-1} (1+r_{sn})x_{sn} - \sum_{s=1}^{n-1} - P + L = -a_n + b_n$$

$$x_{st} \geq 0, \qquad s, t = 1,\ldots, n$$
$$P, L \geq 0.$$

Note that this is a network flow with gains model. The student should be able to draw the network arcs for this problem and the dual problem. Observe that the objective function is to maximize profits if there are any or at least to minimize losses.

## 5.7 Transaction Portfolio Management

The selection of an optimal portfolio of transactions is not solved as of yet. The cash flow, the market instabilities, the introduction of new players, the introduction of new technology, and other exogenous factors would have to be included. The problem considered here is the purchase of contracts from a variety of suppliers. The contracts are risky since the price of electricity may change dramatically. Thus, the price $p_i$ is random. The value of a portfolio with n contracts (x) is:

$$V = \sum_{i=1}^{n} p_i x_i$$

This work assumes that the historical risk estimates are used. Thus, the above forecasting techniques would have to forecast not only the expected price (m) but also the price variance ($\sigma^2$), and the correlation between the various contracts. Note that contracts from the same supplier would have be dependent. Also, contracts from the same marketer would also have similar correlation values.

Let's use the notation:

| | |
|---|---|
| i,j = 1,...n | are the contract indices |
| x=(x1,...,xn) | vector of number of each contract held |
| p=(p1,...,pn) | vector of contract prices at end of auction period |
| po=(po1,...,pon) | present contract prices |
| B | budget for investments |
| r | investment risk parameter |

The formulation of the optimal portfolio then takes a quadratic form that can be solved by LP under specific parameter conditions. The formulation is referred to as a Quadratic Programming (QP) problem. The risk factor (r) determines the location of the portfolio on the efficiency frontier. The resulting formulation is:

$$Max V = \sum_{i=1}^{n} (mp_i)x_i - r\sum_{i=1}^{n}\sum_{j=1}^{n} Cov(p_i, p_i)x_i x_j$$

Subject to:

$$\sum_{i=1}^{n} p_i^o x_i \leq B, \quad x_i \geq 0, \quad \forall i = 1,...,n$$

where Cov is the covariance matrix of the interdependencies of the contracts. Application of the QP extension to LP requires that the second term in the objective function be concave. The equation set is linear.

This is only a partial solution to the contract portfolio selection problem. As with all investments, there is the possibility that the worst case condition will occur. Under such conditions it is necessary to know if the firm can survive the rapid movement of the market without using all of the budget available. The inclusion of such dynamic constraints is the subject of value at risk (VAR). The inclusion of such dynamic limits is beyond this work. However, the application of decision analysis is often used to assess the value at risk.

## 5.8 Risk Management

The key to successful corporations is the management of the risks of doing business. Risk is due to various external causes and to some internal causes as well. The concept of risk management within this work is the use of financial and of physical assets to mitigate unforeseen events and errors in forecasting and operations. The key tool for mitigating such risks is insurance. Insurance may be a contract issued by a bank or insurance company on key individuals within the organization. It may be the use of financial contracts to offset the losses that would occur if the detrimental event occurs. Risk management is a major topic that a complete analysis would justifies a text on this subject alone. The risk management concepts within this work are limited to estimates of the risk taken, estimates of the cost of contracts to insure such losses, and estimates of the information needed to assess such risks.

### *5.8.1 Decision Analysis*

**5.8.1.1 Decision Analysis Introduction.** The first step of decision analysis (DA) is to list viable options for: gathering information, experimentation, and actions [Kaplan, 1978; Bayes, 1763; Bernouli, 1738; Bussey, 1981; Kaufmann, 1968; Lindgren, 1989; Raiffa, 1968]. After the list of events that might occur is generated, then the list is sorted into chronological order. Next, the consequences for each possible state has to be identified as well as the probability of each event. The critical component is to include all costs, benefits and risks. It is always better to do no decision analysis than to do a poor analysis.

The second step in applying decision analysis is to draw the tree. The decision tree includes three types of nodes: decision, natural and result. The decision nodes have connections to natural nodes, one connection for each possible decision. Note that the decision to do nothing is always present and thus must always be included. The natural nodes are connected to other natural nodes or, finally to result nodes. Each connection leaving a natural node identifies the probability (subjective or estimated) of traversing to the next state based on all other probable outcomes for that natural node. The final nodes are result nodes which give the benefit minus the cost for the alternative given the present state defined by all of the probability connections to natural nodes back to the first decision node. All decision trees start with a decision node. All decision trees end with result nodes. The steps to build a tree are summarized as follows:

1. List all possible risk points [n]

2. List all payoffs (benefits-costs) [r]

3. List all payments (tolls) [t]

4. List all decision points in sequence [d]

The next step is to evaluate the tree. The common approach is to assume that all natural nodes are Bayesian (normally distributed). Then the variable to be calculated at each node, the expected monetary value (EMV) is simply a summation of the weighted benefits from the next node (natural or result):

$$EMV_i = \sum_{j=1}^{n} (prob_j * val_j)$$

Where: *i* is the node being evaluated,
*j* is the index of nodes connected to i,
*prob* is the probability of the branch,
*val* is the valuation of the node j.

The alternative selected at each decision node is made by selecting the path, which yields the maximum payoff! The decision tree can be evaluated in table format. This is often done using or with spreadsheet applications. The tree for dispatcher action is fixed by engineering staff.

The next step is to perform a sensitivity analysis. How much do the probabilities have to vary in order to change the decision(s)? At what point is it beneficial to redo the estimates of the subjective probabilities? At what point is more market analysis worthwhile? Such aspects are discussed below.

### 5.8.1.2 Example Applications

*A. Contract Selection*

Consider the basic problem of selecting between two contracts. A dispatcher has the option of selecting contract "A" or contract "B". The initial benefit of the "A" contract has possibly been underestimated. The benefit is now estimated by the subjective probability distribution: \$180,000 @ .2, \$200,000 @ .6, and \$220,000 @ .2. What is the EMV for the "A" contract now? The benefit for the "B" contract is known to be \$200,000 (probability of 1.0). Which contract should be selected? The decision tree is shown in Figure 5-11.

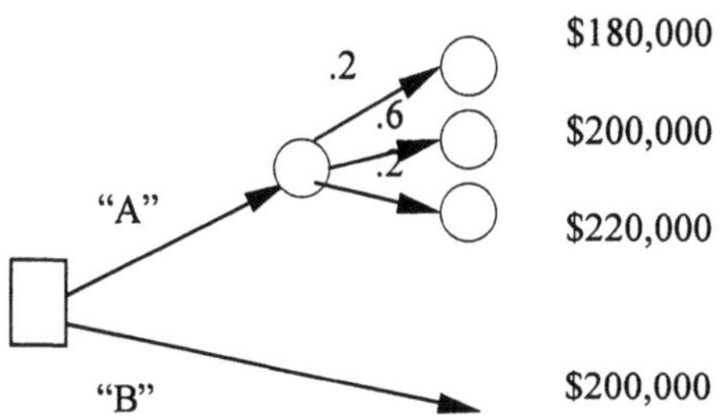


Figure 5-11. Tree Illustration of State Enumeration.

Note that all benefits are shown since all result nodes have values attached. The EMV at the natural node is given by the following calculation:

$$EMV_A = (0.2 * \$180000) + (.6 * \$200000) + (.2 * \$2200000)$$

$$EMV_A = \$200000$$

$$EMV_B = \$200000$$

The decision to be made is simply to compare the EMV for contract "A" with contract "B." Since the EMV for contract "A" is the same, it is not selected, nor is "B." Note that there is not a "do nothing" node. The dispatcher has to select one of the two contracts based on past judgements or call for more analysis of the gathered data. Note that a risk averse dispatcher would select "B" since the outcome is "certain." Such analysis is beyond the scope of this paper.

*B. Reserve Requirement Costing*

Dispatchers in the new environment will have to determine the amount of reserves needed to maintain the security and the reliability of the power system. Assume that the unit availability, the demand duration curve as contracted, and the transaction contracts uncertainty are known for each. How much reserve is needed can be determined by performing a probabilistic production costing as if the industry were still vertically integrated. The example reserve-costing algorithm first linearizes the cost curves by piece wise linear segments. Then, the reserve requirement as required for each contract type (demand or transaction) for the contract duration could be calculated. The contracts would require a given amount of reserves (spinning for the following example). The reserve-costing algorithm solves the optimization problem as a separable linear programming problem for the results shown [ Fahd, 1992; Kumar, 1996].

The DA procedure is to first construct the influence diagram and generate the corresponding decision tree. Determine probability of each path emanating from chance node as described above. That is, use reliability models to compute f(Pg), f(Pl), and f(Tf). Next, calculate production cost f(p) using the LP formulation for all paths of the decision tree. The net payoff or expected cost at the terminal of each path is given as:

Pay off (expected cost) = f(p) + T*Tp + I* Ip

Where: f(p) = production cost
T = number of transactions
Tp = transaction price
I = number of insurance contracts
Ip = insurance price

Next, calculate the expected cost or EMV (including pay tolls due to remedial actions) for all the decision nodes. Finally, choose the transaction (decision node) with the least value of expected cost or EMV.

*C. State Enumeration Method*

In planning fuel supply for the next hour and thus the expected cost of production, the dispatcher is basically acting on the basis of expected monetary value (EMV). In general, one obtains the EMV of a "gamble" with several possible outcomes by multiplying each possible cash outcome by its probability and summing these products over all the possible outcomes.

Tying this with production costing estimation, the structured tree would enumerate all the possible generator status combinations. The first node, as shown in the Figure 5-12, would be the first unit on the priority list with the least production cost from which there are two possible state outcomes for this example: either it is out of service, $s_1 = 0$, or in service, $s_1 = 1$. However, multiple levels of operation can easily be included. Indeed, one industrial application uses 14 levels of operation for accuracy.

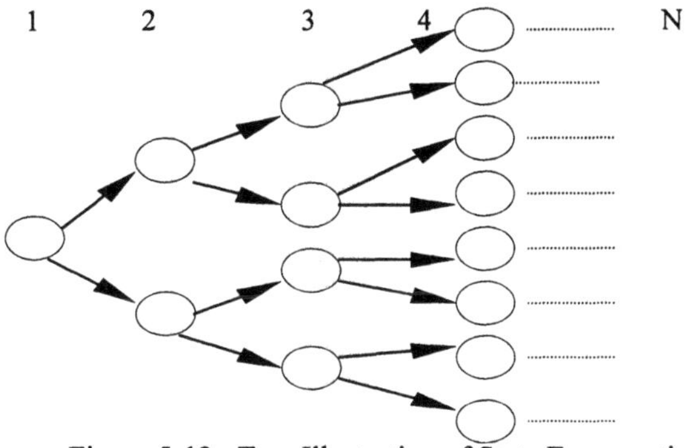


Figure 5-12. Tree Illustration of State Enumeration.

The next stage of the structured tree is the status of generator 2 with $2^2$ possible outcomes. This process is repeated until all possible states are counted. If on one branch the load is satisfied by generator i, then the rest of the N-i branches are deleted (trimmed). With the structured tree built, at most there would be $2^N$ possible state combinations, status = $(s_1, s_2, s_3, \ldots, s_i, \ldots, s_n)$, and their corresponding probabilities ($prob_k$).

Now for each possible state, k, the optimal dispatch is performed on the available units. The dispatch is represented as:

$$z_o = \underset{p_i}{Min}\{\sum_{i=1}^{n} s_o F_i(p_i) + \sum_{j=1}^{m} I_j(f_j) + \sum_{k=1}^{l} C_k(d_k)\}$$

$$\sum p_i + \sum f_j + \sum d_k = 0$$

$$\sum R(p_i) + \sum f_j + \sum d_k \geq RR(S_o)$$

$$p_i \min \leq p_i \leq p_i \max$$

where: $z_o$ = cost of status combination o ($),
$s_0$ = unit status for combination $S_o$,
$F_i$ = operating cost for unit i ($),
$p_i$ = unit output (MWh),
$I_j$ =price of transaction contract j ($)
$f_j$ = transaction flow j (MWh).
$C_k$ =price of customer contract k ($),
$d_k$ = customer demand (MW),
R(..) = reserve contribution (MW),
RR = required reserve (MW).

The expected production cost, EPC, of the structured tree is the sum of the EMVs. The structured tree enumerates all possible generator status combinations, k = 1.... $2^n$:

$S_o = (s_1, s_2, s_3, \ldots\ldots, s_n)$

The procedure's pseudo-code is shown in several steps:

1. k = 1
2. generate a state combination, status, not generated previously
3. calculate the probability of this state, $prob_k$
4. calculate the economic dispatch, $pgen_k$, of the unit state combination
5. if $pgen_k$ is equal to any of the previous $pgen_{k-i}$ go to step 2 else go to step 6
6. calculate EMV
7. if $k = 2^n$ go to step 8 else k=k+1 and go to step 2
8. calculate expected production cost, EPC

*D. Pruned State Enumeration Method*

As the name of this method suggests, the structured tree's branches are pruned to further simplify the tree and the number of calculations needed. Pruning is based on the fractional increase of the EPC when the EMV for all the states with k (k = 0, 1, 2, 3,...,N) units out of service are calculated and added to the EPC. The tree is structured such that the state combinations are enumerated from zero unit outages to N unit outages, i.e. from the most probable (all units available) to the least probable (all units are not available) cases.

This could also be explained by taking a look at the probability distribution function (PDF) of the production cost [Sullivan, 1977, Wang, 1994, Bisat, 1997]. The area under the curve is the expected production cost, EPC. The PDF provides the probability that the EPC will exceed a given production cost. The PDF is stratified into N+1 intervals; each interval corresponds to the EMV contributions of all status combinations corresponding to only k (k = 0, 1, 2, 3,...,N) units being out of service.

So, in essence, the pruned state estimation is determining the area under the curve. The enumeration is stopped at stratum k if the ratio, defined below, falls below a predetermined threshold (ε):

$$ratio = \frac{emv_k}{(emv_k + epc_{k-1})} \leq \varepsilon$$

where, $emv_k =$ sum of the expected monetary value of all states with k outaged units, i.e. the area under the curve of stratum k, $epc_{k-1} =$ expected production cost already determined, i.e. area under the curve of stratums 0 to k-1.

**5.8.1.3 Uncertainties** One of the primary concerns associated with using DA deals with accuracy of costs, benefits and probabilities needed for a good decision to be made? This is obviously an area to philosophize about "engineering judgement." The more accurate the estimates, the better the resulting decision will be. There are many items to be validated for proper application of decision analysis. The uncertainty of payoff and bias is the most obvious. The benefits and the costs must be known (estimated) as accurately as possible. It is all too common to have unknown payoffs which were not included in the analysis. It is very common to have unknown sampling costs for gathering the data to perform an analysis. The primary culprit of such errors isinaccurate reporting. Inaccurate cost estimates, not including all costs of control (load following or regulation), and other "simplifications" have been known to give incorrect decisions. These values are not adjustable by the dispatcher but by the engineering support team.

Some problem characteristics are troublesome for implementation. The chance of extreme loss is often best analyzed with alternative DA techniques. Multiple alternatives, each giving a different degree of insurance against loss or failure, is most often the result of analysis. It is assumed that the outcome is independent of the contract selected for this implementations. Additional exceptions are included to stop the process when warranted.

There are many implementation concerns with dispatcher use of DA. There is a tendency to discardseemingly irrelevant factors, suppressseemingly non-crucial factors, or use incorrect data. The economic payoffs are strongly dependent on the cost and revenue data input. The judgmental probability assignments may not be properly assessed. The multiplicity of alternatives may lead to a very complex analysis. However, the critical time within which the dispatcher must decide when and how to cover reserves is expected to be very short in the new environment. A good approach to avoiding such problems is to treat the DA process as an iterative refinement technique.

**5.8.1.4 Analysis Results**

*A. Sensitivity Analysis*

To perform sensitivity analysis, first insert range of values for forecasted variables most likely to change (i.e. control costs). If radical changes do not change the decision, then the decision is not sensitive to uncertainties of that variable, else it is sensitive. If probabilities are not known, then this is more of an uncertainty analysis, otherwise it is risk analysis.

The step to performing a random sensitivity analysis is to apply one variable at a time, starting with the most significant variable. The most significant impact variables represent the dominant factors: control costs, reserve costs, unit and line availability. The procedure is to plot the dependency of EMV on the variable being varied, use interpolation or iterative refinement to find critical value of variable.

The results of such a sensitivity analysis are often shown on tornado plots. Figure 5-13 shows half a tornado plot with decision limit for dispatcher s.). This is an easy way to identify the variables that give the widest variation in results. The essence is to order variables in decreasing importance (impact). Index values in excess of a target limit would require prohibition of requested action (transaction). Then, it is necessary to examine variables based on value of sampled information for remedial action.

When the tornado plot is outside allowable limits, then the dispatcher would be required to follow other procedures. Note that utility (risk aversion) functions can be used to modify such tornado plots based on management strategy.

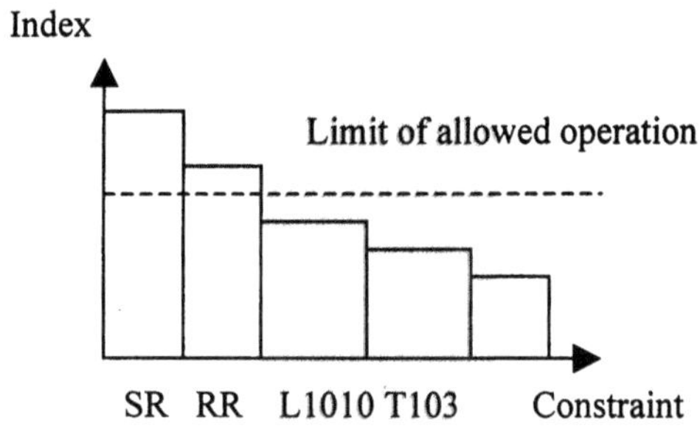


Figure 5-13. Tornado Plot.

Note that SR is the spinning reserve, RR is the ready reserve, L1010 represents the availability of transmission line 1010, and T103 represents the availability of transformer 103

*B. Other Valuations*

Another benefit of sensitivity analysis is finding the value of sampling. The expected value of sample information, EVSI, is based on evaluating the tree for the information needed to change the decision. The typical questions include, at least, the following: How much reserve can be obtained? What is the cost to get it? What is the accuracy of outage information? What are the contract penalties.

The DA method can be used to define the value of information. The measure of information is the probability that a decision-maker's preferred choice will lead to the most desired outcome achievable for a given a specific situation.

**5.8.1.5 Summary** Many applications for decision support systems have been reported in the literature. Some have even been designed and implemented for dispatcher use. However, it has not been necessary, in the past, to include probabilities. Nor has past generation operation required competitive bidding to maintain market share or profits! Other typical applications include: transaction selection, fuel contract selection, hydro scheduling, pumped storage scheduling, maintenance scheduling, etc. DA is potential solution algorithm in place of the traditional scheduling algorithms. New applications will include market share strategies, demand side management, market interconnections, and actual production for both GENCOs and ESCOs. Many DA refinements are not covered in this paper. Such refinements would extend the utility of the approach under more robust conditions.

### 5.8.2 Bidding Based on Decision Analysis

Contract bidding involves decision analysis due to the uncertainty of what the competitors are going to bid. The basic elements of contract bidding includes direct job costs, mark up or return, overhead, and profit. Contract bidding is a multiple objective problem in that maximum profit is desired but it is also necessary to win the award. If management wants many awards to fill capacity, then the price should be low. If management wants maximum profit then capacity used will be low. This constant trade off of high volume vs. high profit is the perennial question. The optimum markup i somewhere in between these two extremes.

If management wants maximum profit then the expected return should be maximized. The average return over N jobs, the expected return over N jobs, and the variance of N jobs, assuming independence of returns, follows.

$$\overline{R} = \frac{1}{N}\sum_{i=1}^{N} R_i$$

$$E[\overline{R}] = \frac{1}{N}\sum_{i=1}^{N} E[R_i]$$

$$V[\overline{R}] = \overline{\sigma}^2 / N$$

If $\sigma^2$ does not grow as a function of N, the variance approaches zero. Then the sample mean value will equal expected value. Strategy should maximize expected return for each contract to result in the largest long-run average return.

Competitor's markup can be considered a random variable based on C, estimate of direct job cost, $X_0$, competitor's bid, and X, competitor's percentage markup.

$$X = \frac{(X_0 - C)}{C} 100\%$$

The exceedance probability is the probability that his competitor's percentage markup exceeds x. Such a distribution is obtained by the analysis of other bids. The value $P_0$ is probability that competitor's bid $X_0$ exceeds the direct job cost C.

$$p(x) = P(X > x) = 1 - F(x)$$

$$p(x_0) = P(X > 0) = P(X_0 > C)$$

The bidder's own percentage markup is *k*. Bidder wins when X>*k*. The probabilities of R = k and R = 0 corresponding to win and loss are given by the following relationship. Percentage return is a random variable with two discrete values (*k*, 0). E[R] is the expected percentage return.

$$P(R=k)=P(X>k)$$
$$P(R=0)=P(X\leq k)$$
$$E[R]=kP(R=k)+(0)P(R=0)=kP(X>k)$$

Consider the simple case of a single competitor. Note that only positive markups are considered. Thus, there are only positive values for x. Assume that the competitor's bids follow the distribution. Then the expected return is easily found.

$$p(x)=p_0e^{-x/\theta}, x\geq 0, \theta>0$$
$$E[R]=kp_0e^{-k/\theta}$$

The optimum markup is found by taking first two derivatives. The corresponding optimal percentage markup and the expected percentage return follows.

$$k^*=\theta$$
$$E^*[R]=\frac{\theta p_0}{e}$$

If there are multiple bidders, the previous can be most easily applied if all competitors have the same distribution form. Assume that all of the bids are independent (no collusion). Then the expected percentage return can be found through the rules of probability theory as follows.

$$p_i(x_i)=p_{0i}e^{-x_i/\theta_i}, x_i\geq 0, \theta_i>0, for\ i=1,\ldots,n$$
$$\prod_{i=1}^{N}P(X_i>k)=\prod_{i=1}^{N}p_i(k)$$
$$E[R]=k\prod_{i=1}^{N}P(X_i>k)=k\prod_{i=1}^{N}p_i(k)$$
$$E[R]=k\widetilde{p}_0e^{-k/\widetilde{\theta}}, \widetilde{p}_0=p_{01}p_{02}\cdots p_{0N}$$
$$and\ \frac{1}{\widetilde{\theta}}=\frac{1}{\theta_1}+\frac{1}{\theta_2}+\ldots+\frac{1}{\theta_N}$$

The optimal percentage markup for the multiple bidders solution is found in a similar fashion.

$$k^*=\widetilde{\theta}$$
$$E^*[R]=\frac{\widetilde{\theta}\widetilde{p}_0}{e}$$

The multiple bidders solution is most interesting if all competitors have the same distribution. Then the optimal solution is dependent on the number of bidders.

$$k^{*} = \frac{\theta}{N}$$

$$E^{*}[R] = \frac{\theta p_0^N}{Ne}$$

The multiple bidder implication is that both optimal percentage markup and maximum expected percentage return are reduced as number of bidders increases. Job return is reduced considerably when competition for the contract is great

The bid error is not symmetric for this distribution. Note that due to the skewness, curve falls off less rapidly from the peak on the high side of k* than on the low side, that the overestimate of k* is less of a loss than an underestimate by the same amount.

The expected value criterion is not always applicable. Consider the normal case of obtaining insurance to remain solvent. Insufficient capital to sustained losses is a major problem for competitive companies. This is a typical insurance problem. The decision is based on subjective preferences. Alternative chosen depends on decision marker's propensity for or aversion to risk. The expected value criterion assumes sufficient capital. An insurance premium is set by finding expected loss, add amount to cover operating expenses and profit.

An alternative to EMV is Risk Profiles and Dominance, especially for strategies. Assume that the mean and variance are sufficient descriptions. Probability distribution completely specified by mean and variance ($\mu$ & $\sigma^2$). Then use simple expressions using $\mu$ & $\sigma^2$ to simply rank decisions. Consider risk curves of these two descriptions.

$$R=f(\mu,\sigma^2)$$

Indifference curves can be used to plot each alternative in two space. Connect alternatives with same R value. Lines connect points equally appealing relative to the criterion established by decision maker. The alias for this graph is "gambler's indifference map." The decision falling on the curve of smaller risk is chosen.

An alternative is to use a utility function to convert the monetary value to a usefulness value. Especially if the actual satisfaction is not proportional to the measure of value. This approach assumes that the value of each consequence representing the satisfaction of the decision maker is tangible and can be expressed directly in a common scale.

Utility of a consequence varies linearly with measure of value expressed in common scale. It is normal to define utility based on psychological assumptions concerning manner in which consequences are valued.

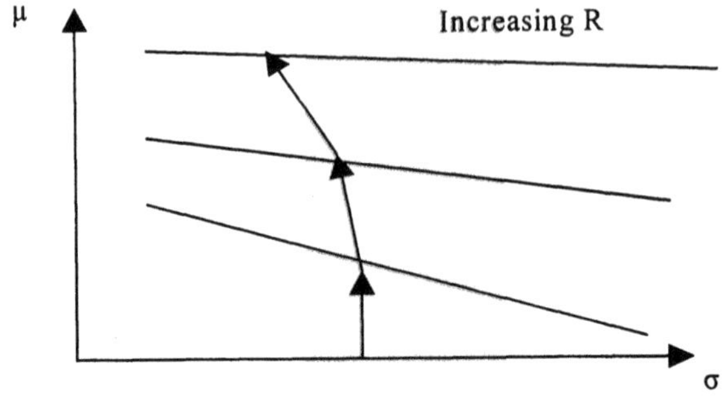


Figure 5-14. Indifference Curve.

Utility functions must satisfy preference properties. Any two consequences ($a_1$ and $a_2$) must have $u_1 > u_2$ IFF $a_1$ is preferred over $a_2$. A gamble "G" resulting in $a_2$ with probability p and $a_1$ with probability 1-p then is a linear combination.

$$u(G) = p\, u_2 + (1 - p)\, u_1$$

Multiple consequences require ordered utilities. Alternative "G" with n possible consequences offering $u_i$ with corresponding probabilities $p_i$ must be a linear combination. Utility functions are typically asset dependent. Consider an utility function that is a function of assets r. The lowest preference is for r = 0. The higher preference is for next level of r. A concave function results that is unique to each decision maker.

Figure 5-15. Utility Function.

Since this is a concave Utility, the alternative "G" with n possible consequences offering $u(r_i)$ with corresponding probabilities $p_i$ by :

$$u(G) = \sum_{i=1}^{n} p_i u(r_i)$$

$$\sum_{i=1}^{n} p_i = 1$$

The relationship of the utility function to EMV is that EMV is linear utility function.

$$E[R] = \sum_{i=1}^{n} p_i r_i$$

$$u(r_i) = k\, r_i, \quad k = \text{constant}$$

Utility functions imbed many implications. The utility scale is totally subjective. It varies from one decision maker to another. It may vary from case to case. A utility function for a company is hard to enforce. Utility values do not reflect value of a time lag between when decision is made and consequence is felt. Additionally, Kenneth Arrow presented the results referred to as Arrow's impossibility theorem. This theorem demonstrate that in general, we cannot write a utility function for a group is all vote equally. The application of von Neumann-Morgenstern utility leads to a very reasonable answer. Specifically, the utility of money is not, in general, a linear function of the amount of money [Luce, 1957]. Utility theory extended to the von Neumann-Morgenstern lottery is a powerful method for creating a value space.

Opportunity loss or regret arises as a measure of value in connection with decision problems including uncertainty. Better decision could be made after the future starts to unfold and decision is reconsidered in retrospect. Loss incurred due to inability to predict, exactly, chance factor outcomes.

Opportunity Loss is difference between value indicated in consequence node and the best that could have been achieved by considering all possible decisions and same outcome. It is a random variable. If alternative chosen produces smallest loss, then, of course, the opportunity loss is zero. Expected opportunity loss (E[L]) represents the long-term average cost that results from having less than perfect information. Computation of E[L] provides information to evaluate risks of each alternative but also for value of information collected and technology developed to reduce uncertainty. Cost of uncertainty E[L] can be described by either a table or a loss function.

A bidding problem opportunity loss would be represented by difference between X and R where X is percentage markup by competitor (continuous variable) and R is the percentage return corresponding to percentage markup k by bidder. R is a discrete variable with values k and 0. One expression for the opportunity loss follows. X is a random variable with known density. Assume that the bidder will use optimum policy, then the opportunity loss follows.

$$l(k,x) = \{^{x-k, \quad for\ 0 \le k < x}_{x, \quad for\ x \le k < \infty}$$

$$L = \{^{X, \quad for\ 0 \le X \le \theta}_{X-\theta, \quad for\ X > \theta}$$

If the density of X is known, then the E[L] can be found as for the above assumed density as follows.

$$E[L] = \int_0^{\theta} x\, f(x)\, dx + \int_{\theta}^{\infty} (x-\theta)\, f(x)\, dx$$

$$F(x) = 1 - p(x)$$

$$p(x) = p_0 e^{-k/\theta}$$

$$f(x) = -\frac{dp(x)}{dx} = \frac{p_0 e^{-x/\theta}}{\theta}$$

$$E[L] = 0.632 p_0 \theta$$

Decision Analysis provides strategies based on subjective probability. Estimated benefits (profit) are required as well as estimated impacts (costs). Logical, consistent, defendable bids can be based on this approach. Additionally, the Value of Information and the Value of Research can also be determined.

The valuation of Futures, Options, and Other Derivatives is often based on decision analysis models. Such details may be found in the next chapter for Futures and Options. Other derivatives are beyond the scope of this work.

## 5.9 Futures Allocations

The valuation of future contracts is based on the expected use of the contract if converted to physical delivery [Kumar, 1996].

### *5.9.1 Allocation of Futures Contract*

The Futures contract offered by the New York Mercantile Exchange enables both buyers and sellers to hedge future production costs up to 18 months into the future. However, such contracts may be delivered only during specific periods of the weekday. The following schedules these contracts, maximizing the benefit to the contract holder taking delivery.

### *5.9.2 Mathematical Formulation*

Under no bankruptcy assumption, market agents are obligated to fulfill the commitments of futures contracts similar to take-or-pay fuel contracts. The futures contract should be modeled as a separate unit (a generator unit for GENCO or a load unit for ESCO) of zero cost in the short-term scheduling problem. The basic framework of allocation of futures contract among the on-peak periods of a specific month can be seen as a profit maximization problem as given below:

Maximize:

$$\sum_{s=1}^{s\max} Pr_s^{*}[(\sum_{i=1}^{n} Pg_{si}) - Pc_s] - \sum_{s=1}^{s\max}\sum_{i=1}^{n} F(pg_{si})$$

Subject to:

$$\sum_{s=1}^{s\max} Pc_s = FC^{o}_{available}$$

$$Pc_{ss\max} \le FC^{o} \quad \forall s = 1,\ldots,s\max and \forall i = 1,\ldots,n$$

$$Pg_{si} \ge Pg^{o}_{i\min} \quad \forall s = 1,\ldots,s\max and \forall i = 1,\ldots,n$$

$$Pg_{si} \le Pg^{o}_{i\max} \quad \forall s = 1,\ldots,s\max and \forall i = 1,\ldots,n$$

$$Pg_{si} - Pg_{s-1,i} \le Ramp_{s,i} \quad \forall s = 1,\ldots,s\max and \forall i = 1,\ldots,n$$

where

| | |
|---|---|
| $Pr_s$ | = Expected price in the sth period |
| $Pc_s$ | = Power generation for futures contract in the sth period |
| $Pg_{si}$ | = Power generation by ith unit in the sth period |
| $F(Pg_{si})$ | = Cost of production of power generation by ith unit in sth period |
| $FC^0_{available}$ | = The unsettled amount of futures contract |
| $FC^0_{smax}$ | = Maximum allowable settlement in one on-peak period |
| $Pg^0_{imin}$ | = Bidding limit for minimum generation of the ith unit |
| $Pg^0_{imax}$ | = Bidding limit for maximum generation of the ith unit |
| $smax$ | = maximum number of periods in forward market simulation. |
| $Ramp_{s,i}$ | = Minimum up/down time ramping limit of the ith unit |

The first and second term of objective function are the total revenue generated and the cost of production, respectively. Thus, the objective is to maximize the profit for a given time horizon. The first two constraints are futures contract allocation constraints. The next two constraints are bidding constraints for minimum and maximum generation capacity limits respectively. (5.29) is operational constraint for minimum up-and down-time ramping limits.

An equivalent formulation can be used by ESCOs to solve the futures contract allocation problem. The operational constraints of specific units (such as hydro generation) should be included in the overall profit maximization problem. The aforementioned framework can be easily modified to accommodate such needs.

### *5.9.3 The Decision Analysis (DA) Approach for Contract Allocation*

Decision analysis can be viewed as a methodology for making decisions with uncertain outcomes. Comprehensive treatments can be found in many volumes [Raiffa, 1968, Hazelrigg, 1996]. Note that the decision analysis method does not compete with other modeling methodologies. Rather, it is complementary because it integrates the results of various models and applies them to decision making. Reference [Howard, 1988] and [Kumar, 1994] describe the application of decision analysis to bulk power marketing. Part of this work led to the development of the framework of transaction selection and evaluation, using decision analysis in a competitive electric market [Kumar, 99]. This approach can easily be modified for application to this model.

The futures contract allocation problem is difficult to solve due to risk and uncertainties associated with fluctuating market prices. In view of these uncertainties, the contract allocation problem is a stochastic optimization problem. However, the DA approach converts the stochastic optimization problem into a set of deterministic optimization subproblems by developing a decision tree as shown in Figure 7-6. The square and oval nodes represent decision and chance nodes, respectively. A decision tree path represents a sequence of decisions after the corresponding price trends are resolved. Thus, each decision tree path has profit associated with the solution of a deterministic optimization subproblem given in (7.2) through (7.6). The solution of these deterministic optimization subproblems are then combined to obtain the solution of the original problem.

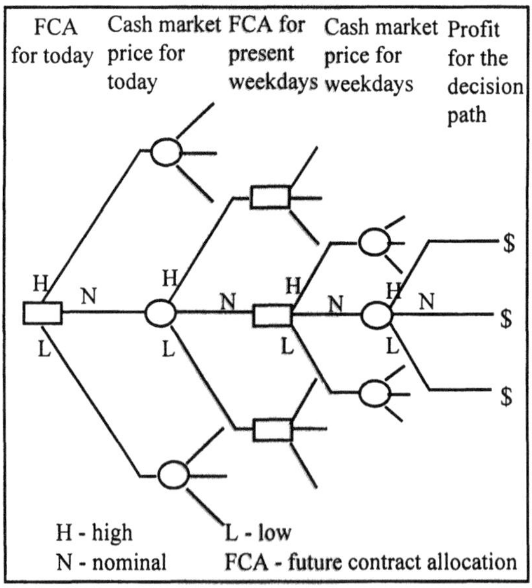


Figure 5-16. Decision Analysis for Futures Contract Allocation.

## 5.10 Auction Market Dynamic Simulation

### *5.10.1 Classical Optimization Approach or Modern Control Theory Approach?*

As noted above, modern control theory can be used to model the competition, the markets and the consumers. Alternatively, the algorithms of optimization given by classical analysis could be used. Both have been used to build a dynamic simulation. However, both must respect the assumption that all players will be profit takers. Specifically, the response of the bidding process will never be an under damped response. Any such bidding strategy would wildly give away profit to the competition or to the consumers. All bidding strategies must be critically damped or, more likely, over damped.

### *5.10.2 Assumptions*

The proposed simulator allows the implementation of futures contracts via forward market simulation. The simulator is developed on the following assumptions.

Assumption 1: The futures market is a monthly market up to 18 months. The forward market is an hourly market for a 1-month period. Figure 5-17 explicitly depicts this time horizon of the different markets.

Assumption 2: The agents are obligated to settle the futures contract via forward market. The allocation of futures contract in forward market simulation is shown in Figure 5-18.

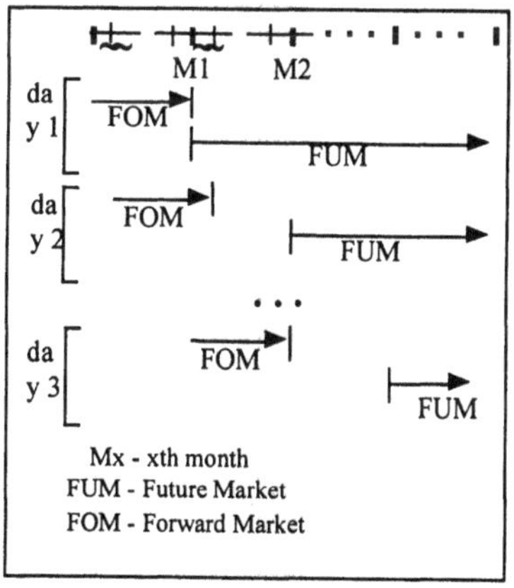


Figure 5-17. Time Horizon of Forward and Futures Market.

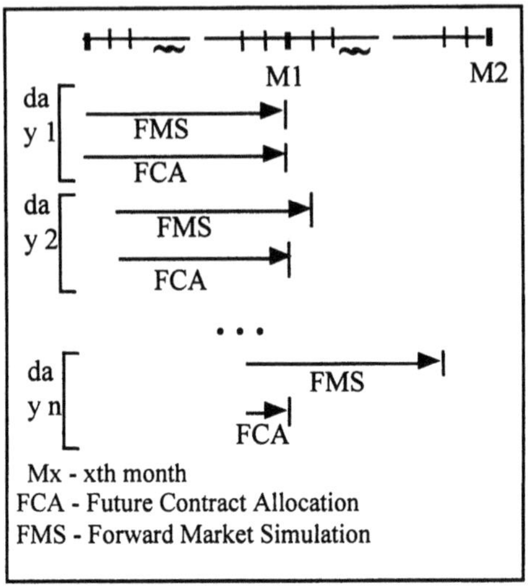


Figure 5-18. Market Simulation and Futures Contract Allocation.

Assumption 3: The overall process of forward market simulation consists of periodic bid development by GENCOs and ESCOs followed by bid matching by auctioneer. This process is repeated several times until the price ESCOvery has occurred in the auction market place. The *price ESCOvery* is defined as the cycle at which the amount of binding contract exceeds a set minimum percentage value. The contracts established at the price ESCOvery cycle are defined as the *closing contracts*. The overall auction process cycle is shown in Figure 5-19. Thus, the major elements of auction markets are (1) rules defining the auction mechanism and contract evaluation, (2) information available to the market agents, (3) GENCO bids, and (4) ESCO bids.

### *5.10.3 Rules of the Auction Market*

Various rules for the market could be accepted as discussed in the previous chapter. The simulator initially reported was based on the following rule. However, changing any of the rules normally results in only small changes to the software if proper software engineering principles are employed (e.g. object oriented design and implementation).

Rule 1: The auctioneer will establish interchange schedules among the participating agents on an hourly basis by using the model developed in previous sections.

Rule 2: The bidding process is performed a fixed number of times within an hour. That is, each agent is expected to submit bids every time the auctioneer requests bids. The auctioneer may decide when the last bids are binding to allow price ESCOvery to occur without excessive gaming of the market. The agents do not know which bid will be binding until the auctioneer declares such conditions.

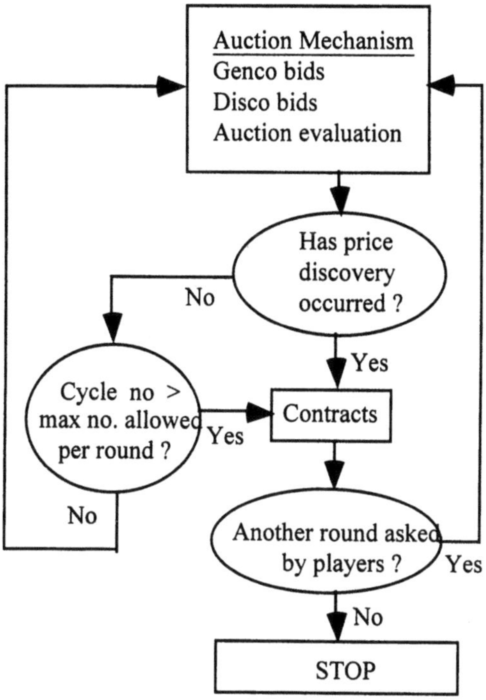


Figure 5-19. Auction Cycle.

However, they do know that there is a maximum number of trials per bidding period. The auctioneer may start another round of bidding if any of the parties wish to bid.

Rule 3: A bid is a specified amount of electricity at a given price. Hence, the agents should determine the price of electricity for the amount they wish to transact (price per block for a given number of blocks).

Rule 4: The agents are obligated to buy or sell the binding bids declared by the auctioneer. Hence, all local operational constraints (such as ramp rate constraint, emission constraint, fuel constraints, minimum downtimes, minimum uptimes, startup procedure curves, etc.) must be considered by the agents while generating bids for the next trading session or round.

Rule 5: The auctioneer would post the following information on an electronic bulletin accessible to all agents.

(1) High bid

(2) Average bid

(3) Low bid

(4) Bids accepted

A chronicle of the above mentioned information would describe the trend of the auction market. Modern control analysis may be used to explain the expected behavior of agents as explained previously.

### *5.10.4 The Overall Scheme for Auction Market Simulator*

The overall scheme for auction market simulator is described in Figure 5-20. Initialization is done on the basis of operating point and contract agreements. The auction mechanism is used to match the submitted bids by the agents.

The agents use future contracts in generating bids to maximize profit by using formulation given in (5.25) to (5.29). The auctioneer ensures that the current day futures contract allocation is feasible and then simulates the forward market for the current month. The auction cycle proceeds until the price ESCOvery occurs. The convergence in forward market is defined as the auction cycle in which the number of closing contracts (as defined earlier) in different periods exceed a set minimum percentage value. At each convergence, the operating points are updated in accordance with the established contracts and the simulation proceeds for the next day.

### *5.10.5 Auction Market Simulation [Kumar, 1996]*

The auction market simulation scheme described has been implemented. The incremental cost curves of generators are modeled as quadratic cost curves given as follows:

$$F(p) = a + b*p + c*p^2 \tag{5.1}$$

The cost curves data used in the simulator are given in Table 5-1.

Revenue curves of ESCOs are also modeled as quadratic curves. However, in this example, these curves are generated such that the data allows enough consumer surplus for facilitating the trade game. The data related to ESCOs are not shown in this report. This approach provides a basis of comparison between auction market outcome and the conventional pool dispatch.

An example of 3-period simulation of forward auction market is used to illustrate the power transaction in the proposed framework. Hydro units being the cheapest always enter the basis of the proposed linear programming (LP) model. Hence, they are ignored for this simulation example. Table 5-2 shows the future contract agreements of the GENCOs. Initial operating points of various units for the three periods are shown in Table 5-3.

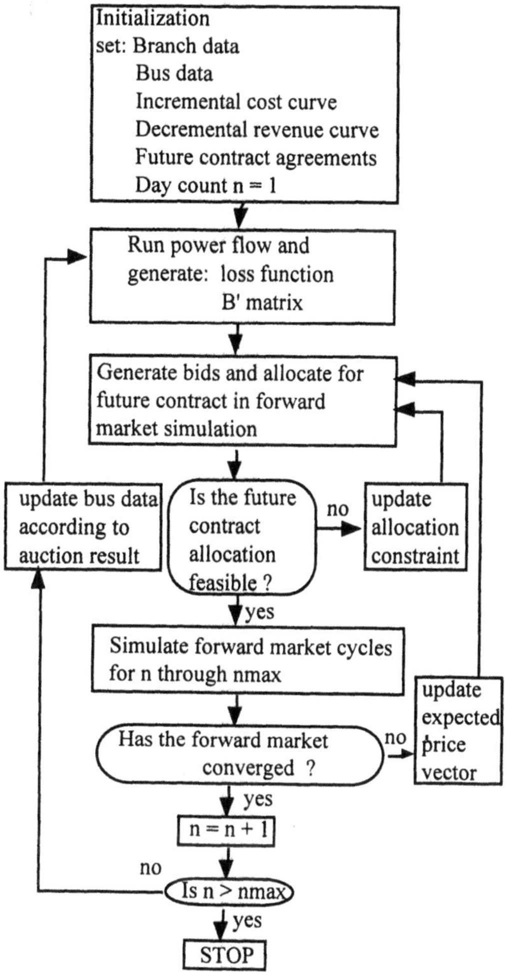


Figure 5-20. The Overall Scheme for Auction Market Simulator.

Table 5-1. Cost Curves of Generators.

| Generator Type | Coefficient A | Coefficient B | Coefficient C |
|---|---|---|---|
| U20 | 78.0 | 7.97 | 0.005820 |
| U76 | 75.9 | 8.69 | 0.001750 |
| U197 | 561.1 | 7.92 | 0.001562 |
| U155 | 173.1 | 7.68 | 0.002394 |
| U350 | 310.0 | 7.85 | 0.008940 |
| U12 | 234.1 | 6.34 | 0.002394 |
| U100 | 342.2 | 8.88 | 0.005572 |
| U400 | 452.1 | 9.71 | 0.002820 |

Table 5-2. Future Contract Agreements in RTS.

| GENCO | Futures Contract MW |
|---|---|
| 1 | 70 |
| 2 | 105 |
| 4 | 45 |
| 4 | 0 |

The simulation process for the agents follows a two-step procedure at every auction cycle. First, the futures contract amounts are allocated among the periods by using the deterministic formulation as described below. Decision analysis treatment is ignored in this example for simplicity. The price estimates are generated by a multi-area economic dispatch program. At the second step, the agents update their operating points on the basis of accepted futures contracts. Then, the bids are generated on the basis of operating points and desired profits. Using these bids, the overall process of auction market simulation for the 3 periods is implemented. The simulation results are summarized in Table 5-4, 5-5 and 5-6.

Table 5-3. Initial Operating Points of Units in RTS.

| Generator Type: GENCO | Period 1 (MW) | Period 2 (MW) | Period 3 (MW) |
|---|---|---|---|
| U20 – 1 | 80.00 | 77.68 | 47.31 |
| U76 – 1 | 165.00 | 165.00 | 165.00 |
| U197 – 2 | 395.53 | 305.45 | 265.00 |
| U155 – 2 | 308.19 | 249.42 | 175.57 |
| U350 – 2 | 73.02 | 57.28 | 50.00 |
| U12 – 4 | 60.00 | 60.00 | 60.00 |
| U155 – 4 | 308.19 | 249.42 | 175.50 |
| U100 – 4 | 125.00 | 125.00 | 125.00 |
| U400 – 5 | 333.00 | 333.00 | 333.00 |

The market clearing price is higher than the lambda pool dispatch. This is due to high value of ESCO bids originally designed to have enough consumer surplus in the system data. Hence, this should not be confused with the high price of transaction. The intent of this example is to show the overall process. The similarity of the proposed approach with the conventional pool dispatch is recognized by observing that all of the market clearing prices are correlated with system lambda. Market clearing prices are in the range of 5% to 15% above the value of system lambda due to randomization in profit desired by the market agents.

Table 5-4 Operating Points of Units After the Binding Contracts in RTS.

| Generator Type: GENCO | Period 1 (MW) | Period 2 (MW) | Period 3 (MW) |
|---|---|---|---|
| U20 – 1 | 80.00 | 80.00 | 80.00 |
| U76 – 1 | 268.70 | 200.42 | 175.00 |
| U197 – 2 | 547.52 | 471.02 | 336.16 |
| U155 – 2 | 310.00 | 310.00 | 269.46 |
| U350 – 2 | 99.57 | 86.21 | 62.65 |
| U12 – 4 | 60.00 | 60.00 | 60.00 |
| U155 – 4 | 310.00 | 310.00 | 269.46 |
| U100 – 4 | 145.00 | 135.00 | 140.00 |
| U400 – 5 | 353.00 | 353.00 | 343.00 |

Table 5-5. Futures Contract Allocation.

| GENCO | Period 1 MW | Period 2 MW | Period 3 MW |
|---|---|---|---|
| 1 | 12 | 26 | 32 |
| 2 | 30 | 35 | 40 |
| 4 | 12 | 18 | 15 |

Table 5-6. Transaction Parameters.

| Transaction Parameters | Period 1 | Period 2 | Period 3 |
|---|---|---|---|
| Market clearing price ($/MW) | 10.58 | 10.41 | 10.13 |
| Lambda for pool dispatch ($/MW) | 9.63 | 9.39 | 8.97 |
| Total amount of sell bids accepted | 292.06 | 360.81 | 314.60 |
| Total amount of buy bids accepted | 285.88 | 353.46 | 304.33 |
| Transmission losses due to transactions (MW) | 6.17 | 7.35 | 10.26 |

### *5.10.6 Bidding Models*

**5.10.6.1 GENCO Bids.** GENCOS would develop bids primarily based on the plant I/O curve. Typically, the incremental cost curve (ICC) of a power plant is monotonically increasing. A piecewise linear ICC is shown in Figure 5-21. To generate a bid, GENCOs need to verify the curve illustrated in the figure. Most of the ICCs are flatter in the lower range of operating points as opposed to much more steeper in the higher range. Thus, more discrete segments are needed as one moves to the higher range of operating points for block bidding.

It is evident that the GENCOs need to generate their ICC accurately. All local constraints such as fuel, emission, and ramp rate must be included in the optimization problem. Moreover, the bid development should be done by using not only the incremental cost curve but also by deciding business strategies with due consideration to moves made by the other agents in the marketplace. Information posted on the electronic bulletin should be properly used to perform a trend analysis and to conjecture performance of other agents.

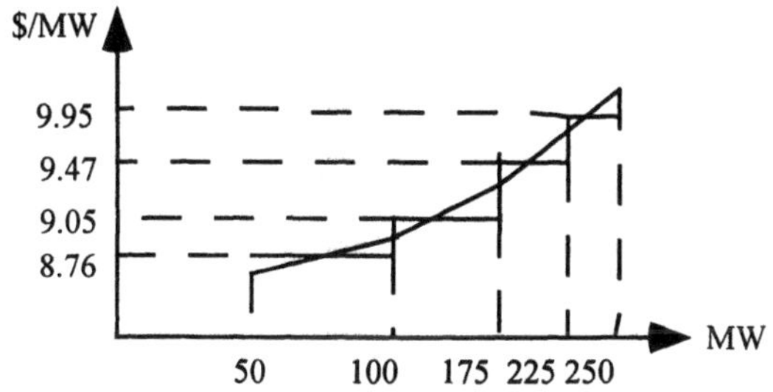


Figure 5-21. GENCO Bids at Different Operating Points.

**5.10.6.2 ESCO Bids** ESCOs would develop bids primarily based on their decremental revenue curve (DRC) as shown in Figure 5-22. At present, the concept of DRC is not very prevalent. But in the future, ESCOs would have to use strategies based on their DRCs to operate efficiently. In essence, the DRC is a marginal revenue curve. Two major components affect the DRC: (1) direct load control [Chu, 1993] and (2) interruptible and curtailable rate programs [Chao, 1983].

The most important element that affects the DRC is the ability to shift the load through different time periods to maximize the overall profit in procurement and delivery. This optimization problem can be solved by dynamic programming using any combination of the following: (a) self generation, (b) storage (cold water, SMES, etc.), and (c) demand side management (DSM) by moving load and losing some). Under interruptible and curtailable rate programs, a ESCO offers a bill ESCOunt to the participating customers in the exchange for the right to curtail their service under prescribed conditions within prespecified monthly frequency of interruptions.

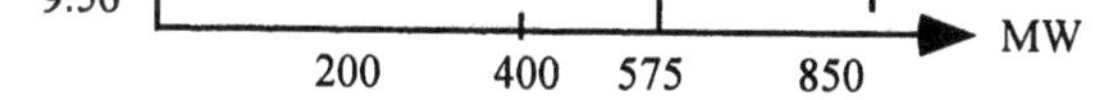


Figure 5-22. ESCO Bids at Different Operating Points.

## 5.11 AGC Simulator in Price-Based Operation [Kumar, 1996a, 1996b]

### *5.11.1 Introduction to Load Following Contracts*

In a vertically integrated industry, AGC operation aims to satisfy the NERC control performance criteria [NERC, 1998] while mitigating growing system problems.

The main objectives of AGC are (1) to hold system frequency at or very close to a specified nominal value, (2) to maintain the correct value of interchange power between control areas, and (3) to maintain each unit's generation at the most economic value. Implementation of AGC schemes is typically in a central location where information pertaining to the system, such as unit MW output, MW flow over tie lines, and system frequency are monitored. Finally, the output of AGC is transmitted to each of the participating generating units to meet the objectives.

Under the new paradigm, AGC operation is accountable to load following contracts described later in this section. However, implementation of these contracts will also meet the NERC control performance criteria as long as the area control error (ACE) is a part of the control objective. This can be done by choosing a suitable market structure and operating mechanism. In such a marketplace, the AGC simulator will need all the information required in a vertically operated utility industry plus the contract data and measurements.

An overview of load frequency control issues in power system operation after deregulation is reported in [Christie, 1995]. The discussion is focused on addressing the operational structures likely to result from deregulation, the possible approaches to load frequency control, and other associated technical issues. The proposed market structure in this work combines features of all the operational structures described in [Christie, 1995].

The focus of this work is to introduce the idea of different load following contracts and provide evidence that the proposed ideas are implementable. Players violating the contractual agreements should be subjected to high penalty. In real-time operation, the contract violation is reflected in higher cost of area regulation requirement. The traditional notion of control area and tie lines is retained in the proposed development.

The new framework requires establishment of standards for the electronic communication of contract data, as well as, measurements among the independent contract administrator (ICA) and the market agents. Increased magnitude of computerized accounting needed by the explosion in the number of transactions is an another technical issue to be solved. In general, a variety of technical regulations will be needed to ensure secure system operation and a fair marketplace.

The growth of transactions between control areas and demands at levels far greater than for classical AGC will require extended algorithms to perform load following in power system operation. One such algorithm with a modified control concept is presented in [Schulte, 1995]. The AGC simulator proposed in this work is capable of using such algorithms. The simulation market consists of three types of transactions as described below.

*Type 1 (Bilateral transactions):* GENCOs and ESCOs negotiate bilateral contracts among each other and submit their contractual agreements to the ICA. The agents are responsible for having a communication path to exchange contract data as well as measurements to perform load following in realtime as shown in Figure 5-23. In such an arrangement, a GENCO sends a pulse to a governor to follow the predicted load as long as it does not exceed the contracted value. The responsibility of the ESCO is to monitor its load continuously and ensure that the load following requirements are met according to the contractual agreement. A ESCO can control its load by using DSM techniques.

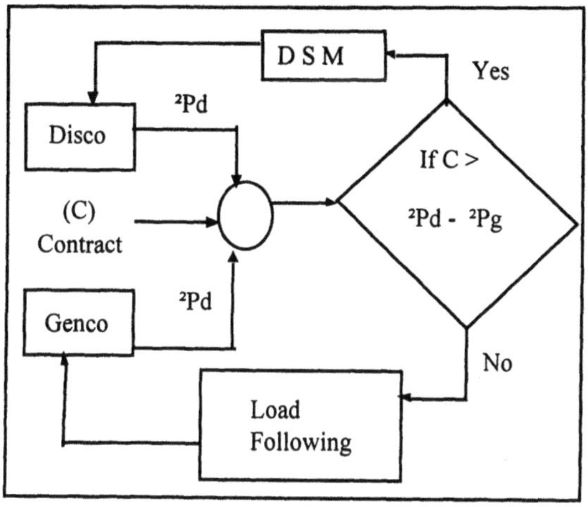


Figure 5-23. A 3-Bus Example System.

### *5.11.2 Example 1: (Transaction 1: Case of Nonbinding Reserve Constraints)*

Bids submitted by the players are shown in Table 5-7. The parameter s1, s2, and Spmax were set to 0.5, 0.25, and 10 MW, respectively.

Table 5-7. Bid Data.

| Bids | Block 1 | | Block 2 | |
|---|---|---|---|---|
| Quote | Size (MW) | Bid ($/MW) | Size (MW) | Bid ($/MW) |
| BUS1(sell) | 25 | 7.5 | 10 | 8.5 |
| BUS2(sell) | 20 | 6.0 | 10 | 8.0 |
| BUS3(buy) | 20 | 9.5 | - | - |

A generic package was used to solve the LP formulation of auction model as described. The results, indicated in Table 5-8, indicate that reserve margin constraints were nonbinding. This result is equivalent to solving the energy and spinning reserve auction markets separately.

The power flow solution was computed with the resulting transaction schedule. Results for the base case and the post state power flow are shown in Appendix D. Results from LP simulation of brokerage system was then compared to that from power flow. Comparisons are shown in Table 5-9. Errors in solution are fairly small and are due to ignoring the reactive power equations and linearization of the system.

Table 5-8. Brokerage Solution.

| Bus No. | Accepted Bids(MW) | Change in $\delta$ | Reserve Allocated |
|---|---|---|---|
| 1 | 1.4 | 0 | 0 |
| 2 | 20 | 0.015 | 5 |
| 3 | 20 | -0.028 | - |

Table 5-9. Verification of Results.

| Variables | Brokerage Solution | Power flow Solution |
|---|---|---|
| $\delta$2(in deg) | 2.12 | 2.13 |
| $\delta$3(in deg) | -6.15 | -5.93 |
| Change in loss(pu) | 0.014 | 0.019 |

### *5.11.3. Example 2. (Transaction 2: Case of Binding Reserve Constraints)*

Bids submitted by the players are shown in Table 5-10. The parameter s1, s2, and Spmax were set to 0.5, 0.25, and 10 MW, respectively. Again, GAMS was used to solve the LP problem disregarding the reactive power and ready reserve constraints. The results, indicated in Table 5-11 indicate that reserve margin constraints are binding. Results are again verified by the post state power flow solution and are given in Table 5-12. The cost of transaction increases as opposed to previous example. This is due to more binding constraints in the system. However, the overall system losses decreased as opposed to Example 1 because LP is intended to minimize the cost of overall transaction rather than the losses only.

Table 5-10. Bid Data.

| Bids | Block 1 | | Block 2 | |
|---|---|---|---|---|
| Quote | Size (MW) | Bid ($/MW) | Size (MW) | Bid ($/MW) |
| BUS1(sell) | 25 | 7.5 | 10 | 8.5 |
| BUS2(sell) | 15 | 6.0 | 10 | 8.0 |
| BUS3(buy) | 20 | 9.5 | - | - |

Table 5-11. Brokerage Solution.

| Bus No. | Accepted bids(MW) | Change in $\delta$ | Reserve allocated |
|---|---|---|---|
| 1 | 6.3 | 0 | 1.25 |
| 2 | 15 | 0.007 | 3.75 |
| 3 | 20 | -0.032 | - |

Table 5-12. Verification of Results.

| Variables. | Brokerage Solution | Power Flow Solution |
|---|---|---|
| $\delta 2$(in deg) | 1.66 | 1.66 |
| $\delta 3$(in deg) | -6.38 | -6.16 |
| Trans. loss(pu) | 0.013 | 0.018 |

The examples (transactions) 1 and 2 demonstrate how the cost of ancillary services are affected by the coupled operational characteristics of energy market and spinning reserves market. In both transactions, the broker receives the same bidding prices for all of the blocks. In the first transaction, reserve allocation constraint is not binding, and hence, bus 2 is contracted for total spinning reserve requirement of 5 MW at the price of 8.0$/MW. In transaction 2, the reserve constraint becomes binding, and accordingly, bus 2 and 3 are contracted for spinning reserve requirements of 3.75 MW and 1.25 MW at the price of 8.0 $/MW and 8.5 $/MW, respectively.

### *5.11.4 Example 3. (Transaction 3: Consideration of a Security Constraint)*

A 3-generator, six-bus test system from [Wood, 1996] is chosen to illustrate the consideration of the security constraint elaborated in Section 5.2.3. The initial condition and the network data are given in [Wood, 1996]. Buses 1, 2, and 3 are considered GENCOs 1, 2, and 3 respectively. Buses 4, 5, and 6 are assumed to be ESCOs 1, 2, and 3 respectively. For simplicity, all bidding is performed in single blocks. Furthermore, the ICA imposes the following security constraint:

$$\text{Lsum}(14, 15) = L_{14} + L_{15} \leq 105 \text{ MW}$$

where: $L_{ij}$ = MW flow on line connecting buses i and j. (5.2)

Since, at the initial condition, the value of Lsum(14,15) is 80 MW, the change in MW flow due to bidding is restricted to 25 MW. Table 5-13 shows the bid data submitted by agents.

Table 5-13. Bid Data.

| Bids | GENCO | | ESCO | |
|---|---|---|---|---|
| Quote | Size (MW) | Bid ($/MW) | Size (MW) | Bid ($/MW) |
| 1 | 45 | 9.20 | 25 | 10.34 |
| 2 | 30 | 9.26 | 30 | 10.27 |
| 3 | 20 | 9.29 | 45 | 10.39 |

First, the brokerage problem is solved without considering the constraint (5.2). The unconstrained solution shows that the change in Lsum(14, 15) is 30 MW; hence, the constraint is binding. Therefore, the problem is resolved with the additional constraint of (5.2) as described in Section 5.2.3. Results for unconstrained and constrained problems are compared in Table 5-14.

Table 5-14. Result for Security Constrained Auction System.

| Problem | GENCO 1 | GENCO 2 | GENCO 3 | GENCO 4 | GENCO 5 | GENCO 6 |
|---|---|---|---|---|---|---|
| Unconstrained | 45.00 | 30.00 | 20.00 | 25.00 | 21.93 | 45.00 |
| Constrained | 37.25 | 30.00 | 20.00 | 25.00 | 30.00 | 29.75 |

Results indicate that the constraint reduces the GENCO 1's and ESCO 3's accepted bids. However, the ESCO 2's accepted bid is increased. In other words, the constraint makes the ESCO 3's bid effectively more profitable. This shows that the consideration of security constraint has direct impact on relative monetary benefits of the market agents. The proposed model is capable of evaluating such impacts.

*Type 2 (POOLCO based transactions) :* Players generate bids (buy and sell) and submit quotations to the ICA. A *bid* is a specified amount of load following at a given price. The ICA binds bids (matching buyers and sellers) subject to approval of the contract evaluation. The bid-matching mechanisms described in [Kumar, 1995, 1996] can be applied to bind the bids for AGC contracts. The objective of matching mechanism is to maximize the number of bid awards for each time period of bid matching. This is equivalent to the classical simulation since the generators are controlled by a single central authority.

*Type 3 (Area regulation contracts):* The ICA obtains contracts with GENCOs to provide area regulation. This is needed because of unscheduled generation and load changes and inconsistent frequency bias existing in the system [Jaleeli, 1992, 1993]. A load change produces a frequency change with a magnitude that depends on the characteristics of the governor and frequency characteristics of system load. All governors respond to this frequency change in the system instantaneously, whether or not they are selected for AGC. The proposed approach defines this governor response as *area regulation contracts.* The cost of area regulation can be allocated among the agents by the ratio of their participation.

### *5.11.5 Classical AGC Scheme*

A classical AGC scheme is shown in Figure 5-24. The scheme consists of a central location where the system data, such as unit output (Pg), tie-line flows (Pt), and system frequency (F) are telemetered to generate raw area control error (RACE).

A filter eliminates inconsequential components of signals in RACE. Subsequently, a processed ACE (PACE) is generated using the load forecast computed by the predictor. The PACE is initially allocated among fast-ramping units to change the desired generation according to ramping participation factors (rpf) of the respective units. Later, PACE allocation is done among the units by calculating economic participation factors (epf) based on an economic dispatch program. Stored energy logic (SEL) computes the potential unit output in response to the desired change in generation. Accordingly, a raise or lower pulse is sent to analog and digital governors by using a pulse code algorithm (PCA) and a communication link, respectively. In the new framework, all the functionalities mentioned above will exist. In this work, the focus is on the construction of ACE and its allocation in the new marketplace.

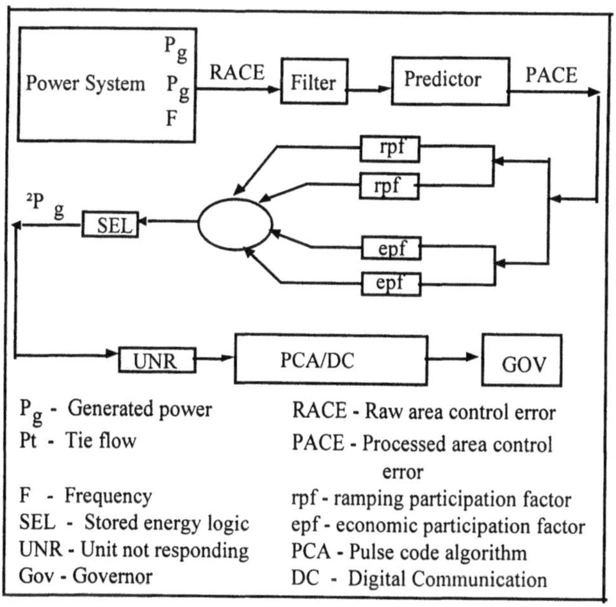


Figure 5-24. A Complete AGC Scheme.

### *5.11.6 AGC Simulator for New Framework*

*Conventional AGC simulator:* A conventional AGC simulator for a two-area system is shown in Figure 5-25. The control error (CE) feedback to a unit on AGC is given below. The ACE is calculated in the mode of tie-line bias control with time deviation correction.

$$CE = ACE = \left[\sum_i Ptie_i - P_{sch}\right] + B^1\left[F - F_{sch}\right] + \int_0 k1 * \left[\sum_i Ptie_i(t) - Psch(t)\right]dt + \int_0 k2 * \left[F(t) - Fsch\right]dt \tag{5.3}$$

where:

| | |
|---|---|
| $Ptie_i$ | = The flow on ith tie line |
| $Psch$ | = The net scheduled interchange |
| $F$ | = System frequency |
| $Fsch$ | = The standard system frequency |
| $B1$ | = The frequency bias (MWsec) |
| $k1$ | = The tie-line bias |
| $k2$ | = The time bias (MW/sec) |

The conventional tie-line bias control concept is correct for a vertically integrated industry, where a single utility company owns one control area and can allocate the CE according to its own wishes. The state space equations of a conventional AGC are given as follows:

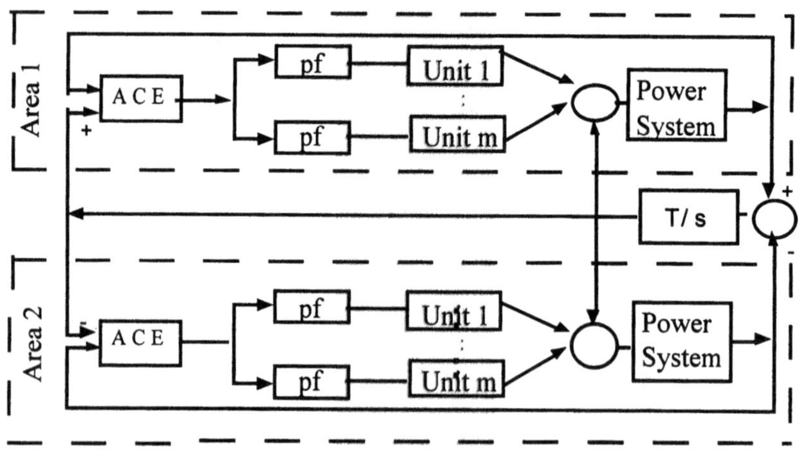


Figure 5-25. Conventional AGC for Two Area System.

*AGC for new marketplace:* In the new framework, many GENCOs and ESCOs will have load following contracts among each other within and across the control area boundaries. In that case, the CE signals will consist of the contract data and measurements among the ESCOs and GENCOs in addition to the area CEs. Such a scheme is shown in Figure 5-24. The modified CE expression for ith GENCO is given by the following.

$$CE_i = ACE^* \, apf_i + \Big[ \sum_{j \in Discos} [cpf_{ji} * DF_j] - GF_i \Big] \tag{5.4}$$

where:

- $ACE$ = The expression given in equation
- $Apf_i$ = Participation factor of ith GENCO in ACE
- $Cpf_{ji}$ = Participation factor of ith GENCO in the total load following requirement of jth ESCO
- $DF_j$ = Total load following requirement of jth ESCO
- $GF_i$ = Total load following generation of ith GENCO

The parameters $DF_j$ and $GF_i$ cannot be measured because they are part of the total load and generation of the ESCO and GENCO, respectively. Hence, they need to be computed as follows:

$$DF_j = DL_j - DC_j \qquad GF_i = GG_i - GC_i \tag{5.5}$$

where:

- $DL_j$ = Net load of ESCO j
- $DC_j$ = Net contracted load of ESCO j
- $GG_j$ = Net generation of GENCO i
- $GC_i$ = Net contracted generation of GENCO i

The parameter $cpf_{ji}$ is set to unity for bilateral transactions. In a POOLCO-based transaction, the value of $cpf_{ji}$ can be determined by the brokering mechanism described previously. The parameter $apf_i$ *is* determined by sorting the GENCOs participating in area regulation contracts in merit order. The state space equations of the modified AGC for new marketplaces are given in Appendix C.

Note that the CE expression given contains the ACE term. Hence, the modified AGC scheme will also satisfy NERC control performance criteria of maintaining the area CE to zero. The simulation results validate this claim. The second term in conservation of energy expression is the component image of the contract quantities in the new marketplace. These component images are continuous, regular, and quiescent. Hence, they are very well suited for generation control as suggested in

[Schulte, 1995]. The AGC scheme shown for the two-area system in Figure 5-26 can easily be extended to multiple area systems.

### *5.11.7 Simulator Features and Capabilities*

The proposed approach is used to develop an AGC simulator for a 3-area test system. The software can simulate bilateral contracts, POOLCO-based contracts, and a combination of both, within and across control area boundaries. The program can simulate any pattern of load changes.

The proposed simulation scheme can be used to study all possible scenarios of contract violation. The approach is based on the premise that the ACE is an integral part of the CE feedback to GENCOs. If the excess demand (equivalent to shortfall of generation from a simulation point of view) is not contracted out to any GENCO, the change in load appears only in terms of area CEs. Hence, the additional demand or the shortfall of generation is shared by all of the GENCOs of the area in which the contract violation occurs.

The details of contract implementation can be used to teach students the issues involved in load following in a price-based operation. More case studies can be presented to encourage thinking about required technical regulations to ensure secure system operation and a fair marketplace. A number of interesting discussions can be motivated by asking questions such as those listed below:

- What are the possible ways in which people can breach the contract?
- How should one detect if somebody is violating the contract?
- What kind of penalty should be designed to prevent such contract violations?
- What should the ICA do to minimize the requirement of area regulation contracts?
- What kind of modifications should the ICA do so that primary generation mechanism can be used to reduce the tie-line oscillations?
- How should one use the enhanced AGC algorithms [Sheble, 1991] to reduce the operational overheads (such as filtering, processing RACE) to meet the increased demand of transactions?
- What modifications should one need to make in control steps for an efficient implementation of contracts?
- What trends should be monitored for technical scrutiny and what kind of correction logic should be applied?

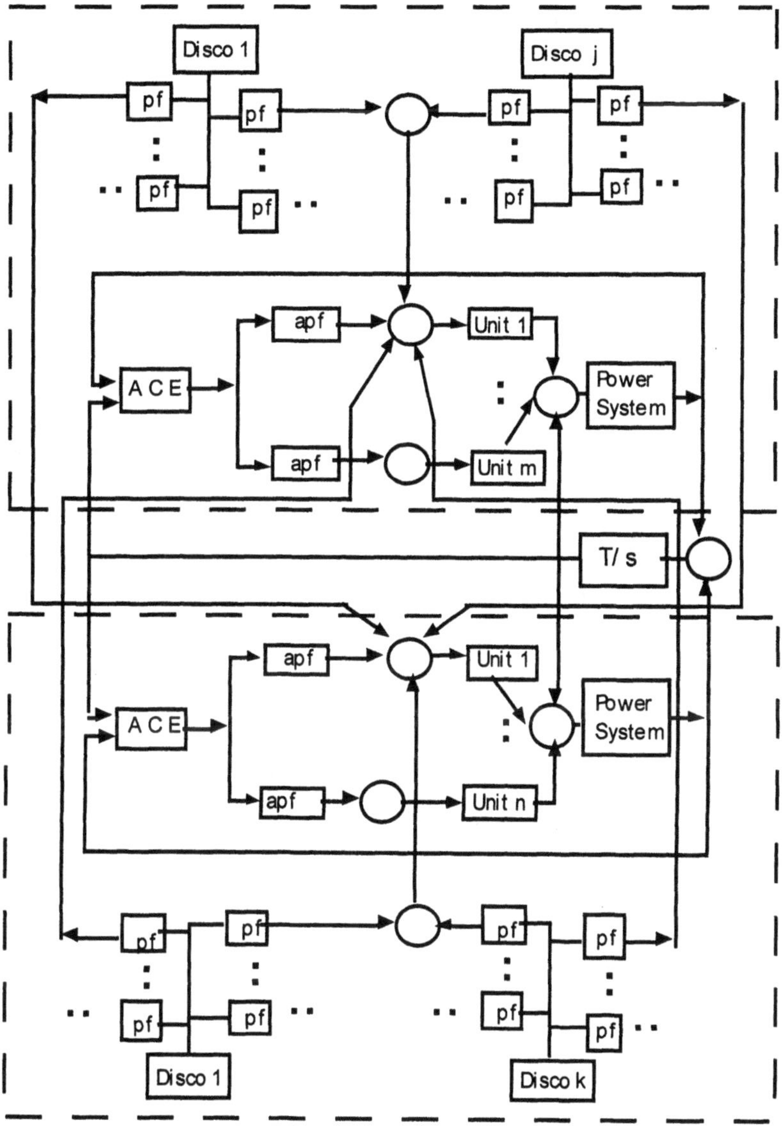


Figure 5-26. AGC For Two Area System in New Market Place.

## 5.12 Advanced Concepts

### *5.12.1 New Environment*

The future environment is not known at this time. However, the impact of heterogeneous products based on reliability or other product differentials (i.e. green power) will require more simulation to determine if the markets are stable. The inclusion of reliability as a product differentiator is critical to the proper price determination for each consumer. The most critical aspect is the rules for the various markets. How many price spikes will occur until a more general commodity market concept is accepted? How many missed profit opportunities or black outs will occur until the markets extend out into the future sufficient for operational and system planning?

### *5.12.2 Advanced Agent Behavior*

Many of the edc $\lambda$ update algorithms are possible methods for generating the computer bid. The interested reader should review the algorithms proposed for edc in the literature.

The double auction winners [Friedman, 1991] of Arizona token exchange demonstrated that the most successful bidding techniques did not use any sophisticated optimization techniques. Indeed, the most successful were those based on simple exponential filtering.

The artificial life techniques (ANN, GA, GP) offer the most complex bidding strategies as documented in [Ilic, 1997].

Additional inter-market simulation would add energy model pointers for fuel markets, hydro markets, emissions markets, and financial markets to provide a complete simulation of the operational planning needed for a competitive environment.

# 6 GENCO OPERATION

*It used to be easy; add generation whenever concerned.*

## 6.1 Introduction

The majority of the research described in this chapter is conducted from the point of view of generation companies (GENCOs) wishing to maximize their expected utility, which is generally comprised of expected profit and risk. Strategies that help a GENCO to maximize its expected utility must consider the impact of (and aid in making) operating decisions that may occur within a few seconds to multiple years.

In the work described here, bidding strategies are developed in an environment in which energy service companies (ESCOs) buy and GENCOs sell power via double auctions in regional commodity exchanges. This power is transported over transmission lines owned by transmission companies (TRANSCOs) at high voltages and over wires owned by distribution companies (DISTCOs) at lower voltages. The proposed market framework by [Kumar 1996a], allows participants to trade via the spot, futures, options, planning, and swap markets. The market simulator developed by Kumar and Sheblé has been adapted to allow computerized agents to trade energy. We use ANN algorithms to find the adaptive bidding strategies for use in a double auction. The strategies may be judged by the amount of profit they produce and are tested by computerized agents repeatedly buying and selling electricity in an auction simulator. In addition to the obvious profit-maximization strategies, one can also design strategies that exhibit other types of trading behaviors. The resulting strategies can be directly used in on-line trading or as realistic models of competitors in a trading simulator. We investigate and discuss methods of minimizing an energy trader's risk using futures and options contracts and through inclusion of risk while judging strategies. We examine unit commitment and discuss how to update it for the competitive environment and its role in developing bidding strategies. We outline methods of automatically inferring other's trading rules and methods of forecasting prices and demand based on historical data. We also discuss how fuzzy logic can be applied in developing bidding strategies. This work facilitates the design of effective trading strategies and profitable portfolios for energy producers.

Within the framework described above, our objective is to maximize the profit of electric energy producers. That is, for a given market, traders will find it helpful to implement bidding strategies in making their bids and offers. Bidding strategies are designed to maximize a trader's utility (a function of risk and profit). We have researched the development of bidding strategies to maximize profit for the spot market using genetic algorithms and GP [Richter, 1997a, 1997b, 1998, 1999]. Although our previous research focused on double auction bidding strategies for the spot market, it is also independently applicable for the futures, options, and forwards markets. We seek to extend our previous work by creating energy trader portfolios that combine the spot market contracts with options and futures contracts with the overall goal of increasing profitability and decreasing risk. In addition, we want to know what strategies others are using, so we investigate inferring trading strategies from historical data.

With neither an obligation to serve nor guaranteed rates of return, the energy trader's objective becomes the maximization of expected profit for his shareholders. In a competitive environment, there may be times when ESCOs may be unable to purchase enough energy for their customers or times when GENCOs may have excess generation. This uncertainty, combined with fluctuating prices and demand, makes profit difficult to predict in any particular scenario. We might then consider a distribution of bids and offers and develop strategies that maximize the trader's expected profit. If a trader uses the strategy long enough, he should get the expected profit associated with that strategy. In the short run, he might see gains or losses very different from the expected profit. This unpredictability means that we consider the strategy risky. The term *risk* can be loosely defined as a measure of the lack of predictability of an outcome associated with a particular decision. Different strategies producing the same expected profits might well have different risks associated with each (see Figure 6-1). Since most traders cannot endure low or even negative profits for long periods, the trader would probably be willing to sacrifice some long-term expected profit in return for reduced risk. Economists use "utility functions" to describe and order preferences. Among other things, a trader's utility should vary directly with actual profit (should be similar to expected profit) and indirectly with risk.

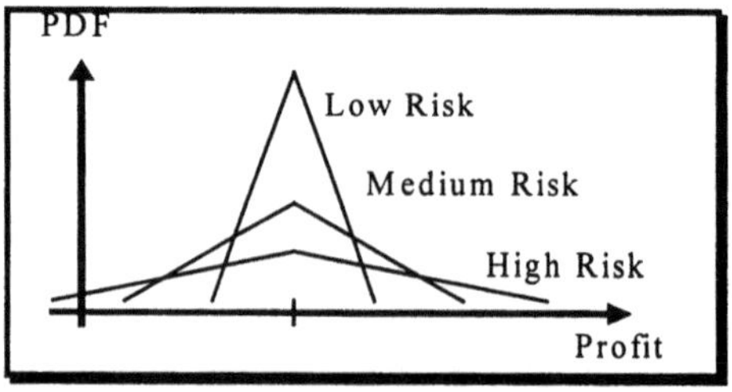


Figure 6-1. Different Risks with the Same Expected Profit.

## 6.2 GENCO Case Studies

### 6.2.1 Output Optimization Strategies

Deregulation in the power industry will have many far-reaching effects for the strategic planning of firms within the industry. One of the most interesting effects will be the optimal pricing and output strategies GENCOs will employ to be competitive in the marketplace while maximizing profits. This case study presents two basic, yet effective means for a GENCO to determine the optimal output and price of their electrical power output for maximum profits.

We will first assume that the switch from the utilities government regulated, monopolistic industry to a deregulated competitive industry will result in numerous geographic regions each consisting of an oligopoly. An *oligopoly* is a market structure with a small number of firms. The market will behave more like an oligopoly than a purely competitive market due to the increasing physical restrictions of transferring power over distances, making it practical for only a small number of generator companies to service a given geographic region.

A GENCO can have one of two positions in an oligopolistic economy. The GENCO will be either a dominant or a secondary participant in the oligopoly. A dominant participant is a GENCO that can service an area with the lowest marginal costs associated with providing the electrical power. This will most likely be the firm that is closest in proximity to the geographic region under consideration.

A secondary participant will be a GENCO that has higher costs, most likely associated with transmission costs, in providing electrical power to a geographic region. A secondary GENCO can service excess demand in an area that the primary GENCO cannot fully service. Any given GENCO will most likely operate as a primary GENCO in their immediate surroundings while acting as a secondary GENCO for geographic locations further away.

For the primary GENCO, a strategy of price leadership is the most effective way of maximizing profits. With the price leadership strategy, the primary GENCO has the opportunity to determine the price of electricity as a function of how much electrical power they are willing to supply to the area. A secondary GENCO does not have enough market share to influence the price of electricity in the area. Therefore, a secondary GENCO must use a market price that has already been determined. The strategy for the secondary GENCO becomes one of optimizing the quantity of electrical power they are willing to supply to an area for a given price.

**6.2.1.1 Review of Economic Theory** To understand the GENCO's market strategies let us review economic theory. In a basic economic supply and demand model, the two primary functions are the supply function and the demand function. Both equations relate price as a function of quantity. In the electric utilities industry, a bid consists of both the price of electricity per megawatt-hour (MWh) and the quantity in MWhs.

The market demand function represents the amount of a commodity that buyers will purchase at a given price. The demand function, on a graph that has price on the vertical y-axis and quantity on the horizontal x-axis, would be a downward sloping curve. The downward slope of the curve makes sense, because as price decreases one would expect the quantity demanded to increase for a given commodity.

The supply function represents the amount of a commodity suppliers are willing to sell at a given price. When we graph the supply function on the same graph as the demand function, it is logical that it would be an upward sloping curve, which means that the amount of the commodity that sellers are willing to supply increases as the commodities price increases. The market supply function represents the total amount of the commodity supplied by all suppliers in the market. In the electric utilities industry, suppliers are made up of individual GENCOs supplying electricity to the market for a given geographical region.

To fully understand how the output and pricing strategies work, we will define two other functions: the marginal cost function and the marginal revenue function. The *marginal cost function* is the additional cost for a single supplier to produce and deliver one more unit of a commodity. The *marginal revenue function* is the additional revenue that a single supplier receives for providing one more unit of a commodity. The marginal revenue function can also be defined as the derivative of the total revenue received with respect to total quantity produced.

The relationship of marginal cost and marginal revenue is of particular interest to the manager of a firm, since profits are maximized when a company's marginal cost equals its marginal revenue. An example showing this is given in Table 6-1 below. For the fictitious commodity output given here, profit is maximized when marginal cost and marginal revenue are both $10.

Table 6-1. Commodity Data for Example.

| Total Quantity | Total Cost | Marginal Cost | Marginal Revenue | Total Revenue | Profit |
|---|---|---|---|---|---|
| 1 | $100 | $100 | $200 | $200 | $100 |
| 2 | $110 | $10 | $100 | $300 | $190 |
| 3 | $111 | $1 | $90 | $390 | $279 |
| 4 | $112 | $1 | $75 | $465 | $353 |
| 5 | $115 | $3 | $50 | $515 | $400 |
| 6 | $120 | $5 | $20 | $535 | $415 |
| 7 | $130 | $10 | $10 | $545 | $415 |
| 8 | $145 | $15 | $5 | $550 | $405 |
| 9 | $165 | $20 | $2 | $552 | $387 |

With a general understanding of terms used in economic theory, we will examine the two strategies that GENCOs can use to maximize profits. We will consider the price leadership strategy for a dominant GENCO operating in an oligopolistic economy. First, we will look at the price leadership strategy in general, apply it to the electric utilities industry, and finally provide an example of how a GENCO can use the strategy to determine its optimal output and price.

**6.2.1.2 Basic Cash Market Cases** The following steps outline how the price leadership strategy is used to set production:

(1) The first step in using the Price Leadership Strategy is to determine the market demand function for the commodity under consideration. It is easiest to have the demand function stated as the quantity demanded as a function of price or

$Q_{MD} = f(P)$

(2) The next step is to determine the supply curve for all competitors, in the oligopolistic industry. This should also be stated as the quantity suppliers are willing to produce as a function of price or

$Q_{CS} = f(p)$

(3) Now we need to determine the residual demand function as the difference between the preceding two equations. This can be calculated by taking the market demand function in (1) and subtracting the supply function of all other competitors in (2) from it.

$Q_{RD} = Q_{MD} - Q_{CS}$

(4) Now we can derive the marginal revenue function from the residual market demand function. First, we state the residual demand curve as a function of Q.

$P = f(Q_{RD})$

Next, we derive the total revenue function from our residual demand function.

$TR = P^*Q = f(Q_{RD}) * Q_{RD}$

Lastly, we derive the marginal revenue function from the total revenue function.
$MR = dTR/dQ$

(5) Determine your company's marginal cost function. This will be the additional cost of producing one more unit of output.

(6) From economic theory, we know that profits will be maximized at the output where marginal cost equals marginal revenue. Set marginal cost function equal to the marginal revenue function to find the optimum quantity to produce, solve for Q.

$MC=MR$

(7) The final step in the Price Leadership Strategy will be to determine the optimal price. To do this, we use our residual demand function since it represents the amount consumers will pay for output produced by our company. We will need to take the optimal quantity calculated in (6) and plug it into the "residual" demand function to solve for the price. If we charged a price higher than this amount, consumers would migrate to our competitors to receive the commodity for a lower price, causing the level of output that we could sell to drop below the optimal output level. If we priced lower than the optimal price, we would not be receiving the full amount that we could charge. This would cut into revenues we could potentially be earning.

**6.2.1.3 Price Leadership Strategy Applied to the Electric Utility Industry** The Price Leadership Strategy applied to a GENCO operating in the electric utilities industry can be outlined as follows:

(1) Determine the market demand. In this case study, we will assume that in the short run, the market demand for electricity will be nearly constant for relatively small changes in price. In other words, the market will be almost indifferent to a small price change in the short run and continue to demand the same quantity of electrical power. This sets the demand curve at some constant quantity regardless of price.

(2) Determine the supply function for all other GENCOs in the area under consideration. Intuition tells us that this function will be exponential, reflecting the increasing transmission costs for surrounding firms to transmit their electrical power to the area under consideration.

(3) The residual demand curve for the GENCOs under consideration will be overall demand minus the supply function for competing firms.

(4) The marginal revenue function can easily be derived from the residual demand function using the definition for marginal revenue. First solve for P, then find total revenue.

$$MR = dTR/dQ$$

(5) Determine the GENCOs marginal cost function. This function is sometimes referred to as the incremental cost function for all units operating as GENCO. Several economic unit dispatch models can be used to determine the incremental cost or marginal cost function for a GENCO.

(6) To determine the output for maximum profits, set the marginal cost function in (5) equal to the marginal revenue function in (4).

(7) Finally, determine the optimal price for the GENCO to set the bid for the electricity that will support the optimal output and maximize profits. This price can be determined by solving the residual demand function for the output determined in (6).

Figure 6-2 illustrates the seven steps presented above.

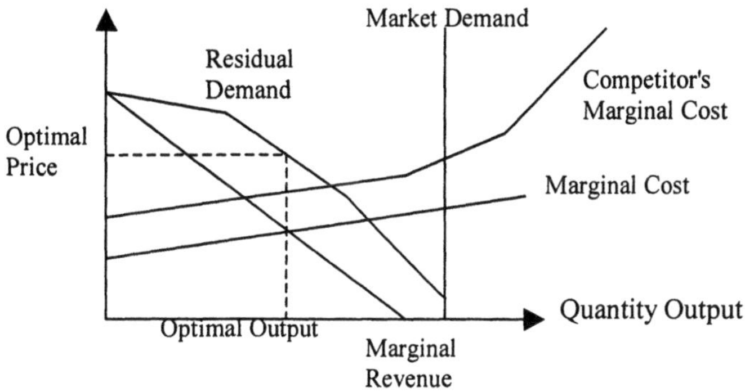


Figure 6-2. Supply Demand Curve Solution.

### 6.2.1.4 Example of Price Leadership Strategy

(1) The market demand for the region under consideration is 1,000 MWh. This is the short-run demand curve and is considered to be a constant demand regardless of price change.

$Q_{MD} = f(P) = 1000$

(2) The competitors combined supply curve for the region under consideration is given as follows:
$Q_{CS} = f(p) = 1000 * (\text{sqrt}(P + 97) - 10)$ E

(3) Residual demand function:

$Q_{RD} = Q_{MD} - Q_{CS} = 1000 - 1000*(\text{sqrt}(P+97) - 10)$

(4) Marginal revenue function:

$Q_{RD} = 1000 - 1000*(\text{sqrt}(P+97) - 10)$

$P = (1*10^{-6})*Q^2 - 0.022Q + 24$

$TR = P*Q$

$TR = (1*10^{-6})*Q^3 - 0.022Q^2 + 24Q$

$MR = dTR/dQ$

$MR = (3*10^{-6})*Q^2 - 0.044Q + 24$

(5) If our GENCOs used only one unit for generation, the marginal cost function would be a continuous function. However, most GENCOs have

multiple unit combinations that may be used to produce the desired power output. Multiple units give a piecewise function that reflects different incremental costs for the different units producing electrical power.

Assuming only one unit for production, our marginal cost curve is given as follows:

$$MC = 0.01 * Q + 2.1875$$

(6) Setting our marginal cost function equal to our marginal revenue function gives us our optimum quantity for output.

$$MR = MC$$

$$(3*10^{-6})*Q^2 - 0.044Q + 24 = .01 * Q + 2.1875$$

$$Q = 413 \text{ or } 17{,}587$$

Since 17,587 MWh is above the market demand we will use 413 MWh.

(7) To determine the optimal price (solve for Q = 413), we will use the residual demand function stated as a function of output in (4).

$$P = (1*10^{-6})*Q^2 - 0.022Q + 24$$

$$P = \$15.08/\text{MWh}$$

**6.2.1.5 GENCO Acting as a Secondary Supplier** For a GENCO operating as a secondary supplier of electricity in a given geographic region, we must use a different strategy to maximize profits. It was assumed that the secondary supplier has no effect on the price of electricity for the geographic region under consideration. Therefore, we will take the price of electricity to be constant. The price of electricity becomes the incremental revenue that a secondary GENCO can receive for selling one additional unit of electricity in a given area. In this way, the price of electricity acts as the marginal revenue function for the secondary GENCO.

We know, according to the laws of economics, that the GENCO can maximize profits by setting its marginal cost function equal to the marginal revenue function. Since the marginal revenue function is the same as the price for a secondary GENCO, we can state that the GENCO will want to produce at the level that equates the marginal cost function of the price of electricity. As before, we can use an economic unit dispatch model to determine the marginal cost function for a GENCO. We would set this function equal to the price and solve for quantity. This quantity will be the optimal quantity that the secondary GENCO must produce to maximize profits for servicing the region.

Consider the following example of a GENCO acting as a secondary supplier. Assume that a secondary GENCO servicing a given region would experience the following marginal cost function that includes transmission costs for servicing that region:

$$MC = (7*10^{-6})*Q^3+(3*10^{-6}) * Q^2 + .001 * Q$$

Furthermore, the price for electricity in the region is given as \$7/MWh. To determine the optimal quantity for the GENCO to supply to the region, we would set the marginal cost function equal to the price, \$7, and solve for Q, giving us a quantity of 99 MWh. For the GENCO to maximize its profits for the region, it would want to supply 99 MWh to that region.

**6.2.1.6 Summary** This case study has introduced two working strategies that a GENCO can use to determine the optimal output (and price) that maximizes profits. The Price Leadership Strategy is used to determine the optimal price and output to use for a GENCO acting as a primary participant in an oligopoly. This same GENCO can also use the optimal output strategy for regions that it serves in which they act as a secondary participant in an oligopoly.

As a final comment, these strategies need to be reevaluated for any change in demand. Since electricity demand commonly changes dramatically over a 24-hour period, prices are usually quoted on an hourly basis. Therefore, it will be necessary to reevaluate the optimal quantities on an hourly basis.

### *6.2.2 Call Option Contracts To Hedge Against Unforeseen Unit Outages*

We have already looked at probabilistic costing and how unforeseen production unit outages can effect the cost structure of a GENCO. Now we will examine how we can use call options as a tool to hedge against these unforeseen outages.

How To Use Call Options As A Hedge Against Unforeseen Outages

When we looked at probabilistic costing, we learned that we could attach a forced outage rate (FORATE) with each unit we have in operation. The forate represented the probability that the unit would incur an unforeseen outage and would be unavailable to produce electrical power. This lack of production can cause a problem if we have already committed to providing the power through contracts such as futures contracts. To meet this commitment, the GENCO would be required to purchase the needed power to cover the commitment.

The purchase can be done on the cash market when the need arises or it can be done by purchasing a call option that would give the GENCO the option to purchase the power if the need arose. The probabilistic cost associated with purchasing the power on the cash market is equal to the forate times the cash price of the electrical power produced by the unit under consideration. However, purchasing the power on the cash market makes it difficult to determine the probabilistic cost since the price of power on the cash market is volatile. If the price on the cash market increased, the GENCO could incur large losses due to a unit outage.

If the GENCO purchased call options to cover the unforeseen outages it would limit the amount it would be required to pay for electrical power if the need arose. The cost of providing this limit would be the cost of the call contracts necessary to cover the possible unit outages. The probabilistic cost of this solution would be the forate times the lesser of the strike price or cash price of the electrical power produced by the unit under consideration, plus the cost of the call options.

**6.2.2.1 Example of Hedging with Call Options** The decision of whether to purchase call options to hedge against unit outages will be determined by a GENCO's risk tolerances and what it thinks will happen to the future price of electrical power. To demonstrate this, let us look at an example. In this example, we will make the following assumptions:

- We have one 300 MW unit for producing power.
- The forate of this unit is determined to be 10%.
- We have committed to providing all 300 MW in a futures contract.
- The current cash price of electrical power is \$20/MWh.
- We do not know if the future price of power will increase, decrease, or remain at \$20.
- Call contracts with a stated strike price of \$20. 5/MWh are currently selling for \$.05/MWh.

To see the effects of using call options as a hedge, we will determine the probabilistic costs for a range of possible cash market prices. The probabilistic costs were calculated without using call options as a hedge and using call options as a hedge.

Table 6-2. Call Impact on Production Costs.

| Cash Market Price | Probabilistic Cost w/o Call | Probabilistic Cost with Call |
|---|---|---|
| \$18 | \$540 | \$555 |
| \$19 | \$570 | \$585 |
| \$20 | \$600 | \$615 |
| \$21 | \$630 | \$630 |
| \$22 | \$660 | \$630 |
| \$23 | \$690 | \$630 |
| \$24 | \$720 | \$630 |
| \$25 | \$750 | \$630 |
| \$26 | \$780 | \$630 |

The table clearly shows the effects of using a call option to hedge against the risk of a unit outage. For lower cash prices on electricity, the use of the call option gives a slightly higher probability cost. However, as the cash price of electrical power increases, the use of a call option results in lower probability costs. The call option limits the maximum price that a GENCO would be required to pay for electrical power if a unit outage occur, limiting the risk to the GENCO.

**6.2.2.2 Using Call Options with Probabilities of Future Electricity Prices** In the above example, the GENCO management would need to decide what they think will happen to the future price of electricity. They would also want to decide if they can financially handle the risk of not using call contracts. This decision can be made easier by forecasting the future price of electricity using models given in the section on the optimized use of electricity futures contracts by GENCOs., If the GENCO wanted to go one step further, it could assign probabilities to the possible future prices of electricity thereby choosing the solution that gives the lowest probability cost.

To demonstrate this, we can add to our earlier example probabilities associated with the range of prices we used. By taking these probabilities, multiplying them by the probabilistic costs already calculated, and summing the products, we can determine two total probabilistic costs: one total probabilistic cost for the GENCO that does not use call options and one total probabilistic cost for the GENCO that does use call options. The results can be found in Table 6-3 below.

Table 6-3. Futures Impact on Production Cost.

| Probability of Cash Market Price | Cash Market Price | Probabilistic Cost w/o Call | Probabilistic Cost with Price Probabilities and w/o Call | Probabilistic Cost with Call | Probabilistic Cost with Price Probabilities and with Call |
|---|---|---|---|---|---|
| .15 | $18 | $540 | $81 | $555 | $83.25 |
| .15 | $19 | $570 | $85.5 | $585 | $87.75 |
| .10 | $20 | $600 | $60 | $615 | $61.5 |
| .10 | $21 | $630 | $63 | $630 | $63 |
| .10 | $22 | $660 | $66 | $630 | $63 |
| .10 | $23 | $690 | $69 | $630 | $63 |
| .10 | $24 | $720 | $72 | $630 | $63 |
| .10 | $25 | $750 | $75 | $630 | $63 |
| .10 | $26 | $780 | $78 | $630 | $63 |
| TOTALS | | | $649.5 | | $610.5 |

Given the fact that the probabilities forecasted an increase in the future cash price of electricity, it is not surprising to see a lower total probabilistic cost for the GENCO using call options. While this demonstrates the usefulness of call options as a hedge, it also shows the importance of accurate forecasts. The ability of the GENCO to control its costs associated with unit outages heavily depends on the accuracy of its price forecasting techniques.

**6.2.2.3. Using Call Options To Cover Multiple Units** In the above examples, above we assumed that the GENCO only operated one unit to produce electricity. In practice, multiple units commonly produce the electricity for a GENCO. Each of the multiple units has its own forate associated with it. If we increase the number of units under consideration, it further complicates the decision of the optimum number of call contracts needed to hedge against unforeseen outages.

In the case study presented on probabilistic costing, we assumed that any additional power that would be needed to cover unit outages could be obtained at a fixed rate. This assumption would be valid for the immediate future. However, if we try to use the same costing methods for periods further into the future, we cannot assume that the electricity prices will remain fixed. Therefore we must make some adjustments to this assumption.

If we used the pruned state enumeration method as our method for probabilistic costing, we could run the model using a forecasted price along with the probability associated with the price. If we did this for every possible price, we could sum the results to obtain a final probabilistic cost. Then we would start again and run the costing model multiple times using prices that reflect the use of call options. Using this method would require an extensive amount of effort and computer resources.

**6.2.2.4 Determining the Optimum Number of Call Option Contracts Needed** Up to this point we have seen the advantages of using call option contracts to hedge against price increases in the event of unforeseen unit outages. We have also looked at the complications involved with using probabilistic costing methods for future time periods given the unknown price of electricity in the future. Now we will look at a method that can be used to determine the optimum number of call contracts needed to hedge against unforeseen outages.

In the process of determining the optimum number of call option contracts needed, we will need to determine the probabilistic outage rate for all units under operation. To do this, we start by taking each unit's forate times the amount of power generated by that unit, giving us the probabilistic outage rate for the given unit. If we sum the probabilistic outage rates for all of the units, we will have the probabilistic outage rate for the entire GENCO.

Next we will want to determine the probability that the call option contracts will be "in-the-money." For a call option contract to be in-the-money, the underlying futures contract prices must exceed the sum of the premium of the call option plus the stated strike price in the call option. If the call contract is in-the-money then the call option contract can be exercised for a profit. If the call option contract is not in-the-money, or is out-of -the-money then the call option contract would be left to expire at an expense equal to the amount of premium invested in the call.

We stated that for a call option contract to be in-the-money the underlying futures contract prices must exceed the sum of the premium of the call option plus the stated strike price in the call option. The probability of this happening will be the probability that the future price of electricity will be more than the sum of the premium of the call option plus the stated strike price in the call option. Forecasting techniques that were discussed in the section on the optimized use of electricity futures contracts by GENCOs, can be used to estimate the probability that electricity prices will be high enough to cause the call option contracts to be in-the-money.

Once we have determined the probabilistic outage amount and the probability that the call option contracts will be in-the-money, we are ready to make the decision on the optimum number of call option contracts we should purchase. The amount of electrical output we would want to cover with the call option contracts would be equal to the probabilistic outage amount times the probability that the futures contracts will be in-the-money. The optimum number of call contracts to purchase can then easily be determined by taking the optimum amount to cover and dividing it by the amount of electricity covered by one call option contract.

**6.2.2.5 Algorithm** Given below is the algorithm for determining the optimum number of call contracts to purchase as a hedge against unforeseen unit outages.

(1) Determine the outage rate for each unit by multiplying the forate of the unit by the production amount of the unit.

(2) Determine the outage rate for the entire GENCO by summing all of the outage rates of the individual units.

(3) Determine the future price of power needed for the call option to be in-the-money by adding the strike price of the call option to the premium for the option.

(4) Determine the probability that the future price of electrical power will be above the price needed for the call option to be in-the-money.

(5) Multiply (2) by (4) to determine the amount of electrical power to cover with call option contracts.

(6) To determine the number of contracts to purchase, divide (5) by the amount of power covered in one call option.

**6.2.2.6 Summary** The above method and algorithm provides a quick, easy way to determine the optimal amount of call contracts needed to hedge against unforeseen unit outages. Using this method does not prevent the GENCO from experiencing losses due to unforeseen unit outages. However, it does limit the amount of the possible losses,.thereby limiting risk to the GENCO.

### *6.2.3 Optimized Use of Electricity Futures Contracts by GENCOs*

Futures contracts allow producers to hedge so that they can limit their losses. All things being equal, a GENCO's profit varies with the price of electricity. Trying to predict the price months in advance so that profit can be known in advance is tricky. Suppose it is April and because of some major decisions (unrelated to insider trading), the board members want to know what the GENCO's profit will be in July.

By considering our fuel contracts and using demand forecasts, we can draw a profit curve based on the price in Figure 6-4. In the figure, this corresponds to the line segment labeled "with no hedge." Not knowing the price means that we have the potential for large losses. The board members do not want to see a line on a graph—they want a simple number. This is where futures hedging comes into play. For the example in the figure, the GENCO can short (i.e., sell non-firm electricity they do not have yet) July electricity with futures contracts. When July arrives, if the spot price is low, they make money on their futures contract and lose on the electricity sold on the spot market. The gain on the futures market offsets the loss in the spot market. If the spot price in July is high, then the electricity sold on the spot market yields a profit while the futures contract will produce an offsetting loss. The result is that the net profit is much more predictable due to the hedge, and we can give the board members the information they were requesting.

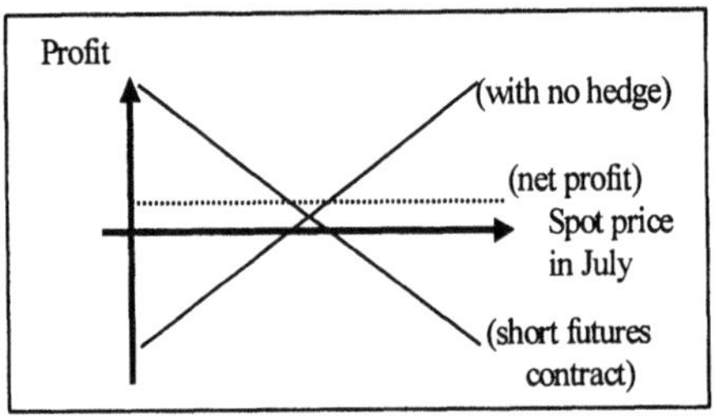


Figure 6-4. Hedging with Futures and Contracts.

This case study will focus on the optimized use of futures contracts in the deregulated electric utilities marketplace. Specifically, the study will concentrate on the optimum quantity a GENCO should commit to selling in the futures market. To begin the case study, we must know a few things about the futures contracts.

In general, a futures contract is a legal agreement that binds two participants to the future sale of a commodity for an agreed-upon price at an agreed-upon time and location. The legal agreement has two sides with one side representing the seller and one side representing the buyer. Either side can sell their obligation, or position, to another party at any time before expiration if they choose.

Futures contracts trade in standardized units on a highly visible, competitive open auction market or exchange. To trade effectively, the underlying market must meet three broad criteria: the prices of the underlying commodities must be volatile, there must be a diverse, large pool of buyers and sellers, and the underlying physical products must be homogeneous.

Each market participant understands that prices are quoted for products with precise specifications, delivered to a specified point, over a specified period of time. In reality, standard deliveries of most futures contracts rarely take place, in the case of energy contracts, less than 1%. However, because existing futures contracts provide for physical delivery, they ensure that any market participant will be able to transact for physical supply and that the futures prices will be truly representative of cash market values. Therefore the prices displayed on the trading floor reflect the marketplace's collective valuation of how much buyers are willing to pay and how much sellers are willing to accept for a given commodity.

The electric utilities market is divided into three regions: eastern U.S., western U.S., and Texas. However, current futures contract trading (1998) is only available for the western U.S. (California). This region encompasses a large geographical area with a number of major suppliers representing virtually every available source of electric generation: oil, coal, natural gas, nuclear, hydro, and geothermal. The suppliers active in this wholesale market are a diverse group of investor-owned utilities, municipally owned utilities, co-generators, government power authorities, and power marketers.

At the time of this writing, futures contracts for electricity are only traded on the New York Mercantile Exchange (NYMEX). Trading was initiated on March 29, 1996, when the NYMEX launched two electricity futures contracts: one based on delivery at the California/Oregon border and the other at the Palo Verde switchyard in Arizona. Both sites are major market centers. Given the rapid development of the power markets in this region, the NYMEX felt it was prudent to offer contracts representing both delivery sites to ensure the futures market will coincide with the industry's need for a price reference and risk management tool.

**6.2.3.1 NYMEX Electricity Futures and Options Contract Specifications** Some of the contract specifications for futures contracts on electrical power as stated by NYMEX are as follows:

Trading Unit: 736 MWh delivered over a monthly period.
Trading Hours: COB 10:30A.M. -3:30P.M.;Palo Verde 10:30A.M. -3:25 P.M. for the open outcry session. After-hours trading will be conducted via the NYMEX electronic trading system Monday through Thursday, 4:15 P.M. to 7:15 P.M.
Trading Months: 18 consecutive months.
Price Quotation: Dollars and cents per MWh.
Minimum Price Fluctuations: $.01 per MWh ($7.36 per contract).
Maximum Daily Price Fluctuation: generally $15.00 per MWh ($11,040 per contract) for the first two months.
Last Trading Day: Trading will terminate on the fourth business day prior to the first day of the delivery month.
Delivery Rate: 2 MW throughout every hour of the delivery period (can be amended with mutual agreement of the buyer and seller).
Delivery Period: 16 on-peak hours: hour ending 0700 prevailing time to hour ending 2200 prevailing time. (This can be amended at the time of delivery with mutual consent of the buyer and seller.)
Scheduling: Buyer and seller must follow Western Systems Coordinating Council scheduling practices.
Exchange of Futures: For, or in connection with, physicals (EFP). The commercial buyer or seller may exchange a futures position for a physical position of equal quantity by submitting a notice to NYMEX. EFPs may be used to either initiate or liquidate a futures position.
Position Limits: 5,000 contracts for all months combined but not to exceed 350 in the last three days of the delivery month or 3,500 in any one month.

Trading Symbol: Palo Verde: Ky, California/Oregon Border: MW

**6.2.3.2 Who Participates in Futures Trading and Why** Market suppliers, or generator companies, trade on the exchange to lock in buyers of the electricity they will produce for a guaranteed price they can sell at. On the demand side, participants use futures contracts as a way of locking in the price they will have to pay in the future for a supply of electricity. Both sides can use the futures markets as a means of hedging against unwanted risks brought from possible price fluctuations. Speculators and investors also participate in futures contracts as a way of investing their funds in hopes of making a profit from the changes in the price of the futures contracts.

The price for electricity futures contracts is determined much the same as the cash or current price of electricity. As defined in classical economics, the price for a commodity is determined by equating the market demand function with the market supply function. With futures contracts, estimates must be used to determine the supply and demand functions.

Estimates for the demand function can be attained by analyzing historical demand data for the time of day and time of year under consideration. Estimates for the supply function can be derived from analyzing individual GENCOs servicing the area. It is particularly critical to examine the market of the fuels used by these GENCOs. If GENCOs in the area primarily use coal as a fuel, higher coal costs will result in higher costs associated with producing electricity.

**6.2.3.3 Strategies for Generator Companies Participating in Futures Market** GENCOs need to have two questions answered before participating in the electricity futures market: What should the future price of electricity be and what should my position as a GENCO be?

To answer the question of what the future price of electricity should be, a GENCO will need to estimate the future demands and supplies of electricity for a given region. We have already discussed how these estimates can be obtained in the previous section. However, the estimates will not always match those of the market. The result will be that we will estimate the future price of electricity to be greater than or less than the amount stated in the futures contract. The strategies that we will look at assume that we are 100% certain that our estimates are correct. Further studies could be done on using probabilities of the estimates we are using.

If we believe the future price of electricity will be greater than the futures contracts state, we would generally be unwilling to participate in selling electricity in the futures market. In other words, we believe that we would be able to sell our electricity at a higher price in the future cash market than is stated in the futures contract market.

If we believe the future price of electricity will be less than the futures contracts state, we would want to participate in selling electricity in the futures market. We now need to calculate the optimal amount that we will want to commit to sales in the futures markets. First, we present a strategy for doing this that assumes our participation in the futures market will not be significant enough to affect the price of the futures contracts. Then, we present a strategy that assumes our position can affect the price of the futures contracts.

**6.2.3.4 Strategy for Participation that Will Not Affect Futures Prices** If we assume that our participation will not affect the price of the futures contracts, the scenario becomes much the same as a secondary GENCO deciding how much to electricity sell on a cash market. The price of the futures contract becomes our marginal revenue function. Our marginal cost function can be derived by using the same methods we used in the optimization strategies for cash prices. However, we will need to estimate what these costs will be in the future.

As economic law dictates, our profits will be maximized by allowing marginal revenue equal marginal cost. Since the marginal revenue function is the same as the price of the futures contracts, we let our marginal cost function equal the price of the futures contracts. Solving for the quantity gives us the optimal quantity that the GENCO should commit to selling in the futures market to maximize profits given that our participation will not affect the price of the futures contracts.

**6.2.3.5. Strategy for Participation that Will Affect Futures Prices** If we assume that our participation in the futures market will affect the price of the futures contracts, we can no longer use the price of the futures contracts to represent our marginal revenue function. Assuming that the more we sell on the futures market, the less we will receive for the futures contracts gives us a demand function. This *demand function* is the demand for futures contracts. From this demand function we can derive a marginal revenue function.

The marginal cost function can be determined using methods previously discussed. Once we have the marginal revenue and marginal cost functions, we will set these equal and solve for quantity. This quantity is the optimal quantity that we will want to sell in the futures market. The price that we will be able to sell this quantity for in the futures market will be determined by the demand function of the futures market.

**6.2.3.6. Summary** This case study defined a futures contract in the electric utilities marketplace. We also gave some contract specifications that apply to these futures contracts on electricity. If a GENCO determines that it is profitable to participate in the selling of power through futures contracts, it has two strategies that it can use to determine optimum participation. The first strategy will give the optimum quantity to sell in the futures market, assuming that its participation is not influential enough to change the price of the futures contracts. The second strategy will determine the optimum quantity to sell in the futures market, assuming that its participation does influence the price of futures contracts. The following algorithm summarizes the strategies for a GENCO wanting to participate in the futures market.

**6.2.3.7 Algorithm**

(1) Use estimated future demand and supply functions for electricity to determine an estimated future price of electricity.

(2) If the estimated future price of electricity is more than the going futures prices, the GENCO should, not sell in the futures market. Goto (12).

(3) Will the GENCOs participation be enough to affect the pricing of futures contracts in the market? If yes, goto(7). If no, goto(4).

(4) Determine the GENCO's marginal cost function.

(5) Set the marginal cost function equal to the price of the futures contracts.

(6) Solve for quantity. This is the optimal quantity the GENCO should sell in the futures market. Goto (12).

(7) Determine the GENCO's marginal cost function.

(8) Determine the demand function for the futures market.

(9) Derive the marginal revenue function from the demand function in (8).

(10) Set marginal revenue equal to marginal cost.

(11) Solve for quantity. This is the optimal quantity that the GENCO should sell in the futures market.

(12) Stop.

### 6.2.4 Options on Electricity Futures Contracts

Futures options contracts can also be used to reduce risk. Consider the GENCO that wants to maximize its profit and reduce its risk. One alternative is that the GENCO pays a premium for an options contract that gives it the right to sell (short) electricity at the strike price. (If the price was higher than the strike price, the GENCO would let the option expire). Figure 6-5 shows how the option contract can be used to hedge profit. Notice that when the price is low, the GENCO can exercise the option and have a futures contract, as in the previous example, to offset its losses in the spot market. When the price is high, the GENCO has no obligation to sell at the strike price; the net profit is the profit from the electricity produced by the GENCO and sold on the spot market minus the premium paid for the options contract. The GENCO has limited the amount of money that it can lose but can still reap the benefits of a high price in July. Another alternative using swaps would be a short call.

The subject of this case study is option contracts on electricity futures. The case study begins by defining option contracts in general. It then presents the usage of option contracts within the electric utilities marketplace. Finally, the case study shows how to use the Black and Scholes option pricing model to determine the value of a call option on an electrical futures contract.

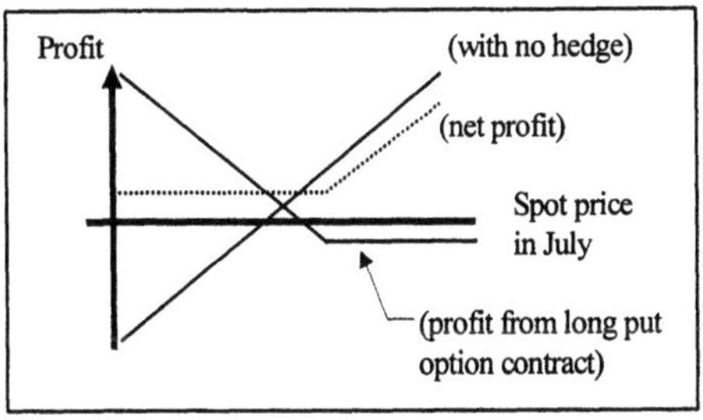


Figure 6-5. Hedging with Futures Options Contracts.

An *option* is a contract that allows the buyer of the contract to purchase or sell an underlying security for a stated price. The buyer pays a one-time payment or premium to have this option of buying or selling the underlying security. The name "option" comes from the fact that the buyer of the contract is never obligated to exercise their right to buy or sell. The life of the contract lasts until the contracts stated expiration date.

The seller of the contract is obligated to purchase or sell the underlying security for the stated price if the option is exercised by the buyer. The seller receives the premium payment for taking on the risk of having to buy or sell the underlying security at the stated price no matter what the going rate is on the market for the underlying security.

The two types of option contracts are named calls and puts. A *call* gives the contract buyer the right, but not the obligation, to buy the underlying security at a specified price (the strike or exercise price) for a specified period of time. A *put* gives the contract buyer the right, but not the obligation, to sell the underlying security at a specific price for a specified period of time. If prices do not move in a direction that makes exercising of the contract profitable, the options buyer will simply let the contract expire and forfeit the premium paid.

"Long" denotes ownership; "to go long" means to purchase the item in question. In the figure, long indicates that the trader has purchased the option and now has the right to buy (call) or the right to sell (put) the future. A trader who writes the option is "short"; "to go short" is to sell the item in question. Assume that the item in question is a MWh of electricity. In the long call diagram, the long trader has paid a premium (e.g., $1) to the option writer for the call option. This call option gives the trader the right to buy a MWh for the strike price (e.g., $7). If the price goes above the strike price plus the premium (e.g., $8), the trader has made a profit. The long trader has reduced risk by limiting his losses to the premium. In the diagrams labeled "short," we see what happens from the option writer's point of view. He receives the premium for assuming the risk and is obligated to sell the MWh at the strike price even though the market price is higher. The diagrams labeled "put" show how the put works. The long put trader pays a premium to lock in a maximum price (exercise price) that he will have to pay for the MWh. The short put trader takes that premium in return for promising to sell the MWh for that same exercise price.

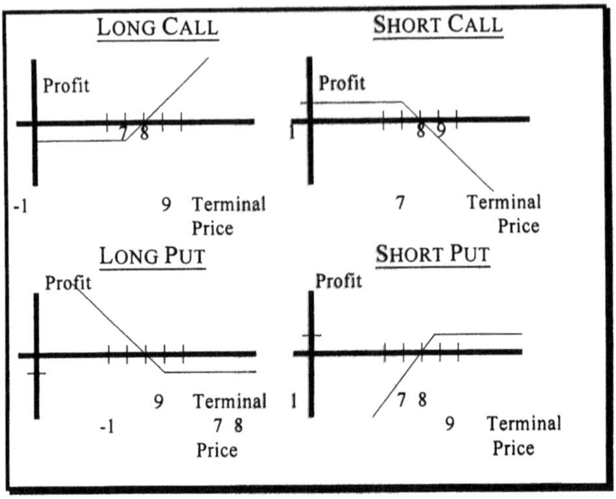


Figure 6-6. Methods of Using Options.

**6.2.4.1 Uses of Option Contracts** Call options permit the buyer of an option to purchase the underlying security at a stated price, allowing the buyer to protect himself against the risk of a price increase. Buyers of electricity who incur the risk of rising electricity prices, can purchase calls to limit their exposure, putting a cap on costs while retaining the ability to take advantage of falling prices.

Put options allow the option buyer to sell the underlying security at a stated price. This allows the seller of the put option to protect him against the risk of a price decrease. Producers of electrical power who incur the risk of falling electricity prices can purchase puts to limit their exposure, putting a floor below which revenues cannot fall while retaining the ability to take advantage of rising prices.

The sellers of both types of options immediately generate cash flows for themselves from the receipt of premiums. These cash flows are the return for investors willing to accept the risk of price fluctuations in the utilities market. Furthermore, the options contracts can be used by power marketers and integrated utilities who can be exposed to rising and falling prices. They do this by simultaneously buying calls and selling puts (or buying puts and selling calls) to protect their market position while generating cash flow.

The nature and flexibility of the options market allows participants to engage in strategies as aggressive or conservative as their tolerance for risk and their budgets will allow. The NYMEX Division electricity options contracts complement the futures contracts and provide yet another hedging instrument for market participants to increase their flexibility in managing their business risk. By using options alone or in combination with futures contracts, strategies can be found to cover virtually any risk profile, time horizon, or cost consideration.

**6.2.4.2 Valuation of Call Option Contracts** Several different theories and methods are used by analysts to determine the value of options. However, the Black and Scholes option pricing model tends to be the most widely used method for pricing options, both in industry and academia. This option pricing model stems from the original model that the two professors developed in 1973 to price European call options on nondividend paying stocks. Since the behavior of options on electrical futures is not the same as European options, the original model has been altered to encompass the changes in behavior. However, the underlying principals of their model will still be used in the pricing model for options on electrical futures.

The Black and Scholes formula for pricing call options like those used in the utilities market. We will only discuss the use of this formula for call contracts; however, a similar calculation is used for put contracts as well.

$$C = SN(d_1) - Ke^{-rT}N(d_2)$$

$$d_1 = ln(S/Ke^{-rT})/(v*sqrt(T)) + (1/2)v*sqrt(T)$$

$$d_2 = d_1 - v*sqrt(T)$$

where: C = Value of the call option.

N(x) = The cumulative normal distribution function. Use of a table to determine this amount is helpful. These tables can be found in almost any statistics text.

r = The continuously compounded interest rate. A good estimate for this is the stated interest rate on 90-day Treasury bills.

K = Strike price of the call option. The *strike price* is the stated price for which the underlying futures contract can be purchased. Note that the higher the strike price, the less valuable a call option since the strike price represents a higher cost of exercising the call or purchasing the futures contract.

S = The current price of the underlying futures contracts.
T = Time to maturity of the option stated as a fraction of a year. The longer the time to maturity, the higher the price of the option. This reflects the increased chance of having the option be in-the-money and thus profitable.

v = The annualized standard deviation of the price of the underlying futures contract. This is the measure of volatility of the price of futures contracts. A database of historical prices on futures is the easiest way to estimate a measure of volatility. A more volatile underlying security means a more valuable option contract. This also reflects the increased chance of having the option be in-the-money and thus profitable.

**6.2.4.3. Summary** Option contracts can be used as a valuable tool in hedging against risks in a deregulated utilities market. In a later case study, we expand on one of these uses.

To take advantage of option contracts, it is essential to determine the contract's value. Because of differing estimates, market prices on option contracts may or may not be the same value calculated by a potential investor. Therefore, a GENCO should always determine the value of an option contract before investing. The calculated value may suggest a good buy or a bad buy in the options market.

## 6.3 Building Fuzzy Bidding Strategies for the Competitive Generator

The advent of power plants deregulated has generated a need for a wide variety of software tools as shown by the preceding case studies. As discussed in General Operations Planning, each company now has to operate as a profitable entity without support from the transmission company, the distribution company, or the energy services company.

Although some research has been conducted on bidding strategies, little has been performed for the electric industry. [Rajan, 1997] developed a sub-optimal bidding technique that could provide a lower bound for profit in single-shot bidding. However, this work predicts that multi-shot bidding (with opportunities to adjust bids and offers to reach price discovery) is a more probable, competitive alternative. Much less work has been done on bidding strategies for this type of auction.

Developing bidding strategies with evolving trading agents for the deregulated electric utility industry is a new field of research. Bidding strategies are historically based on heuristic algorithms developed from expert judgment and experience. Most companies use heuristic algorithms that can best be described by knowledge-based systems and fuzzy logic. Apart from the electric utility industry, interest has grown in recent years for using evolving, or adaptive, agents to simulate trading behavior. Research with adaptive agents has proved to be a useful means of exploring trading markets outside of the electric industry. [LeBaron, 1997] used evolving agents to learn to play financial markets. [Tesfatsion, 1995] described research in which trading agents decide who to trade with based on an expected payoff. [Ashlock, 1995] used genetic programming (GP) combined with a finite state automata to play a classic academic game called Divide the Dollar, which involved bidding behavior and strategies. [Ashlock, 1997] used the same game to study kinship effects and concluded that when evolving strategies are used by buyers and sellers, unless they come from separate populations, collusion is likely to occur. [Andrews, 1994] used a game based on a double auction to verify that genetic search is useful. They showed that GP-based agents actually do learn, and they compared the performance of the GP-based strategies to those developed using simulated annealing. In addition, they showed that at the beginning of the genetic algorithm, it is possible to use a less rigorous fitness test than is needed in later generations. While their findings may be useful to the genetic algorithm community, their experiments leave room for further improvements in strategy building. Previous work by this author in Chapter 6 of the work by M. Ilic, F. Galian, and L. Fink [1998] summarizes the artificial life approach.

Some research has been conducted on bidding strategies for electric systems in other countries. Bidding strategies were analyzed [Finlay, 1995] for the restructured power pool of England and Wales system and it was mathematically shown that an optimal bidding strategy exists for its bidders. Finlay's work differs from that describe here mainly in that his objective was not to maximize the profit of the individual generation companies, and the actual system is different from those proposed in the U.S. Hence it is not directly applicable to our scenario.

In this section, we build on previous research on building bidding strategies for electric utilities in the competitive environment. The previous research is briefly reviewed. The deregulated market-place is defined and modeled. Fuzzy logic is included to make bidding strategies adaptive. Four methods for building bidding strategies that use fuzzy logic and/or genetic algorithms are discussed and outlined. Economical inputs are fuzzified for use in determining a generator's bid. Methods of tuning and searching for the optimal rule are discussed. We also discuss how an agent using the bidding strategies can compare them on the basis of profitability.

### *6.3.1 Introduction*

Economists have developed theoretical results of how markets are supposed to behave under varying numbers of sellers or buyers with varying degrees of competition. Often the economical results pertain only when aggregating across an entire industry and require assumptions that may not be realistic. These results, while considered sound in a macroscopic sense, may not be helpful to a particular company that does not fit the industry profile but is trying to develop a strategy that will allow it to remain competitive.

Generation companies (GENCOs) and energy service companies (ESCOs) that participate in an energy commodity exchange must learn to place effective bids in order to win energy contracts. Microeconomic theory states that in the long term, a hypothetical firm selling in a competitive market should price its product at its marginal cost of production. The theory is based on several assumptions (e.g., all market players will behave rationally, all market players have perfect information) which may be true industry wide but not be true for a particular region or firm.

Chapter 1 describes the deregulated marketplace to be considered during this research. A description of the authors' previous research on evolving bidding strategies for generation companies using genetic algorithms appears in [Richter, 1997a]. We build the foundations for strategy development on the preceding basic ideas. The next section provides the basics of fuzzy logic and examines how the economic inputs of ESCOs and GENCOs might be fuzzified to build better bidding strategies. The following section outlines the models that we are using and the research that we are currently pursuing to build better bidding strategies.

### *6.3.2 Fuzzy Bidding*

The field of fuzzy logic was made popular by Lotfi Zadeh during the 1960s. Fuzzy logic provides a methodical means of dealing with uncertainty and ambiguity. It allows its users to code problem solutions with a natural language syntax with which people are comfortable. In fact, people regularly use fuzzy terms to describe things or events. For instance, if asked to describe a person, we might use terms like "pretty tall," with a "big nose" and "somewhat overweight." These terms can be defined differently by different people. A certain amount of ambiguity or uncertainty is associated with any description involving natural language terms such as these. Most of the things we deal with daily in this universe are ambiguous and uncertain. "The only subsets of the universe that are not in principle fuzzy are the constructs of classical mathematics". [Kosko, 1992]

Fuzzy logic allows us to represent the ambiguous or uncertain with membership functions. The membership functions map the natural language descriptions onto a numerical value. Membership to a particular description or class is then a matter of degree. For instance, if we define a person's height as described in Figure 6-7, then we can see that a person who is 6 feet in height is tall with a membership value of one. This membership value is also known as a truth value. In the same figure, we can see that a person who is 5 feet 9 inches is tall to a lesser degree but at the same time also short to a certain degree.

Using similar reasoning, we might say that electrical demand is high in a region if it goes above 100 MW and normal if it is between 50 MW and 75 MW. What if the demand is 90 MW? Using traditional logic, we would classify it as neither high nor normal. However, using fuzzy logic, we might find that this demand is actually both high and normal, each to a certain degree (based on its membership function). Similarly we could have fuzzy membership functions for other inputs like fuel costs, risk aversion, level of competition, etc.

Once defined, these inputs can be used in a set of fuzzy rules. For instance, a simple rule might be as follows:

- IF demand is HIGH, then bid should be HIGH.

where a high bid would be defined using another membership function. Multiple input conditions can be considered by combining rules with the "and" and "or" functions. For example, a rule might be as follows:

- IF demand is LOW AND risk aversion is HIGH, THEN bid should be LOW.

Although it may not be necessary, we could have an output for all combinations of inputs. A three input fuzzy rule system where each input is broken into five classifications might be represented as in Figure 6-8. Each small square contains the output of a rule on how to bid relative to cost. Since some conditions are very unlikely to occur, each of these squares need not have an output. In addition, a particular input may be classified in more than one square at a given instant. In the figure, the letters V, L, H, C, and N stand for very, low, high, cost, and normal, respectively. The output of the rule states how to bid with respect to generation cost. We could have more or fewer inputs, and we could use different classifications.

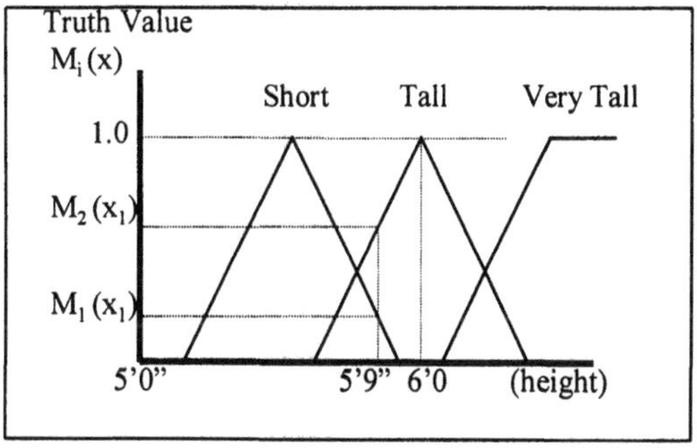


Figure 6-7. Fuzzy Membership Functions.

Figure 6-9 shows fuzzy system architecture. The inputs are fed into the rule base. The output (i.e., the bid values in the example) of each rule can be classified by a fuzzy membership function in the same manner as the inputs. The output of each rule may be assigned a certain weight, depending on how important we determine that rule or corresponding input to be. We can then sum the weighted output of the rules and determine an overall fuzzy output. However, when we place the bid, we cannot say, "bid high." We need a way to convert the fuzzy output to a single number. This is called the defuzzification process.

According to [Kosko, 1992], defuzzification means to round off a fuzzy set from some point in a unit hypercube to the nearest bit-vector vertex. Practically speaking, defuzzification has been done by using the mode of the distribution of outputs as the crisp output or by the more popular method of calculating the centroid or center of mass of the outputs and using that as the crisp output. The fuzzy centroid, $\overline{B}$ , can be calculated as follows:

$$\overline{B} = \frac{\sum_{j=1}^{p} y_j m_B(y_j)}{\sum_{j=1}^{p} m_B(y_j)}, \qquad \text{where } B = \sum_{k=1}^{m} w_k B_k^{'}$$

Where: B is the output distribution that contains all information,

$m_B(y_j)$ is the membership value of $y_j$ in the output fuzzy set B. See Figure 6-.5.

### *6.3.3 Comparing Bidding Strategies*

This section compares approaches that we are taking in developing bidding strategies. First, we will be manually generating fuzzy bidding rules using expert knowledge. Secondly, we will search for good rule sets from a limited search space. With a small number of inputs and a limited number of weighting, we can do an exhaustive search of all rules and determine the best possible rule. (The best rule is the one whose use results in the largest amount of profit for its user.) Thirdly, if we increase the number of fuzzy inputs, increase the number of membership functions describing the inputs, and allow more flexibility with the weighting,

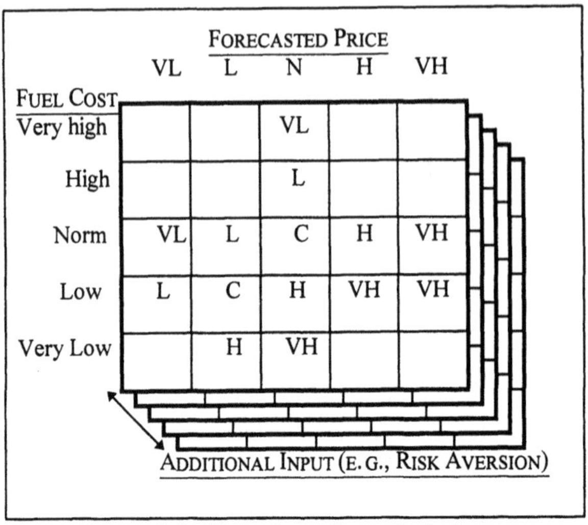


Figure 6-8. Three Input Fuzzy Rule Set.

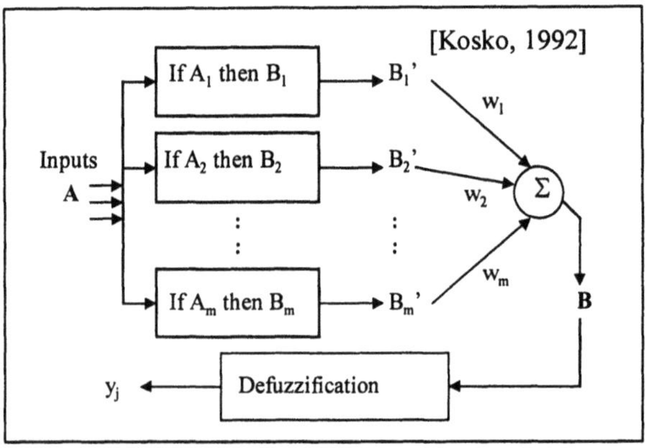


Figure 6-9. Fuzzy System Architecture.

perhaps it becomes desirable to use an optimization algorithm to search for the optimal rule rather than do an exhaustive search. Finally, we will attempt to use a technique developed by [Richter, 1995] to extract, from a historical database containing the bidding details of an auction, the rules others used to develop their bids.

The research described here builds on the techniques used by [Richter, 1997a]. To measure the performance of the bidding rules created in each method described below, a group of GENCOs will compete to serve the electrical demands of the ESCOs. Electricity buyers will be aggregated into a single large ESCO. See Figure 6-10. TRANSCOs and transmission constraints will not be directly considered, but can be accounted for after the fact if desired.

### 6.3.4 Manually Generating the Fuzzy Rule Sets

If we consider only a limited number of fuzzy economical inputs, (e.g., expected price, risk aversion, and generating costs) then it is possible to generate rules manually with expert knowledge from power traders. We can transform the rules-of-thumb used by experienced power traders into a fuzzy rule base. We can also use theoretical economics to influence the rule sets that we construct. If we have fuzzy inputs, each divided into classifications, then we could need as many as 125 rules in each rule set (one for each square in Figure 6-8). Each of the rules can be weighted according to its importance. If any weighting is allowed, we have infinite possibilities.

### 6.3.5 The Searching for the Optimal Fuzzy Rule Set

To reduce the amount of time spent tuning the rule sets, we can predefine a structure and allow a computer program to search through the possibilities to find the optimal rule set. If we predefine each of the three inputs by five fixed ranges and only allow discrete rule weightings (e.g., 0.0, 0.1, 0.2, ..., 1.0), then there are a finite number of permutations to investigate. A possible indication of optimality would be obtained by having an agent use each of the possible rule sets while engaging in a fixed set of trial auctions, competing with a set of agents who had evolved to play the described market [Richter, 1997a] as is performed by dynamic programming.

### 6.3.6 Using ANNs to Extract Expert System Bidding Rules from Historical Data

The authors have investigated the use of ANNs and other artificial intelligence techniques to search through large databases in order to learn the expert system rules that can be used to reproduce the historical results. Previously, this technique was successful in finding load characteristics. Based on extensive records, the software is able to determine what factors characterize the underlying structure. Similarly, a database of trading data could be fed into the software (which would require tuning and some restructurization) to determine what bidding rules traderes were using. Determining the rules that other electricity traders and brokers are using could benefit those who wish to gain a competitive edge in the deregulated market.

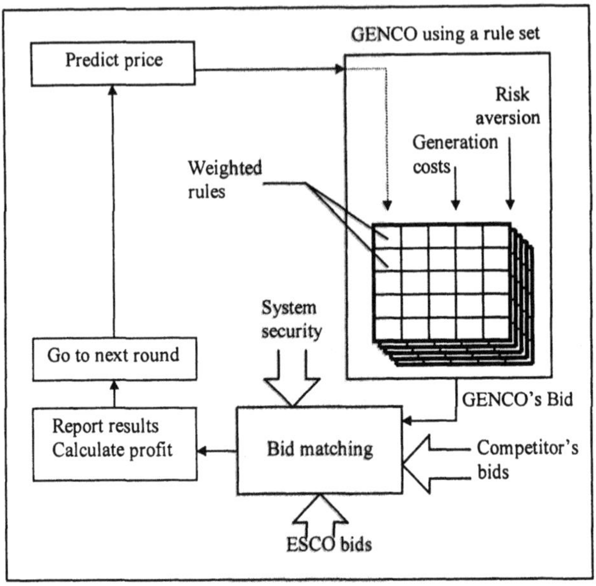


Figure 6-10. Using the Rule Set.

## 6.4 Profit-Based Unit Commitment For The Competitive Environment

As the electrical industry restructures, many of the traditional algorithms for controlling generating units are in need of modification or replacement. Previously used to schedule generation units in a manner that minimizes costs while meeting all demand, the unit commitment (UC) algorithm is one of these tools that must be updated. A UC algorithm that maximizes profit will play an essential role in developing successful bidding strategies for the competitive generator. Simply bidding to win contracts is insufficient; profitable companies require bidding that results in contracts that, on average, cover the total generation costs. No longer guaranteed to be the only electricity supplier, a GENCO's share of the demand will be more difficult to predict than in the past. Removing the obligation to serve softens the demand constraint. A price/profit-based UC formulation that considers the softer demand constraint and allocates fixed and transitional costs to the scheduled hours is presented in this chapter. In addition, we describe a genetic algorithm solution to this new UC problem and present results for an illustrative example. We discuss the implications of allocating the fixed costs via different methods and look at ways of handling uncertainty in price and load forecasts.

### *6.4.1. Introduction*

The U.S. electric marketplace is in the midst of major changes designed to promote competition. No longer vertically integrated with guaranteed customers and suppliers, electric generators and distributors will have to compete to sell and buy electricity. The stable electric utilities of the past will find themselves in a highly competitive environment. Although some states (e.g., California) are already operating in a restructured environment, a standardized final market structure for the rest of the U.S. has not yet been fully defined. The authors believe that regional commodity exchanges, in which electricity contracts are traded, should play an important role.

Previous sections and publications [Kumar, 1996a, 1996b, Sheblé, 1994b, 1996b] have described a framework in which DISTCOs, GENCOs, ESCOs, and TRANSCOs interact via contracts. The contract prices are determined through an auction. Electricity traders make bids and offers that are matched subject to the approval of an independent contract administrator (ICA) who ensures that the system is safely operating within limits.

Operating within such a framework, traders will create and implement bidding strategies to make their bids and offers. These bidding strategies might be designed to limit the traders' risk, to maximize profit, or some combination of both. Recent publications [Richter, 1997a, 1997b, 1998b] have reported research that uses genetic algorithms and GP to evolve bidding strategies that maximize profit for the spot market. In [Richter, 1998a], the authors investigated managing an energy trader's risk and profitability by combining spot market contracts with options and futures. For simplification, the previous work, described in other chapters, avoided the UC problem by ignoring startup and shutdown costs, minimum uptimes and downtimes, ramp rates, etc. In this chapter, a profit-based UC is considered and its implication for bidding strategies is discussed.

Previous researchers in [Maifeld, 1996] developed and implemented a genetic-based UC algorithm. Their algorithm was able to consistently find multiple good unit commitment schedules in a reasonable amount of time. Unlike other UC solution techniques, it used true costing. These attributes remain desirable qualities. We have updated this algorithm for the price/profit-based competitive environment and provide results of its use on some illustrative examples.

The remainder of this section is organized as follows. The first part provides a brief description of the UC problem and formulation and highlights modifications needed for the competitive environment. The second part describes a LaGrangian relaxation (LR) algorithm solution to the updated UC problem. The next section discusses implications of the updated UC on bidding strategies. The next section outlines a method for making the strategies more robust under conditions of uncertain demand and prices. The final part provides some conclusions and identifies areas of future work.

### *6.4.2. Updating UC*

For the vertically integrated monopolistic environment, UC is loosely defined as the scheduling of generating units to be on, off, or in standby/banking mode such that costs are minimized and constraints like demand and reserves are met. Considering inputs like variable fuel costs, startup and shutdown parameters/constraints of each power plant, and crew constraints adds to the complexity of the problem. The schedules are valued based on their costs. To determine the cost associated with a given schedule, an economic dispatch calculation (EDC) where each of the non limit constrained operating units is set so that their marginal costs are equal, must be performed for each hour under consideration. One possible way to determine the optimal schedule is to do an exhaustive search. Exhaustively considering all possible ways that units can be switched on or off for a small system can be done, but for a reasonably sized system it would take too long. Solving the UC problem for a realistic system generally involves using methods like LR, dynamic programming, or other heuristic search techniques. Many references for the traditional UC can be found in [Sheblé, 1994d] and in Wood and Wollenberg [1996].

In the past, demand forecasts advised power system operators of the amount of power that needed to be generated. In the future, bilateral spot and forward contracts will make part of the total demand known a priori. The remaining part of the demand will be predicted as in the past. However, the GENCO's share of the remaining demand may be difficult to predict since it will depend on how its price compares to that of other suppliers. The GENCO's offer price will depend on its prediction of its share of this remaining demand as that will determine how many units they have switched on or in banking mode. The UC schedule directly affects the average cost and indirectly the offering price, making it an essential input to any successful bidding strategy.

In the past, utilities were obligated to serve their customers. This was translated into a demand constraint that ensured all demand would be met. For the UC problem, this might have meant switching on an additional unit just to meet a remaining MW or two. With the obligation to serve gone, the GENCO can now consider a schedule that produces less than the predicted demand. They can allow others to provide that 1 or 2 MW that might have increased their average costs (they might not have secured that contract for which they would have had to compete).

Demand forecasts and expected market prices are important inputs to the profit-based UC algorithm; they are used to determine the expected revenue that affects the expected profit. If a GENCO comes up with two potential UC schedules each having different expected costs and different expected profits, it should take the one that provides for the largest profit, which will not necessarily be the one that costs least. Since prices and demand are critical in determining the optimal UC schedule, price prediction and demand forecasts become crucial. Takriti, Krasenbrink, and Wu [1997] present a description and a stochastic solution of the UC problem that considers spot markets. Their research differs in that they choose to minimize costs rather than maximize profits.

The existence of liquid markets gives energy trading companies an additional source from which to supply power. To the GENCO, the market supply curve can be thought of as an additional pseudo-unit to be dispatched. The supply curve for this pseudo-unit represents an aggregate supply of all of the units participating in the market at the time in question. The price forecast essentially sets the parameters of the unit. This pseudo-unit has no minimum up time, minimum down time, or ramp constraints; there are no direct startup and shutdown costs associated with dispatching the unit. As described later, each GENCO is responsible for its own unit's transitional costs, which must be recovered through adjustments in its offering price. The offer should roughly be equivalent to the marginal cost of the unit at the hour considered shifted by some amount to account for profit and for transitional costs.

The liquid markets that allow the GENCO to schedule an additional pseudo unit, also act as a load to be supplied. The total energy supplied should consist of previously arranged bilateral contracts and bilateral or multilateral contracts arranged through the markets (and their associated reserves and losses). While the GENCO is determining the optimal unit commitment schedule, the energy demanded by the market (i.e., market demand) can be represented as another DISTCO or ESCO buying electricity. Each entity buying electricity should have its own demand curve. The market demand curve should reflect the aggregate of the demand of all buying agents participating in the market.

Mathematically, the traditional cost-based UC problem has been formulated as follows [Sheblé, 1985]:

$$\text{Minimize F} = \sum_{n}^{N}\sum_{t}^{T}(C_{nt} + MAINT_{nt})\cdot U_{nt} + SUP_{nt}\cdot U_{nt}(1-U_{nt}) + SDOWN_{nt}\cdot(1-U_{nt})\cdot U_{nt-1}$$

subject to the following constraints:

$$\sum_{n}^{N}(U_{nt}\cdot P_{nt}) = D_t \qquad \text{(demand constraint)}$$

$$\sum_{n}^{N}(U_{nt}\cdot P\max_n) \geq D_t + R_t \qquad \text{(capacity constraint)}$$

$$\sum_{n}^{N}(U_{nt}\cdot Rs\max_n) \geq R_t \qquad \text{(system reserve constraint)}$$

As we redefine the UC problem for the competitive environment, the demand constraint changes from an equality to less than or equal (we assume that the buyers purchase reserves per contract) relationship, and the objective function shifts from cost minimization to profit maximization.

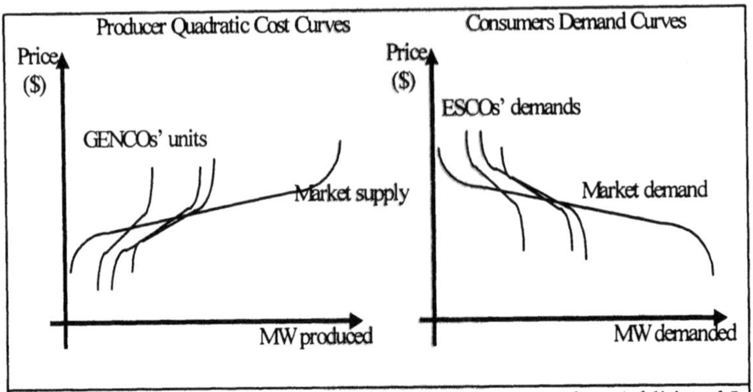


Figure 6-11. The Market as an Additional Generator and an Additional Load.

$$\text{Max } \Pi = \sum_{n}^{N}\sum_{t}^{T}(P_{nt}\cdot fp_t)\cdot U_{nt} - F \qquad \text{(revenue - costs)}$$

subject to:

$$\sum_{n}^{N}(U_{nt}\cdot P_{nt}) \le D_t' \qquad \text{(new demand constraint)}$$

$$Pmin_n <= P_{nt} <= Pmax_n \qquad \text{(Capacity limits)}$$

$$| P_{nt} - P_{n,t-1} | <= Ramp_n \qquad \text{(Ramp rate limits)}$$

where:

$U_{nt}$ = up/downtime status of unit n at time period t
($U_{nt}$ = 1 unit on, $U_{nt}$ = 0 unit off)

$P_{nt}$ = power generation of unit n during time period t

$D_t$ = load level in time period t

$D'_t$ = forecasted demand w/ reserves for period t

$fp_t$ = forecasted price for period t

$R_t$ =system reserve requirements in time period t

$C_{nt}$ = production cost of unit n in time period t

$SUP_{nt}$ = startup cost for unit n, time period t

$SDOWN_{nt}$ =shutdown cost for unit n, time period t

$MAINT_{nt}$ = maintenance cost for unit n, time period t

N = number of units

T = number of time periods

$Pmin_n$ = generation low limit of unit n

$Pmax_n$ = generation high limit of unit n

$Rsmax_n$ = maximum contribution to reserve for unit n

Maximizing the profit is not the same as minimizing the cost. Since we no longer have the obligation to serve, the GENCO may choose to generate less than the demand. This allows a little more flexibility in the UC schedules. In addition, our formulation assumes that prices fluctuate according to supply and demand. In the past, engineers assumed that if they could levelize the load curve, they would be minimizing the cost. When maximizing profit, the GENCO may find that under certain conditions it may profit more under a nonlevel load curve. The profit depends not only on cost but on revenue. If revenue increases more than the cost does, the profit will increase.

It was necessary to redefine EDC because it is now price based rather than used to minimize costs. Where the old EDC ignored transition and fixed costs to adjust the power level of the units until they each had the same incremental cost (i.e, $\lambda_1=\lambda_2=...=\lambda_i=...=\lambda_T$), our new EDC attempts to set $\lambda$ equal to a pseudo price (i.e., produce until the marginal cost equals the price). This pseudo price is the hourly forecasted price modified to account for transition and fixed costs. One way to accomplish that is shown in the following formula,

$$\lambda_t = fp_t - \frac{\sum_t \sum_n (transition \quad costs) + \sum_t \sum_n (fixed \quad costs)}{\sum_t^T \sum_n^N P_{nt}}$$

which results in a \$/MWh pseudo price. Other allocation schemes that adjust the marginal cost/price according to the time of day or price of power would be as easy to implement and should be considered in building bidding strategies. Other allocation schemes are discussed in previous sections. Transition costs include startup, shutdown, and banking costs, and fixed costs (present for each hour that the unit is on), would be represented by the constant term in the typical quadratic cost curve approximation. For the results presented later in this chapter, we approximate the summation of the power generated by the forecasted demand.

## 6.5. Price-based UC Results

The UC was run on a small system so its solution could be easily compared to a solution by exhaustive search. Before running the UC, the GENCO needs to obtain an accurate hourly demand and price forecast for the period in question. Developing the forecasted data is an important topic but beyond the scope of our analysis. For the results presented in this section, the forecasted load and prices are taken to be those shown in Table 6-4. In addition to loading the forecasted hourly price and demand, the UC program needs to load the parameters of each generator to be considered. We are modeling the generators with a quadratic cost curve (e.g., $A + B(P) + C(P)^2$ ). The data for the 2-unit case is shown in Table 6-5.

In addition to the 2-unit cases, a 10-unit, 48-hour case is included in this chapter to show that the LR works well on larger problems. While dynamic programming quickly becomes too computationally expensive to solve, the LR scales up linearly with number of hours and units. Figure 6-12 shows the costs and average costs (without transition costs) of the generators as well as the hourly price and load forecasts for the 48 hours. The data was chosen so that the optimal solution was known a priori. The dashed line in the load forecast represents the maximum output of the units.

Table 6-4. Forecasted Demand and Prices (2 Gen Case).

| HR | Load Forecast | Price Forecast |
|---|---|---|
| 1 | 285 MWh | 25.7 $/MWh |
| 2 | 293 MWh | 23.06 $/MWh |
| 3 | 267 MWh | 19.47 $/MWh |
| 4 | 247 MWh | 18.66 $/MWh |
| 5 | 295 MWh | 21.38 $/MWh |
| 6 | 292 MWh | 12.46 $/MWh |
| 7 | 299 MWh | 9.12 $/MWh |
| 8 | 328 MWh | 8.88 $/MWh |
| 9 | 326 MWh | 9.12 $/MWh |
| 10 | 298 MWh | 8.88 $/MWh |
| 11 | 267 MWh | 25.23 $/MWh |
| 12 | 293 MWh | 26.45 $/MWh |
| 13 | 350 MWh | 25.00 $/MWh |
| 14 | 350 MWh | 24.00 $/MWh |

Table 6-5. Unit Data for 2 Generator Case.

| | Generator 0 | Generator 1 |
|---|---|---|
| Pmin (MW) | 40 | 40 |
| Pmax (MW) | 180 | 180 |
| A (constant) | 58.25 | 138.51 |
| B (linear) | 8.287 | 7.955 |
| C (quadratic) | 7.62e-06 | 3.05e-05 |
| Bank cost ($) | 192 | 223 |
| Start cost($) | 443 | 441 |
| Stop cost($) | 750 | 750 |
| Min up (hr) | 4 | 4 |
| Min down (hr) | 4 | 4 |

In the schedules shown in Table 6-6, it may appear as though minimum up and down times are being violated. When calculating the cost of such a schedule, the algorithm ensures that the profit is based on a valid schedule by considering a zero surrounded by ones to be a banked unit, and so forth.

### 6.5.1 UC and Bidding Strategies

UC will remain an important tool in the new environment. Although customers are no longer guaranteed, bilateral contracts will ensure that the GENCO knows the majority of its load ahead of time. An accurate forecast of the remaining demand and hourly prices will be important inputs for solving the UC problem. Once the UC schedules are generated, they will be of little use to the GENCO unless it can actually win customers from competitors at the price that it assumed in determining the UC schedule. For this reason, the UC schedule becomes an important input to the hourly bidding strategy builder.

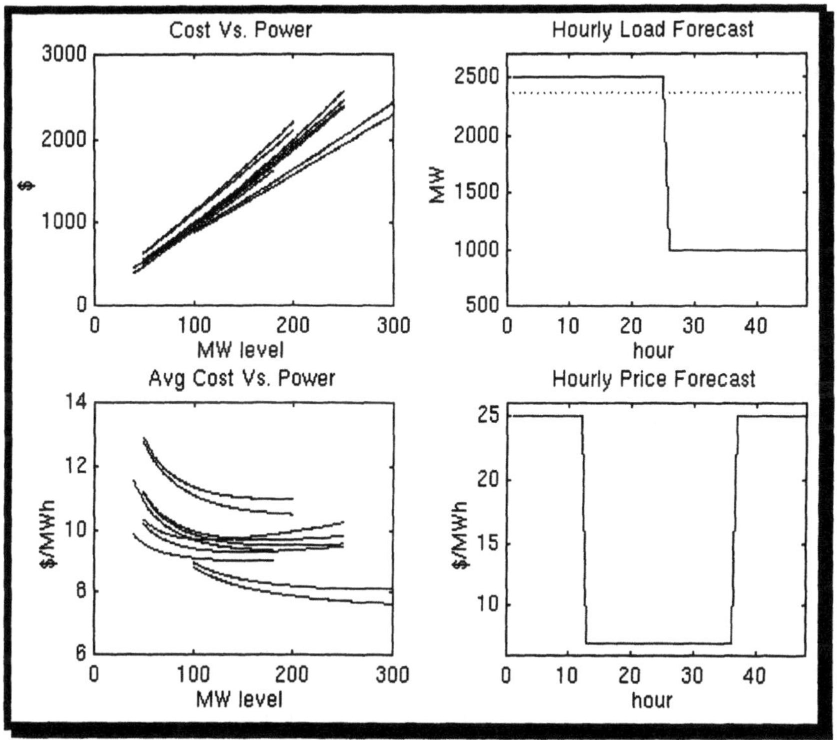


Figure 6-12. Data for 10 Unit, 48 Hour Case.

When using the newly formulated UC program, solutions with more profit are valued more highly than those with less profit. Since forecasts of load and prices may vary widely from the actual prices and loads, we may wish to reduce or increase the amount of power that we are generating as we get closer to the time of production. A schedule that allows a wider range of possible power generation levels without switching additional units on or off is more valuable that a rigid schedule. So the primary measure becomes the profit level, but the flexibility of the schedule also plays a part in judging which schedule is better.

One can begin to think of an entire UC schedule as being a bidding strategy or as having a very close connection to hourly bidding strategies. Another area in which this is evident is the method of assigning the transition and fixed costs associated with the schedule. We described a scheme that allocated these costs in proportion with the number of MWs being produced at any given period. Allowing other factors (e.g., time-of-day, day-of-week, a time of increased competition) to influence this allocation could provide additional profits. In a market in which one is submitting an entire schedule, this can be very important. One might shift peak period costs to offpeak hours to win the bid that may be decided on the price of a particular time of day. However, in a market framework, this is less important. One can argue that allowing utilities to shift costs to offpeak hours to win a bid, while their total cost is higher, may result in a lower total social welfare and that a market framework that awards utilities contracts based on a certain hours price/variable costs and then are guaranteed to have their startup and shutdown costs covered is suboptimal.

Table 6-6. The Best UC Schedules Obtained.

| | solution for 2 unit, 10 hour case |
|---|---|
| **Unit 1** | 1111100000 |
| **Unit 2** | 0000000000 |
| **Cost** | $17,068.20 |
| **Profit** | $2,451.01 |
| | **solution for 2 unit, 12 hour case** |
| **Unit 1** | 111111000011 |
| **Unit 2** | 000000000000 |
| **Cost** | $24,408.50 |
| **Profit** | $4,911.50 |
| | **solution for 10 unit, 48 hour case** |
| **Unit 1** | 111111111111000000000000000000000000111111111111 |
| **Unit 2** | 111111111111000000000000000000000000000000000000 |
| **Unit 3** | 111111111111000000000000000000000000000000000000 |
| **Unit 4** | 111111111111000000000000000000000000000000000000 |
| **Unit 5** | 111111111111000000000000000000000000000000000000 |
| **Unit 6** | 111111111111000000000000000000000000000000000000 |
| **Unit 7** | 111111111111000000000000000000000000111111111111 |
| **Unit 8** | 111111111111000000000000000000000000000000000000 |
| **Unit 9** | 111111111111000000000000000000000000111111111111 |
| **Unit 10** | 111111111111000000000000000000000000111111111111 |
| **Cost** | $325,733.00 |
| Profit | $676,267.00 |

## 6.6 Handling Uncertainty in Price and Load Forecasts

After the solution is found, what will happen if the forecasts are different. Suppose the price at hour 100 was 10% higher than anticipated. At present, the UC user can modify the forecasts and re-run the algorithm. However, the possible set of cases that might be of interest could be large. Re-running the algorithm for each case could be time consuming, even if the algorithm starts from the near optimal solution found before making the change to the forecast.

When forecasting loads and prices, the farther the time period is in the future, the more likely it is that the forecast will deviate from the expected value. So, along with the expected price and load for each time period, our industrious engineer in the forecasting department decides to give us these expected values along with error bars that indicate how confident he is of their precision (see Figure 6-13). How can we use this additional information?

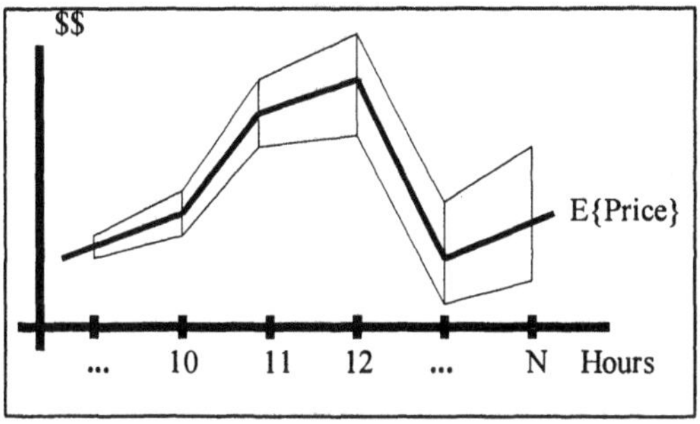


Figure 6-13. Expected Prices with Error Bars.

We propose to use this additional information as follows. Rather than use the profit calculated directly by using the expected profit and load curves, we assume that there is a distribution (probably Gaussian) of values about that data point. We use Monte Carlo sampling to determine and expected profit. We draw a number of data points from the distribution of profits at that hour and a corresponding number of data points from the distribution of load forecasts at that hour. These will be used in determining the expected profit of that particular unit commitment schedule. During each generation of the LR UC solution, we monitor the variability of each solution by examining the duality gap. As the UC schedule solutions come closer to the optimal, we increase the amount of demand expected. With no proof that we are going to get to the optimal, how do we know when to begin increasing the number of Monte Carlo sample points? The variance of the solutions is a good indicator of when to increase the number of data points.

The result of the above should be a schedule that would provide the largest expected profit. The method described above is expected to increase the computational requirements of the LR-UC, but we still expect reasonable solution times because the LR-UC performs quite rapidly.

## 6.7 Improving Bidding Strategies Through Intelligent Data Mining

### *6.7.1 Overview*

The author has investigated the use of artificial intelligence techniques to search through large databases in order to learn the expert system rules that can be used to reproduce the historical results. We have developed software using this technique that is being used to develop standardized treatment methods for hospital patients receiving medical care [Richter, 1995]. A database of trading data could be fed into the same software (which may require some restructurization for this application). The data mining algorithm (DM) would find a number of bidding rules that achieve the same results as those being used by the traders who generated the bidding data. Determining the rules that other electricity traders and brokers are using could benefit those who wish to gain a competitive edge when participating in the deregulated market. Such a tool would also be welcomed by regulating agencies who wish to ensure that the markets are efficient and fair.

### *6.7.2 Introduction*

Many industries (e.g., power systems, commodity trading) have problems that require highly specialized knowledge to solve or diagnose. These problems have typically been addressed by human experts who have had much training and have domain specific knowledge. This knowledge is the fundamental ingredient of an expert's problem-solving abilities. Systems equipped with the appropriate knowledge have been shown to demonstrate expert level performance in many applications. Despite the success, the current state of expert system technology suffers from some serious limitations. One of these limitations is that the development of an expert system, for the most part, remains an art. While tools and methodologies have emerged to provide considerable help, the process of representing and refining the knowledge used by the domain expert remains ill-defined and time consuming. Application of expert systems still remains restricted to fairly narrow, self-contained problem domains, and performance typically degrades sharply as the system approaches the boundaries of human knowledge. With traditional computer-based expert systems, there is little ability to adapt or reorganize knowledge as performance requirements change over time. The greatest potential for removing these limitations lies in the area of machine learning. This chapter presents a technique for solving these problems by considering an adaptive learning strategy based on ANNS.

### *6.7.3 Applications of Expert Systems*

Many industries have applications for expert systems. For instance, in the healthcare field, many areas might be improved through the use of an expert system. Many doctors would welcome an expert system that could suggest a diagnosis or prescribe tests based on symptoms and history of an admitted patient. On occasion, differences in doctors' training, opinions, and methodologies could lead them to diagnose the same problem differently. If an expert system were implemented, it could reduce the number of these inconsistent diagnoses. An expert system would be useful in diagnosing patients more quickly, which would be especially helpful in trauma centers where correct decisions must be made immediately. Often the expert system can provide a tool for analyzing or diagnosing a problem that in the past had no analysis tool available. A number of expert systems have already been implemented successfully in the healthcare field.

Other fields like electrical power systems have applications that have also benefited from expert systems. In recent research, expert systems have been used to help the operator in control of electrical power systems to maintain safe operating conditions by controlling real-time data acquisition and numerical algorithms (load flow, stability analysis) [Germond, 1992; Holland, 1975]. In fact, many areas within the electric utility industry have been studied and have had papers published on expert system usage. Some of these areas include distribution planning, system design, feeder configuration, substation considerations, planning, operations, transmission planning, operations, load forecasting, generation commitment scheduling, load shedding, security assessment, and line overload alleviation.

If the limitations of the expert system development discussed in the introduction were removed, a much more widespread implementation of expert systems would be seen. Currently, research is being conducted that explores the use of expert systems using genetic algorithm based machine learning. The data and experimental results presented in this work are based on research in the healthcare field but are directly applicable to other industries.

### *6.7.4 Overview of Machine Learning*

*Machine learning* is the automatic improvement in the performance of a computer system over time, as a result of experience. Some of the advantages of an expert system based on a learning algorithm over the traditional type are that the system with the learning algorithm

- Delivers more accurate solutions
- Covers a wider range of problems
- Obtains answers less expensively

Machine learning can be applied to almost any domain, but in practice the greatest successes have been related to classification tasks.

For a system to learn, the following components are needed: system rules, a performer, a critic, a learner, system inputs, and system outputs. (See Figure 6-14.)

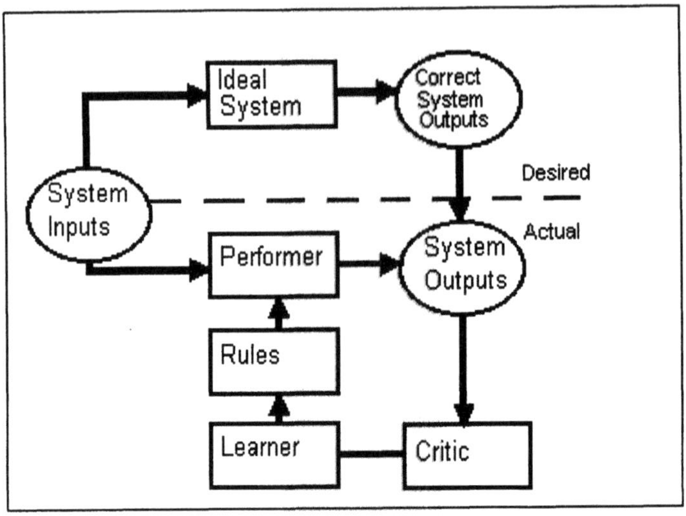


Figure 6-14. A General Learning Structure.

- Rules: a set of information structures that encode the system's present level of expertise.
- Performer: a task algorithm that uses the rules to guide its activity.
- Critic: a feedback module that compares actual results with those desired.
- Learner: a mechanism that uses feedback from the critic to amend the rules.

The problem of learning may be thought of as a search through the entire space of possible descriptions of the problem to find the correct classification structure. In anything other than trivial problems, the number of possible descriptions becomes enormous, and to use a completely enumerative search technique would be computationally unfeasible using today's computer technology. This research proposes to use a genetic algorithm to learn the expert system rules to correctly classify a system's inputs.

### *6.7.5 Expert Systems*

The building of critical decision criteria can be accomplished by expert systems. In general terms, an *expert system* is a classification system that uses a production system of rules and messages to represent a compact, computationally complete knowledge base. The algorithm builds the rule and message system entirely into an if-then-else program construction. These statements allow the user to easily incorporate the problem specific criteria into the rule base.

The expert system rules are set up in the form of a decision tree. Each node in the tree represents an if-then-else operator, and each branch represents the path of the decision made at the previous node. The number of possible outcomes doubles at the end of each branch. The doubled solution can be considered a parallel set of if-then-else nodes. In each iteration of the program, one parallel set of nodes, beginning with the farthest left node, is analyzed against the if-then-else rule at the node. In the following iteration, the program tests the next set of parallel if-then-else nodes in the tree along with all the previous parallel sets. Eventually, each rule is tested against each node and branch in the tree.

Typical inputs in our research are patient diet information, symptoms, and preliminary test results. These inputs are presented to a rule at the decision nodes. For example, if the patient has high blood pressure, then the node chooses to perform some action; if not, it will choose to perform some other action.

When the branches have been tested against the rules once (i.e., one generation has been completed), the program repeats the cycle of testing the nodes and branches. The program continuously repeats the entire process until no changes need to be made in the decision path. At this point, the criteria have been met, the program ends, and the solution is presented [Mann, 1990; Martin, 1988].

## 6.8 Bidding Strategies for LR Based Power Auction

### 6.8.1 Introduction

Many GENCOs have already adopted the LR based auction for trading analysis. LR-based auction is a decomposition algorithm that formulates the auction problem as UC problem and uses LR to find the solution. The LR-based auctions and other types of auctions are illustrated and compared in [Dekrajangpetch, 1997]. [Hao, 1997], and [Jacobs, 1997] discussed the objective functions of power pool-type auctions. The discussion is on cost minimization versus consumer payment minimization. [Post, 1994] gave a thorough explanation of auctions.

LR has many advantages over other methods used for solving UC problem, e.g., the computational requirement of using LR varies linearly with number of generation units ($N$) and stages ($T$) while the computational requirement of dynamic programming (DP) varies exponentially with $N$ and $T$, $(2^N-1)^T$. Nevertheless, LR has some weaknesses when it comes to convergence. The solution found by LR might not be feasible or near optimal if LaGrangian multipliers have not been updated properly.

The difficulty in updating LaGrangian multipliers has led to problems that occur in implementing LR-based auctions. The problems stem from selecting identical or similar generating units. These units can prevent LR from finding an optimal or even a feasible solution. In addition, the solution found may be inequitable to similar units because the decision to alter data made by the dispatcher is heuristic. This will result in contested auctions. The details of these problems are illustrated in detail in [Dekrajangpetch, 1998b]. [Johnson, 1997] showed the effects of variations in the near optimal solution from LR-based UC to the profits of units in the competitive electricity market.

Such details were originally noticed by Virmani et al. [Virmani, 1997] who observed some implementation aspects of LR while applying it to realistic and practical UC problems and also discussed handling the identical generating units by committing them as a group or adjusting their heat rates slightly to make them distinct and then committing them separately. There will be many independent power producers (IPP) or GENCOs in the competitive market with the most recently developed gas turbine units and therefore identical or similar generating units will be prevalent. This prevents us from handling the identical units as they were handled in [Virmani, 1997]. Adjusting the heat rates cannot be used due to the contractual nature of the bid. Since so many units are expected to be similar, the solution found by committing as a group may not be the optimal solution for the system.

LR is an optimization technique that decomposes the main complex mathematical programming problem into simple subproblems that are additively separable by relaxing the hard constraints, e.g., coupling constraints. Each subproblem is coupled through common LaGrangian multipliers, one for each period. Each subproblem is solved separately. The LaGrangian multipliers at each iteration are updated until a near-optimal solution is found. The quality of the solution is characterized by the "duality gap." The *duality gap* is the spread between the primal and the dual objective function values. The larger the gap the more uncertain the quality of the solution. LR has been successfully applied to the UC problem. The UC problem is a large-scale, mixed-integer nonlinear programming problem [Sheblé, 1992b; Wood, 1996]. The UC problem is more complex due to the incorporation of various hard constraints (e.g., ramp rate constraints, minimum uptimes, minimum downtimes, emission constraints, pond level constraints of pump storage units, etc.). The LR algorithm is successful since a LaGrangian multiplier updating procedure has been suitably developed to converge efficiently with a subsequently very small duality gap. [Fisher, 1981] reviewed three approaches for updating LaGrangian multipliers: the subgradient method, column generation techniques of the simplex method, and multiplier adjustment methods. Among these methods, the subgradient method is promising and widely used in UC. Some of the developed algorithms and LaGrangian multiplier updating procedures can be seen in [Virmani, 1989; Merlin, 1983; Zhuang, 1988; Guan, 1994; Peterson, 1995; and Gjengdal, 1996].

LR has many advantages over other methods used for UC, e.g., the computational requirement of using LR varies linearly with number of generation units ($N$) and stages ($T$) while the computational requirement of DP varies exponentially with $N$ and $T$, $(2^N - 1)^T$.

Nevertheless, LR has some weaknesses when it comes to convergence. The solution found by LR might not be feasible or near optimal if LaGrangian multipliers have not been updated properly.

This work focuses on how to change unit data to obtain an advantage while using LR as an auction method. The authors suggest alternative strategies based on previously published problems with selection by unit commitment and subsequent dispatch by economics. The auction scenario considered is one sided and the objective function is cost minimization. The suggested strategies can be applied to both uniform and discriminating pricing. Uniform pricing means every GENCO gets paid the same price while discriminating pricing means each GENCO gets paid corresponding to its bid.

Section Two describes LR-based auctions used in various places. Section Three explains the formulation, the algorithm, and the LaGrangian multiplier updating procedure for LR used in this work. The subgradient method is used for updating LaGrangian multipliers. The notation used in this work is also presented. Section Four outlines the implementation problems of LR. The problems are divided into two categories: problems with identical units and problems with similar units. The results of testing with four generator systems are described. Section Five illustrates the strategies used in submitting bids for GENCOs to gain an advantage over competitors. The procedure of changing unit parameters to gain advantage based on the strategy is also illustrated. Section Six outlines sensitivity analysis results on the four-unit system. The sensitivity analysis is performed on three parameters of two peak units, linear and constant coefficients of unit cost function, and startup costs. This section shows the percentage of difference between individual parameters of these two units that will result in the optimal solution while fixing other parameters. Section Seven suggests parameter changes for real-time maximum generation. Section Eight summarizes how to change unit parameters based on the strategy in Sections Five and Seven and the sensitivity analysis results in Section Six. Section Nine presents future work. Section Ten presents conclusions of this research and other methods proposed to implement auctions.

### *6.8.2 LaGrangian Multiplier Update*

This section outlines the formulation, algorithm, and LaGrangian multiplier updating procedure for LR used in this work. The notation used in this work is also described in this section. This work uses the formulation and algorithm of LR for UC described in Merlin et al. [Merlin, 1983], except for the following simplifications. The spinning reserve constraints have been neglected. The minimum-up and minimum-down time constraints are neglected. The fuel cost is assumed to be a quadratic function. The criterion for stopping is reached when duality gap is less than or equal to 0.026 per unit. Another criterion added to the algorithm is that LR will terminate when number of iterations exceeds 100. A rather large number, 100, is used for the small studied system because the cases studied are those in which LR has difficulties in converging to the optimal solution. The subgradient technique is used for updating LaGrangian multipliers.

The notation used in this work is as follows:

$P_i^t$ power produced by unit i at stage t
$P_i^{min}$ minimum capacity of unit i
$P_i^{max}$ maximum capacity of unit i

| | |
|---|---|
| $a_i$ | quadratic coefficient of fuel cost of unit i |
| $b_i$ | linear coefficient of fuel cost of unit i |
| $c_i$ | constant coefficient of fuel cost of unit i |
| $stup_i^t$ | start-up cost of unit i from stage t-1 to t |
| $load^t$ | demand at time t |
| $\lambda^t$ | LaGrangian multiplier at time t |
| $\lambda$ | vector containing $\lambda^t$ from t=1 to t = T |
| $iter$ | number of current iterations |
| $N$ | number of generating units |
| $T$ | number of stages |

Each lambda is updated according to [Zhuang, 1988].

$$\lambda^t = \max[\lambda^t + \frac{pdif^t}{(\alpha + \beta * iter) * norm(pdif)}, 0] \tag{6.1}$$

α and β are constants and $pdif^t$ can be defined at (6.2),

$$pdif^t = load^t - \sum_{i=1}^{N} P_i^t \tag{6.2}$$

and so *pdif* is a vector containing $pdif^t$ from t=1 to *T*. *norm(pdif)* is the Euclidean norm. $P_i^t$ here is calculated from DP, not from Economic Dispatch.

The values of α and β are determined heuristically. The general guidelines for selecting their values are explained in [Fisher, 1981]. In this work, the values used can be divided into two categories according to sign of $pdif^t$ as follows:

Category 1: $pdif^t > 0$: α=0.02, β=0.05.
Category 2: $pdif^t \leq 0$: α=0.5, β=0.25.

α and β when $pdif^t \leq 0$ are rather big, and bigger than those are when $pdif^t > 0$, to make LR converge suitably for the cases studied.

### *6.8.3 Problems in Implementation*

The system under investigation here is composed of four generating units (modified from [Wood, 1996]) and these units are committed for four stages. Startup cost is not incorporated in Cases A or B because the illustration of the problem is clear without it. Cases C and D include startup cost.

Different starting λ can cause LR to find different solutions when the range of the optimal λ is small. Thus, in each section, multiple starting λ were used for each experiment. These starting values are what cause LR to not find the best or any solution. All of the starting λ used in this work are summarized in Table 6-7. Each of the starting λ is composed of four elements, one for each time period. These four elements are ordered from the first to the fourth stage.

The implementation problem can be separated into two main categories.

Table 6-7. Reference Notation For λ.

| Notation | λ |
|---|---|
| $\lambda_a$ | [12.5 12.5 12.5 12.5] |
| $\lambda_b$ | [6 6 6 6] |
| $\lambda_c$ | [7.7 9.8 16.3 14.2] |
| $\lambda_d$ | [9 9 9 9] |
| $\lambda_e$ | [6 6 12.5 12.5] |
| $\lambda_f$ | [6 6 12.5 6] |
| $\lambda_g$ | [6 12.5 12.5 6] |

*A. Problem: Identical Units*

There are two primary effects when identical units exist. The first is that LR may find only suboptimal solutions. The second is that LR may be unable to find any feasible solutions.

(1) *Finding Only Suboptimal Solutions*

The generating unit data to demonstrate this problem is shown in Table 6-8. Unit one is identical to unit four. Unit three is the least expensive unit, and units one and four are the most expensive units. System loads are shown in Table 6-9. This data constitutes Case A.

The solution found by LR is shown in Table 6-10. The solution found by LR is not the optimal solution. The solution found from LR is the same as the optimal solution at stages 1, 2, and 4 but is different from the optimal solution at stage 3. Either unit 1 or 4 is selected to generate at 500 MW at stage 3 for the optimal solution, while both units 1 and 4 are selected to generate at 250 MW at stage 3 for the solution found by LR. The total cost of the optimal solution is $20,162.75. This is less expensive than the cost of the solution that LR found, $20,412.75. The difference in the cost is more pronounced when the startup costs are considered. This is because two units, units 1 and 4, are turned on at stage 3 for the solution found by LR, while only one unit, either unit 1 or 4, is turned on at stage 3 for the optimal solution.

Table 6-8. Case A: Generating Unit Data.

| Unit(i) | $a_i$ | $b_i$ | $c_i$ | $P_i^{min}$ | $P_i^{max}$ |
|---|---|---|---|---|---|
| 1 | 0.002 | 10 | 500 | 100 | 600 |
| 2 | 0.0025 | 8 | 300 | 100 | 400 |
| 3 | 0.005 | 6 | 100 | 50 | 200 |
| 4 | 0.002 | 10 | 500 | 100 | 600 |

Table 6-9. Case A: Load Data.

| Stage | 1 | 2 | 3 | 4 |
|---|---|---|---|---|
| Load | 170 | 520 | 1100 | 330 |

Table 6-10. Case A: Solution.

| Stage(t) | Unit 1 | Unit 2 | Unit 3 | Unit 4 |
|---|---|---|---|---|
| 1 | 0 | 0 | 170 | 0 |
| 2 | 0 | 320 | 200 | 0 |
| 3 | 250 | 400 | 200 | 250 |
| 4 | 0 | 130 | 200 | 0 |

The problem arises because LR uses DP to find the optimal solution for the subproblems. Identical or very similar units must have the same optimal states for DP to find the best solution. This is why LR cannot find the optimal solution that selects either unit 1 or 4 at the third stage. This means that the solution found by LR may not be the least expensive nor the best for the whole system when identical or very similar units exist.

(2) *Not Finding any Feasible Solutions*

Units 1 and 4 are still identical units to demonstrate this problem. The system load at the third stage is changed to be between the summation of $P_i^{min}$ of units 1, 2, 3, and that of units 1, 2, 3, 4. In addition, $P_i^{max}$ of units 2 and 3 are reduced so that only selecting units 2 and 3 cannot meet the load at the third stage. The purpose of changing data in this way is to force only either unit 1 or 4 to be selected at the third stage. The loads at other stages are reduced to accommodate the decreased total maximum capacity. The generating unit data is the same as in Table 6-2, except that $P_i^{max}$ of units 2 and 3 are changed to 150 and 80, respectively. The load data is shown in Table 6-11. This constitutes Case B.

Three starting values for λ have been used to demonstrate the importance of the initial guess for LR. After running 100 iterations for each starting λ, LR could not find any feasible solutions. The reason is that to cover the load at the third stage at the lowest cost requires units 2 and 3 to be selected. Units 1 and 4 can only be committed in two possible combinations of states; both units are either selected or not selected. The case in which both units are not selected cannot occur because the summation of $P_i^{max}$ of units 2 and 3 is less than 340. The case in which both units are selected cannot occur because the summation of $P_i^{min}$ of units 1, 2, 3 and 4 are larger than 340. Note that a big primal objective function value is used for the stage without enough committed generation.

This example points out another disadvantage of using LR for auctions when identical and very similar units exist. Not only does LR not find the real optimal solution, but it is also sometimes difficult for LR to even find a feasible solution.

A concluding remark is based on the economic interpretation of the LR iterations. If an energy market is considered, the LR algorithm proposes a sequence of hourly prices (λ) to buy energy from GENCOs. GENCOs independently plan their output power in response to the price sequence, meeting their respective constraints. This results in a surplus of power in some hours and deficit of power in some other hours. The LR algorithm balances demand by modifying the sequence of prices. A reasonable procedure is to modify prices proportionally to their corresponding mismatches (subgradient). This procedure is repeated until convergence in prices is attained. These prices are in turn implemented. A reserve market working in a similar fashion as the energy market can also be implemented.

Table 6-11. Case B: Load Data.

| Stage | 1 | 2 | 3 | 4 |
|---|---|---|---|---|
| Load | 80 | 210 | 340 | 350 |

Thus, identical units will be jointly selected or not selected. However, it is not possible to select some of them while the rest are not. This produces two problematic behaviors. First, it is possible to miss the minimal solution if it requires that some of the identical units be selected and not the rest. Second, it is possible not to find any feasible solutions. This happens whenever the selection of all identical units in a given hour produces an infeasible solution. Alternatively, if not selecting all the identical units in a given hour makes it impossible to supply the demand.

Rules to solve the problems with identical units may be constructed to make identical units sufficiently dissimilar. However, this can not necessarily always *preserve fairness*. One rule, for instance, is to penalize each company (unit) in a rotating fashion. However, such rules to preserve *fairness for every* unit are very difficult to construct.

(3) *Problem: Multiple Optimal Solutions*

The data to demonstrate the next problem are shown in Tables 6-12 and 6-13, Case C. Unit 1 is *similar* to unit 4. They are peaking units. This demonstration includes startup costs for units 1 and 4.

Two approaches were used. One used different starting λ and the other changed the order of the unit data as it is fed to the program (alternating between the two peak units, units 1 and 4). LR is run for two unit data input orders, unit order 1 2 3 4 and 4 2 3 1, and for each unit data input order, five starting λ are used. LR is run 100 iterations for each case. In 100 iterations, LR may find the optimal solution more than once because LR is run for a fixed number of iterations instead of running until the duality gap is satisfied is to find out if different optimal solutions are found.

Table 6-12. Case C: Generating Unit Data.

| Unit | $a_i$ | $b_i$ | $c_i$ | $P_i^{min}$ | $P_i^{max}$ | $stup_i$ |
|---|---|---|---|---|---|---|
| 1 | 0.002 | 10 | 500 | 100 | 600 | 3300.7 |
| 2 | 0.0025 | 8 | 300 | 100 | 400 | 0 |
| 3 | 0.005 | 6 | 100 | 50 | 200 | 0 |
| 4 | 0.002 | 9.88 | 542 | 100 | 600 | 3324.7 |

Table 6-13. Case C: Load Data.

| Stage | 1 | 2 | 3 | 4 |
|---|---|---|---|---|
| Load | 170 | 520 | 1100 | 1000 |

The result is easily explained. The unit data input order does not affect solution, i.e., unit order 1 2 3 4 and 4 2 3 1 give the same solution. The optimal solutions found by LR in all different starting $\lambda$ are the same. When LR found optimal solutions more than once, they are still the same as shown in Table 6-8.

Actually there are two optimal solutions for this data. One is what LR found (shown in Table 6-14). The other is shown in Table 6-15. The optimal $\lambda$ of the solution in Table 6-14 is used to test if LR will find the other optimal solution. This does not happen.

Table 6-14. Case C: Optimal Solution.

| Stage(t) | Unit 1 | Unit 2 | Unit 3 | Unit 4 |
|---|---|---|---|---|
| 1 | 0 | 0 | 170 | 0 |
| 2 | 0 | 320 | 200 | 0 |
| 3 | 0 | 400 | 200 | 500 |
| 4 | 0 | 400 | 200 | 400 |

Table 6-15. Case C: Alternate Optimal Solution.

| Stage(t) | Unit 1 | Unit 2 | Unit 3 | Unit 4 |
|---|---|---|---|---|
| 1 | 0 | 0 | 170 | 0 |
| 2 | 0 | 320 | 200 | 0 |
| 3 | 500 | 400 | 200 | 0 |
| 4 | 400 | 400 | 200 | 0 |

Various starting $\lambda$ and two different unit data input orders were used to obtain these results. Only one optimal solution is discovered. This optimal solution is the one in which LR selects unit 4 at the third and fourth stages while unit 1 could have been selected and would have provided the same total cost, $30,801.2. Thus, this is unfair to unit 1.

Many new installations are using similar generating units. Therefore, using LR as an auction method may be inequitable to some generation companies. LR might not select these units, even though these units can provide the same total cost as the units originally selected.

### *6.8.4 Parameter Changes to Achieve Desired Dual Solution*

The formulation for power pool-type auctions is the same as that of UC. Based on [Dekrajangpetch, 1997], the dual decomposable problem is shown in (6.3) to (6.5). The minimization in (6.5) is performed for each unit separately and is subject to individual constraints.

$$Max \quad dobj(\lambda^t) \tag{6.3}$$

where

$$dobj(\lambda^t) = \sum_{t=1}^{T} \lambda^t load^t + \sum_{i=1}^{N} d_i(\lambda^t) \tag{6.4}$$

$$d_i(\lambda^t) = \min_{u_i^t, P_i^t} (\sum_{t=1}^{T} [F_i(P_i^t)u_i^t + stup_i^t - \lambda^t P_i^t u_i^t]) \tag{6.5}$$

Suppose there is one single peak period ($t$) in total $T$ periods. There are two similar peak units that are not selected in all other nonpeak periods. These two units only have a chance to be selected for the peak period. Assume the dual solution, $P_1^t$ is equal to $P_1$ if unit 1 is selected for the peak period ($u_1^t=1$). Assume the dual solution , $P_2^t$ is equal to $P_2$ if unit 2 is selected for the peak period ($u_2^t=1$). Then the functions $d_1(\lambda^t)$ and $d_2(\lambda^t)$ can be shown as (6.6) and (6.7).

$$d_1(\lambda^t)=F_1(P_1)+stup_1^t-\lambda^t P_1 \tag{6.6}$$

$$d_2(\lambda^t)=F_2(P_2)+stup_2^t-\lambda^t P_2 \tag{6.7}$$

Assume that the optimal dual LaGrange multiplier at the peak period is $\lambda^{t*}$. If unit 1 is selected, it is because $d_1(\lambda^{t*})$ is less than zero, which corresponds to the value of optimal $\lambda^{t*}$ in (6.8). Note that zero is the value of $d_1(\lambda^{t*})$ when unit 1 is not selected. On the contrary, if unit 1 is not selected, it is because $d_1(\lambda^{t*})$ is greater than zero, which corresponds to the value of optimal $\lambda^{t*}$ in (6.9). Equations for unit 2 are similar to (6.8) and (6.9). Only is subscript $1$ changed to $2$.

$$\lambda^{t*} > [F_1(P_1)+stup_1^t] / P_1 \tag{6.8}$$

$$\lambda^{t*} < [F_1(P_1)+stup_1^t] / P_1 \tag{6.9}$$

This study can be separated into four cases based on the unit selection: only unit 1 selected, only unit 2 selected, both units selected, and neither unit selected. The optimal dual LaGrange multipliers at the peak period, $\lambda^{t*}$ , for these four cases are described in (6.10, 6.11, 6.12, 6.13) respectively.

$$[F_1(P_1)+stup_1^t] / P_1 < \lambda^{t*} < [F_2(P_2)+stup_2^t] / P_2 \tag{6.10}$$

$$[F_2(P_2)+stup_2^t] / P_2 < \lambda^{t*} < [F_1(P_1)+stup_1^t] / P_1 \tag{6.11}$$

$$\lambda^{t*} > max( [F_1(P_1)+stup_1^t] / P_1, [F_2(P_2)+stup_2^t] / P_2) \tag{6.12}$$

$$\lambda^{t*} < min( [F_1(P_1)+stup_1^t] / P_1, [F_2(P_2)+stup_2^t] / P_2) \tag{6.13}$$

From (6.10) and (6.11), we see that the unit with the lower ratio of the dual optimal total production cost and startup cost to the dual optimal power is selected. Although both units can be selected at the same time according to the optimal $\lambda^{t*}$ in (6.12), this is not desired for a GENCO for two reasons. First, a GENCO does not know the value of the optimal $\lambda^{t*}$ when the LR algorithm stops. The algorithm might stop at the optimal $\lambda^{t*}$, which is not in the range of (6.12) and (6.13) because the objective function is good enough and the demand and other constraints are satisfied. If this occurs, there is a chance that a GENCO's unit will not be selected. Second, a GENCO would prefer that only its unit is selected rather than sharing the power sale with other units. Thus, a GENCO should develop strategies so that there will be a greater chance for its unit to be selected than other units.

The strategy is that a GENCO should submit a bid that has low total cost to power C/P ratio. However, this will result in low revenue for a GENCO. Thus, a GENCO should modify the submitted bid to have low C/P ratio on the peak portion of the power but high C/P ratio on the low power portion. This technique of bid modification will allow a greater chance for a GENCO's unit to be selected and keep a GENCO from having low revenue. Note that this technique will avoid the case that neither unit is automatically selected.

Although the derivation above is based on two similar peak units, it can be applied to any number of similar peak units and it is still true. The purpose of using two units simplifies the explanation. The derivation can also be used with multi-peak periods.

Case C of section four is used to illustrate this concept. Units 1 and 4 are similar units. The C/P ratios of both units are shown in (6.14) and (6.15). Unit 4 has higher C/P ratios than unit 1 for almost the entire range of production except the range from 550 MW to 600 MW. The result for Case C is unit 4 is selected for periods 3 and 4. This is because the dual optimal power for units 1 and 4 lies in the range for which unit 4 has lower C/P ratio and the optimal $\lambda^{t*}$ is between the optimal C/P ratios of units 4 and 1. For example, one set of the optimal LaGrange multipliers at the third and fourth periods ($\lambda^{3*}$ , $\lambda^{4*}$) is (16.7924, 12.7253) \$/MWh. Corresponding to the values of ($\lambda^{3*}$ , $\lambda^{4*}$), the dual optimal values of power for both units 1 and 4 are 600 MW at the third period and 600 MW at the fourth period. The corresponding optimal C/P ratios of unit 1 at the third and fourth periods are the same and the value is 17.5345 \$/MWh. The corresponding optimal C/P ratios of unit 4 at the third and fourth periods are the same and the value is 17.5245 \$/MWh. It is evident that the optimal C/P ratio of unit 4 is lower than that of unit 1 and this is why unit 4 is selected instead of unit 1.

$$C/P\ (unit\ 1) = 0.002 * P^t_1 + 10.00 + 3800.7/P^t_1 \tag{6.14}$$

$$C/P\ (unit\ 4) = 0.002 * P^t_4 + 9.88 + 3866.7/P^t_4 \tag{6.15}$$

Actually, the real dispatch power at the third and fourth periods of either unit 1 or 4 are 500 MW and 400 MW. At these levels of power, the C/P ratios of unit 1 are 18.6014 at the third period and 20.3018 at the fourth period while the C/P ratios of unit 4 are 18.6134 at the third period and 20.3468 at the fourth period. It can be seen that the C/P ratios of unit 4 are higher than the C/P ratios of unit 1 at the real dispatch power at both the third and fourth periods. However, the selection of unit is based on the dual problem and thus the comparison of C/P ratios is based on the dual solution (dual power) although the dual power is not the real generating power for units.

The strategy for submitting bids to have a greater chance to be accepted has been illustrated above. Next the procedure of adjusting bid parameters will be described. The parameters considered in this work are divided into three groups: quadratic coefficient ($a_i$), linear coefficient ($b_i$), and constant cost and startup cost ($c_i$+$stup_i$). Note that group three has two parameters. Thus, the resulting of change considers the total effect on the two parameters together.

As explained above, this is not what a GENCO desires because of low revenue. If these three groups of parameters are lowered individually, this will lower the C/P ratio for the whole production range. The numerical example to be shown demonstrates what happens when only linear cost is lowered. Two peak units are used for illustration. Unit 1 of Case C is used and thus unit 1 has C/P ratio as (6.14). Unit 4 has the same parameters as unit 4 of Case C except that its constant cost ($c_4$) is changed to be 500 and its startup cost ($stup_4$) is changed to be 3300.7. The C/P ratio of unit 4 is shown in (6.16).

$$C/P\ (unit\ 4)=0.002*P^{t}_{4}+9.88+3800.7/P_4^{t} \tag{6.16}$$

If two of the three groups of parameters are changed simultaneously, the three possible strategies are as follows.

Strategy 1: lower quadratic coefficient ($a_i$) and increase linear coefficient ($b_i$)
Strategy 2: lower quadratic coefficient ($a_i$) and increase constant cost and start-up cost ($c_i$+$stup_i$)
Strategy 3: lower linear coefficient ($b_i$) and increase constant cost and start-up cost ($c_i$+$stup_i$)

These three strategies are desired because they result in low C/P ratio on the peak portion of the power but high C/P ratio on the low power portion. The produced power for the unit having the reduced C/P ratio for each of the three strategies is shown in (6.17), (6.18), and (6.19), respectively. These formulas are useful for bid modification because they tell the range of peak power portion in which a unit's C/P ratio is lower than other units.

$$P = -\ (b_1-b_2)\ /\ (a_1-a_2) \tag{6.17}$$

$$P=sqrt\ (-\ [(c+stup)_1-(c+stup)_2]\ /\ (a_1-a_2)\ ) \tag{6.18}$$

$$P = -\ [(c+stup)_1-(c+stup)_2]\ /\ (b_1-b_2) \tag{6.19}$$

Units 1 and 4 of Case C provide a good illustration/example of Strategy 3. Unit 4 has a lower linear coefficient than unit 1, and unit 4 has higher constant cost and startup cost than unit 1 ($b_4$=9.88, $b_1$=10.00, $(c+stup)_4$=3866.7, $(c+stup)_1$=3800.7). The power level at which the C/P ratio for unit 4 falls below that of unit 1 is $P = -(3800.7-3866.7)\ /\ (10.00-9.88) = 550.00$ MW.

### 6.8.5 Sensitivity Analysis

This section summarizes the results of sensitivity analysis in [Dekrajangpatch, 1998b]. Sensitivity analysis is performed by using the generating unit data and load data of Case A as Case D.

Case D consists of a sensitivity analysis for each of the cost parameters of units 4. The linear and constant parameters of the production function are varied. The startup cost is also varied. The startup cost data is $3,000 for both units 1 and 4. The procedure varies each of these parameters of only unit 4 in the amount of -10% to 10% of the original value, in increments of 1%. Three starting λ are used for implementing the result of varying each parameter. The sensitivity analysis results depend on the subgradient updating procedure used.

The results show that the optimal solution can be found only if there is a difference in the parameters. If two or more units have similar values, then it is hard for the algorithm to select between the two. The algorithm can find the optimal solution with only 1% difference when either varying linear cost coefficient or startup cost for all three starting λ. When varying constant cost, 4% difference is needed for the algorithm to find the optimal solution for one starting λ while 1% difference is needed for other two starting λ. The reason the algorithm needs 4% difference for one starting λ can be understood if the updating procedure is examined. The optimal value of λ cannot be reached by the updating algorithm from a value of one starting λ. The problem exists primarily at the peak demand level. At this level of operation, the optimal solution cannot be found. This problem is that the range of optimal λ of the peak period is small. Thus, if the vector λ is not updated properly based on the system data and the starting value, LR cannot converge to the optimal solution.

The other sensitivity analysis result is to find the number of iterations needed to find the optimal solution versus percent change of each parameter. The result is that varying the constant cost requires more iterations than varying either of the other costs and varying the startup cost requires more iterations than varying the linear cost. In addition, the sensitivity analysis result above shows that the constant cost parameter requires 4% difference while only 1% difference is needed for the linear coefficient and startup cost. It can then be implied that for the same percent change of each parameter between units 4 and 1, the order from the most difficult to the least difficult for LR convergence is constant cost, startup cost and linear cost, respectively. In other words, the resulting cost is least sensitive to the constant cost and most sensitive to the linear cost.

### 6.8.6 Parameter Changes for Maximum Generation

Section Five suggests the strategies for adjusting parameters to enhance the chance of a unit being selected. The unit selection is performed in the dual problem. After the units are selected, the amount of power to be supplied by the units is decided in the primal problem by economic dispatch calculation. Thus, a GENCO should be concerned about this in adjusting its parameters.

The strategy for a GENCO to have higher real-time generation is to submit the bid with low incremental cost. The incremental cost of a unit is $2*a_i*P_i+b_i$. Thus, a GENCO should submit the bid with either a low quadratic coefficient ($a_i$) or a low linear coefficient ($b_i$) or both.

### *6.8.7 Summary of Parameter Changes*

This section summarizes how to change unit parameters based on the strategy in Sections Five and Seven and the sensitivity analysis result in Section Six. Three strategies for enhancing the chance of a unit being selected are presented in Section Five. Not only does a GENCO desire a unit to be selected, a GENCO also wants its unit to produce as much power as it can. Thus, a GENCO should submit the bid with either a low quadratic coefficient ($a_i$) or a low linear coefficient ($b_i$) or both based on the strategy in Section Seven. Combining the strategies in Sections Five and Seven, we see that Strategies 2 and 3 are dominant and should be used by a GENCO. Strategy 1 is not dominant because its increased linear coefficient ($b_i$) will in turn increase the unit's incremental cost.

When a GENCO adjusts parameters, a GENCO should always be concerned about the sensitivity of each parameters. If a sensitive parameter is adjusted, the unit's selection is more risky.

The strategies proposed in this work are for peak units because LR has convergence problems during peak periods. For base-loaded units, strategies just involve tradeoffs between lowering bids and increasing profits.

### *6.8.8 Summary*

Problems in implementing an auction with LR when identical or similar units exist in the system have been known. For nearly identical units, a GENCO can always force unit selection to be dispatched for a price advantage. Additionally, a GENCO can always force more generation dispatch after unit selection. This makes LR a biased auction method that would favor the gaming GENCO. For similar units, the optimal solution may be directed by a sophisticated GENCO. Any subset of similar units can be used to force alternative optimal solutions. This is inequitable to the units not in the chosen subset that actually can provide an alternative optimal solution.

Because the dispatcher has to use heuristic selection, there is no obviously fair solution to these problems. The auction procedure should be separated. This can be considered a decentralized unit commitment as suggested by [Sheblé, 1996b, Sheblé, 1994b, Kumar, 1996a, Kumar, 1996b, Sheblé, 1994e, Sheblé, 1994c, Kumar, 1997] and as implemented in Spain [Debs, 1998, Olero-Novas, 1998]. Instead of submitting cost models to the ICA, GENCOs submit period (hourly) bids, which are composed of prices and quantities to the ICA. ESCOs also submit similar bids to the ICA. Then, the ICA can use other auction methods that are not based on heuristic rules [Sheblé, 1996b, Sheblé, 1994b, Kumar, 1996a, Kumar, 1996b, Sheblé, 1994e, Sheblé, 1994c, Kumar, 1997]. Interior point linear programming is an example of a method that is not based on heuristic rules [Dekrajangpetch, 1998c].

# 7 ESCO OPERATION

*The customer is always right if we make more money.*

"The customer is always right, at least in the eyes of the customer. It is the customer who has the needs we wish to fulfill to maximize the profit received by our management and shareholder. The customer is number one." Such phrases are very different from the relationship between the regulated monopoly and the public. Almost all of the engineers I have met thought of the public trust. Indeed, most utilities espoused the desire to satisfy the everyday customer.

A competitive market is entirely different. Highlighting this difference is the role of the intervenor, now becoming the energy service company (ESCO), between the various components of the electric energy industry and the customer. The ESCO has to maintain a portfolio of agreements with every GENCO, TRANSCO, DISTCO, and customer to procure electricity, ship it efficiently, and deliver it to the customer. The portfolio risk management requires the ESCO to maximize profits for the ESCO shareholders, obtain the least costly supply, and ensure that transportation costs are minimum and the delivered product meets all contract requirements at the customer's site. The product must be supplied when promised with the proper quality and reliability.

The major tool in balancing the supplies with the demands for the ESCO is Demand Management and Risk Management. *Demand side management* reduces the variations in demand to even out the requirements of the suppliers, mitigates the limitations of the transportation network, and minimizes the control requirements of the suppliers. This should enable negotiation with suppliers to achieve the least cost procurement possible. *Risk management* reduces the variations in supplies, the uncertainty of demand, and the uncertainty of transportation such that procurement can be obtained at the best moment in time, specifically at the best price.

## 7.1 Determining Value in a Deregulated Market

The fright of many companies has turned to flight. The first part of this chapter examines the value of a company in the business of providing electric energy to the public.

### *7.1.1. Introduction*

This work describes processes and methodologies for determining value in a deregulated market. The processes require measurement, appraisal, and assessment for each component functioning to meet market demand. They provide the ability to successfully compete in current markets while expanding market share. The objective is to ascertain the real, potential, or unrealized value of each element in an open, competitive market.

Deregulation directly effects the competitive process within any given market area. A free market environment will have the effect of creating virtual markets restricted only by available supply and transport capacities as opposed to regulated service areas with specified boundaries. Entities that are marketing power need to have a clear understanding of a market area's boundaries, volume requirements, price elasticity, profitability, product needs (commodity and unique), customer classifications, and specific customer services. This knowledge provides a basis for understanding the key elements of current demand, recognizing and capitalizing on latent demand, and using projected demand in the futures markets to stabilize pricing and ensure profitability.

### *7.1.2. Case Study*

The following is an example of a process and the methodologies that might be used in assessing the acquisition value of a G&T Cooperative as shown in Figure 7-1. This example is used to illustrate the concepts of the analysis and therefore should not be considered complete.

**7.1.2.1 Background** One of the premises of the methodology is that the current market served and its underlying demand is a major component of the value of the cooperative as a whole. In addition, it is assumed that the purchaser is primarily interested in acquiring market share. In other words, understanding the value of the market (current and potential) is the key to understanding the acquisition value of the cooperative.

The primary components to be valued consist of

- The Business (Balance Sheet, Cash Flow, Operating Expense, etc.)
- Asset Valuation (Physical Plant, Contracts, Transmission, etc.)
- Economic Components (Supply and Demand).

**7.1.2.2 Supply** *For any power producer*, the market area is restricted to the distance that power will carry across the grid in any direction. The endpoints of the boundary describe a polygon that is considered the trade area for the producer. All consumers that reside within that trade area are considered within the "reach" of the producer. Those consumers that are current customers are the producer's "share" of the market. Any part of the trade area that overlaps another producer's trade area creates a market of increased competition.

*For any point on the grid*, supply is available from any producer within carrying range. This defines the supply boundaries for that point. Current and latent production capabilities along with marginal costs per kilowatt-hour (kWh) for each level of production can be estimated for each producer within range. These estimates assist in projecting the effect of incremental profitability per kWh on equity value (share price, capital credits, etc.). Determining the minimal profitability acceptable for each producer by category (i.e., industrial, commercial, consumer (or by type of power)) enables informed price negotiation at each point on the grid. In other words, for any given category, product, or service, once you know the acceptable price range for each producer, you can comparison shop on the futures market.

The ability to supply the market demand for power at a competitive price includes

- Production capacity
- Production utilization
- Operational efficiency
- Contracts for resources (current and future opportunities)
- Covenants & restrictions
- Other resource purchase opportunities
- Reselling latent capacity

Within any organizational environment there are opportunities for expense reduction, operating efficiencies, maximizing asset utilization, etc. all of which will have an effect on valuation. The unrealized value of an organization includes the value of operational and production opportunities yet to be implemented.

The presently required FERC data, which would assist in this analysis, is the available transfer capability (ATC). Unfortunately, these data are not sufficiently robust to adequately predict the transfer capability under any conditions other than the present loading without any new contracts added and without proper trend and statistical analysis. It is also possible to analyze transfer reach based on previous interchange filings with FERC. Such transfer reach is an estimate of the market range presently available under regulated conditions. The price of acquiring such market has no historical data to predict what the price will be for each future time period in a deregulated transmission market. However, analysis of neighboring companies and other competitors provide sufficiently accurate estimates since the transmission will be priced to fully utilize the equipment. If the transmission is regulated in the new environment, then such forecasts will not be necessary. Fortunately, deregulation cannot occur instantaneously. Instead, phases of deregulation have to occur for the orderly transition of companies, or at least their assets, into the competitive environment.

The required ancillary services for transmission capability also determine the market reach for each competitor. The devices presently used for voltage control, power factor improvement, transformer control of voltage or flow, etc. provide the capability for market reach. It is a combination of transmission equipment and ancillary service equipment needed to provide the capability to increase market share. The reach of the transmission system is a key component of the market share maintained by any competitor attached to the system. Open access on the transmission system requires orderly use of the transmission grid under proper operational and safety guidelines.

**7.1.2.3 Demand** For a given a market area (trade area), demand is an aggregate of the needs of all consumers. Measuring demand potential begins by understanding the composition of consumers contained within the area.

One method is to identify all commerce within the reach of the trade area. Industry, estimated sales volumes, number of employees, etc can then classify each commercial site. These classifications are in turn applied against power usage estimates by industry type and size of facility. The usage estimates are also compared to historical demand for the region and projected against growth estimates, economic factors and other criteria. This process eventually results in a time series demand profile for the specific area of study. These data, in turn, provide projections for timing and pricing on futures trades and as a basis for forecasting profitability. The result is a measure for revenue stabilization as desired by equity markets.

A demand analysis should, at a minimum, forecast and account for consumer and commercial growth projections, understand industry to industry relationships, recognize the effect of changing demographics, and allow for changes in competition.

Using the demand profile, a market strategy for the trade area can be established which might include selectively targeting specific industries, product specialization, customer retention and acquiring market share from competitors. The market strategy will help to determine volume, pricing and timing when purchasing power.

In addition to determining the demand for the producer's market area, the same techniques can be applied to determine demand requirements and market potential in any competitor's trade area. This is particularly useful for acquiring market share in areas where the competitor's trade area overlaps that of the producer, and in competitive bidding for power.

**7.1.2.4 Customer Profile** The customer base can initially be classified into three categories: industrial, commercial, and consumer. The categories are then segmented and indexed by demand, utilization, profitability, and penetration to indicate the performance of any individual segment against the norm as shown in Figure 7-2.

This assessment is the basis for understanding the behavior of categories and their segments pertaining to utilization and profitability. These measures can then be applied to historical local, regional, and national trends. A trend analysis for power utilization indicates the revenue potential of any category segment in the market. In addition, these profiles can be used in forecasting potential behavior as regulatory changes are implemented.

**7.1.2.5 Locate Customers** This process involves determining the exact location of each customer. The information is then used to analyze the geographic distribution of specific types of customers. These data are valuable in understanding market distribution, areas of concentration of similar customers (their utilization, profitability, etc.) and recognizing market opportunities. The distribution of customers against a given market area indicates the penetration and share of the market being served.

**7.1.2.6 Trade Area Development** The method for developing trade areas as described in the Supply section of this document depicts the trade area *reach* or total market available. Trade areas are usually developed by plotting customer records and drawing a perimeter around (some percentage of) those customer locations. The area contained within the perimeter is the primary trade area. The trade area, as described in this analysis, is based upon transfer capability and the capacity of production facilities. These constraints define the available market geography. The market, therefore, is the aggregation of all customers (current and potential) contained within the trade area boundary along with inherent demographic information, purchase behavior patterns, realized and potential revenue streams. Customers currently serviced are the market *share* of that trade area. Individual trade areas analyses specific to demand, usage, profitability and customer categories are then developed.

**7.1.2.7 Calculate Market Share** *Market share* is the percentage of the market potential that is being captured (serviced). This is a measurement that is built from information generated by the trend analysis, customer profiles, and trade area *reach*. Within any category segment, the demand for power being satisfied by an entity is its *share* of the aggregate market for that category segment. Calculating *share* within a market provides a measure of the market potential that is not being directly serviced and indicates additional revenue opportunity.

**7.1.2.8 Segment Targets: Growth, Unserved, Underserved** Segment targets are those sections of a category that provide opportunity for additional revenue. They provide target areas that can be aggressively pursued to acquire market share from competitors, exploit latent demand, or capture new and emerging markets. Ranking segments by potential allows conservation of marketing resources while focusing efforts on segments with a high probability of revenue (profitability). Understanding market potential is another aspect of valuation. Market *reach* is a measure of revenue potential, market *share* is a measure of current revenue capability, and market potential is the revenue stream most profitably acquired within the market *reach*.

**7.1.2.9 Competitor Analysis** Understanding your competition includes recognizing where and how they may compete in your market. Calculating their potential to underbid your pricing is essential in maintaining market share. Plotting their trade area reach and overlaps, measuring their ability to serve a market category and estimating their efficiencies all provide insight to their constraints on pricing. It is a measure of the threat they impose on your market share and an indication of your ability to capture their market share in any given category segment (or geography).

**7.1.2.10 Attrition Analysis** This is a calculation of attrition by customers at different price levels. A multi-tiered analysis can be implemented based upon competitive pricing, minimal margin pricing, loss leader, reselling power purchased from outside (including blended rates), etc. Once the segmentation analysis is complete, insight into purchase behaviors and loyalty factors can be inferred. Profitability measures will help to direct which customers may be comfortably abandoned and which customers must be retained.

**7.1.2.11 Distribution Realignment** Once legislative actions are in effect, lingering constraints or restrictions can be assimilated into the assumptions of a market analysis. The effect of these constraints and restrictions are to impose artificial barriers in the market. However, as these changes are implemented, risk can be minimized through alternate strategies. The models enable the development of alternate strategies by allowing changes in the assumptions and providing "what if" analysis capabilities.

**7.1.2.12 Optimize Pricing Strategies** Optimized pricing for specific target markets can be calibrated using competitive analysis, category segment demand, and attrition analysis. These strategies can include volume discounting, the effect of utilization cycles, variable pricing, etc. Pricing strategies are a powerful tool in valuing the market under different scenarios to provide a deeper understanding of value in an acquisition atmosphere.

**7.1.2.13 Specific Target Marketing** Distinguishing specific markets for targeting is the result of all of the analyses previously described. Understanding where these markets exist and their potential value is also a measure of latent value within the organization. Market trends (growth, decline, utilization, etc.), pricing strategies and knowledge of the competition provide the elements for success in customer retention and maximizing revenue potential.

**7.1.2.14 Spot Market Strategies and Implementation** Much higher levels of confidence can be assumed with spot market strategies in an environment where the markets are redefined, market reach and share calculated, competitors analyzed, attrition measurements taken, target markets identified, and pricing strategies developed. These strategies can work in conjunction with market strategies (targeted segment acquisition, etc.), measured market share (trend analysis, attrition prevention tactics), and pricing strategies to provide the timing and pricing for bids on the spot market.

**7.1.2.15 Cost/Profitability** Determining the price range per kWh required by each producer indicates the minimum price they can accept and allows "comparison shopping." Forecasting the demand of an area provides the volume requirements for any level of geography, industry or customer within the trade area at different points in time. This information, combined with the profit goals for the purchasing firm, helps to establish the price ranges for bidding. The objectives are to maximize profitability while submitting successful responses to competing bids in the spot market.

**7.1.2.16 Contract Structuring** The viability of the terms and agreements within any existing or future contracts can now be applied against different market environments. A strategy for exploiting the maximum value can be developed along with tactics that minimize risk and loss.

### *7.1.3 Minimizing Price Volitility and Market Risk*

The following represents a model designed to use information outputs of the previously described analyses, Figure 7-3.

Supply and market demand information is input to generate various expense and revenue data sets. The data are then processed using nonlinear program techniques to provide a range of optimized results under varying expense and revenue scenarios.

Expense data is a combination of the price mixes of kWh costs for different qualities of power, costs associated with transmission capabilities, variances in utilization, effects of weather, etc. This information is processed through futures trading simulation of the spot markets to account for competitive bidding and resource pricing. The futures market pricing results are then fed back into the pricing model to provide resource cost forecasting information.

Revenue data is processed against varying market information such as levels of penetration, competition, growth, attrition, etc. In addition, different mixes of industrial, commercial and consumer revenue streams and profitability are used to simulate the effect of actions such as marketing programs, pricing strategy implementation, and competitive response.

The results are to provide the ability to understand and target realistic goals and to develop the strategies and tactical implementation plans that will achieve those goals.

### *7.1.4 Market Analysis In A Deregulated Environment*

Deregulation will directly affect the competitive process within any given market area. A free market environment will have the effect of creating virtual markets that are restricted only by available supply and transport capacities as opposed to designated areas with specified boundaries. Entities that are purchasing power for resale need to have a clear understanding of a market area's boundaries, customer classifications, product needs (commodity and unique), volume requirements, price elasticity, profitability, and specific customer services. This knowledge will provide them with a basis for understanding the key elements of current demand, recognize and capitalize on latent demand, and use projected demand in the futures markets to stabilize pricing and ensure profitability.

### *7.1.5 Consumer Payment Processing in a Deregulated Environment*

Deregulation will directly affect the competitive process within any given market area. A free market environment will have the effect of creating virtual markets that are restricted only by available supply and transport capacities as opposed to designated areas with specified boundaries. Entities that are selling power need to have a clear understanding of a market area's boundaries, customer classifications, and specific customer demands for products and services. Customer services include all types of payment processing for all transactions. The knowledge required to provide effective, efficient customer services includes understanding the composition of consumers within the market and key elements of demand. In addition, providers must have the ability to forecast customer service and payment processing demands in future markets, retain and acquire customers, penetrate new markets, stabilize pricing, and ensure profitability.

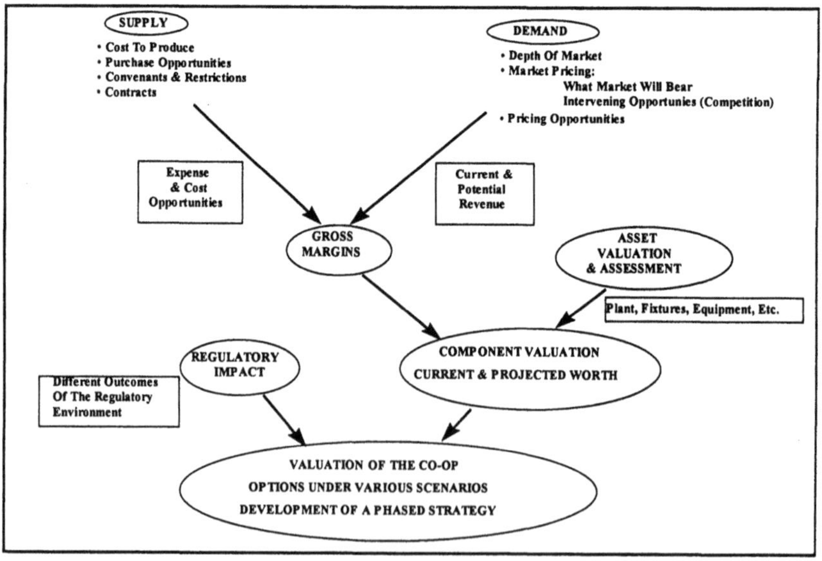


Figure 7-1. Valuation Process Chart.

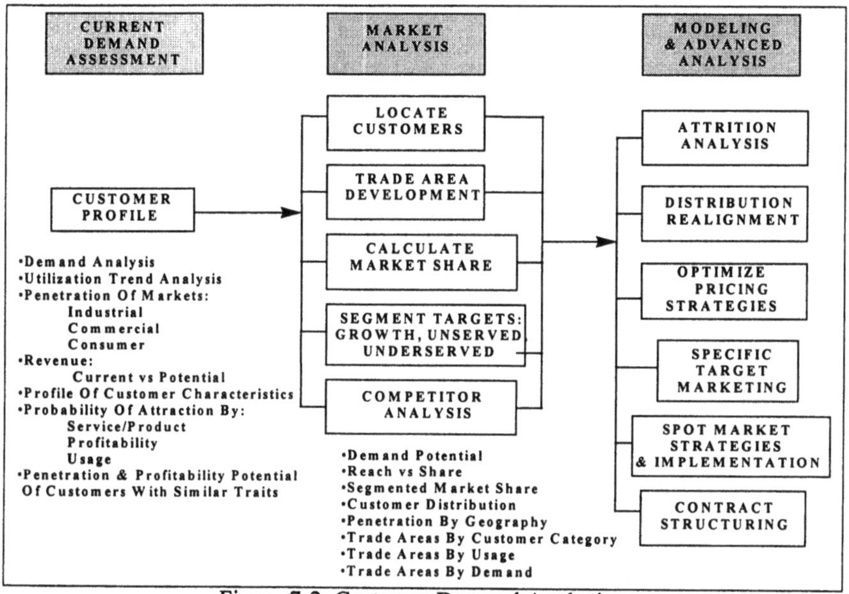


Figure 7-2. Customer Demand Analysis.

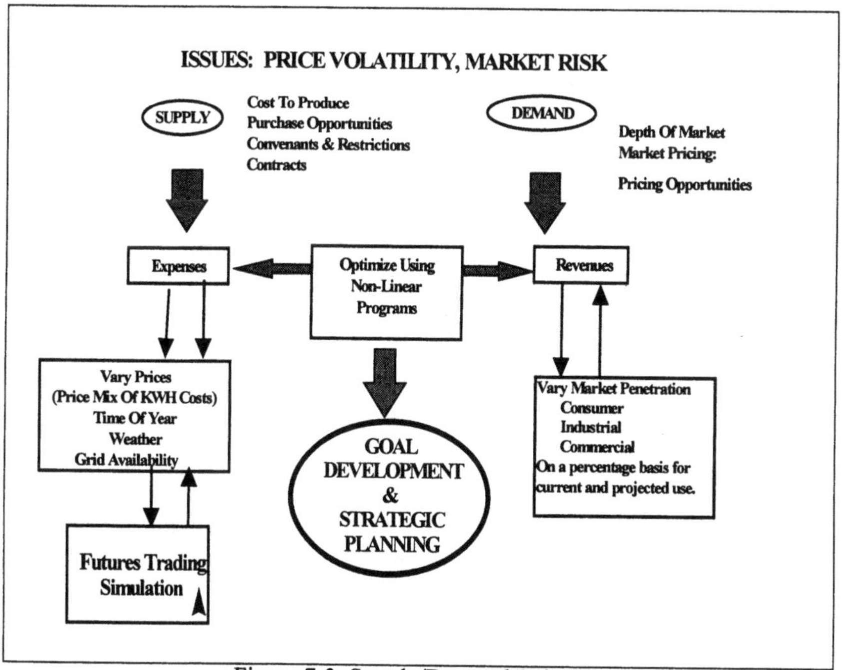


Figure 7-3. Supply/Demand Balance.

### 7.1.6 Summary

This work outlines the process for company valuation in a competitive environment. The intent is to refocus the valuation from a regulated return to market share of customers. The focus of the industry now must turn to selling services instead of electrons.

## 7.2 Cost-based Load Management Using LP

### 7.2.1 Introduction

Load management, introduced in the 1970s, is aimed at reducing the operating cost while maintaining the reliability of the electric power network. Generally, load management can be categorized into the following sections: direct load control (DLC), which allows the utilities to shed remote customer loads unilaterally; indirect load control, which allows customers to control their loads independently according to the price signals sent by the utilities; and storage capacity, which allows both utilities and customers to store energy during the off-peak/low cost session and consume during the peak/high cost session. This work examines only the cost-based DLC algorithm using linear programming.

Selectively grouping the customers' load (LP), the utilities are then capable of offering incentive to respective customers for direct control over selected loads. Various algorithms, dynamic programming primarily [Chen, 1995; Cohen, 1987; Hsu, 1991; Lee, 1984], have been developed to reduce the system peak, operating cost, or spinning reserve. However, while one approach [Chen, 1995] failed to recognize the fact that the maximum controllable load varies from one to another period, another [Hsu, 1991] ignores the load-dependent payback pattern. Linear programming, a relatively inexpensive, powerful method, is used by Kurucs et al. [Kurucz, 1996] to reduce the system peak. However, the algorithm failed to consider all possible control duration. Thus, if the load/cost zig-zagged over time, the resulting solution of the approach may not be optimal. Furthermore, while Kurucs tried to reduce system peak, Le et al. [Le, 1983] pointed out that resulting system production cost of flattening the load might not be the lowest. Thus, the approach does not include all appropriate costs.

This work introduces the cost-based DLC scheduling using the LP algorithm. A relatively inexpensive, powerful approach, the developed algorithm includes various benefits of others. First, rather than determining amount of energy to be controlled during a certain period, which is a rather indirect approach, the new algorithm tries to decide the number of groups to be controlled during a certain period. Second, instead of "two-phase" strategy [Kurucz, 1996], the developed algorithm is capable of including all possible control durations. Third, including the nonfixed maximum controllable load and payback pattern over time, the algorithm resolves the two problems through determining groups to be controlled. Introducing paybacked energy that depends on the load during the controlled period resolves the varied payback pattern. Finally, by changing the constant parameters of the objective function and constraints from period to period, the LP algorithm considers nonlinear cost function and yet obtains integer solutions that are optimal.

### 7.2.2 Nomenclature

$F_j(*)$: system production cost at period j
$SC_j$: system cost at period j
: system load level at period j
$Pn_j$: uncontrollable load level at period j
$Pc_j$: system controllable load level at period j
$P_{i,j}$: load available for DLC at period j by customer/load type i
$R_i$: rate charged on controllable load of customer/load type i ($/MW)
$pr_j(*)$: average cost function at period j in $/MW, reflecting fixed, operating, and maintenance cost.
$x_{i,j}$: number of groups of customer/load type i that are controllable at period j
$x_{i,j,k}$: number of groups of customer/load type i that are undergoing load shedding at period j of length k
n: maximum number of customer/load type
m: maximum period under study
$\bar{k}_i$: maximum length of period for load shedding for customer/load type i
$\alpha_{i,j,k,s}$: payback ratio at period (j+k+s-1) for load shedding at period j of length k on customer/load type i
q(i,j,k):the maximum payback period for load shedding at period j of length k on customer/load type i
u(*): unit step function where

$$u(*) = 1 \qquad \text{if } * \geq 0$$
and
$$u(*) = 0 \qquad \text{if } * < 0$$

c: per megawatt cost of conducting load shedding
$\underline{g}_i$: maximum divisible groups of customer/load type i
P: maximum load that can be increased at any period for the price/cost to remain in the feasible range
$\underline{P}$: minimum load that can be decreased at any period for the price/cost to remain in the feasible range

### 7.2.3 Model

Dividing the load forecast into two sections the amount of power available for control from one to another period is obtained. Distinguishing the controllable load of one customer/load type to another is then categorized in more detail. A lower rate structure is offered to the customers for controllability over their load, denoted $R_i$. The system production cost at any period is calculated as follows:

$$SC_j = F_j(Pt_j)$$
$$F_j(Pn_j + Pc_j)$$

$$= \frac{F_j\left(Pn_j + Pc_j\right)}{\left(Pn_j + Pc_j\right)}\left(Pn_j + Pc_j\right)$$
$$= \frac{F_j'\left(Pn_j + Pc_j\right)}{\left(Pn_j + Pc_j\right)} Pc_j \tag{7.1}$$

since the noncontrollable loads are fixed during any period, they may be deleted from the function. Let

$$pr_j = \frac{F_j'\left(Pn_j + Pc_j\right)}{\left(Pn_j + Pc_j\right)} \tag{7.2}$$

be the average cost function. Then, the simplified function is as follows:

$$SC_j = pr_j Pc_j$$

where

$$Pc_j = \sum_{i=1}^{n} P_{i,j} \tag{7.3}$$

Divide the control choice of each customer/load type by $g_i$, at any period:

$$SC_j = \frac{pr_j\left(Pc_j - \frac{x_{i,j}\sum_{i=1}^{n} P_{i,j}}{g_i}\right)}{g_i}\sum_{i=1}^{n} x_{i,j}P_{i,j}$$
$$= \frac{pr_j'(x_{i,j})}{g_i}\sum_{i=1}^{n} x_{i,j}P_{i,j} \tag{7.4}$$

Then, redefine:

$$pr_j = pr_j'(x_{i,j}) \tag{7.5}$$

The average cost function is transformed into a function of the number of controllable groups, $x_{ij}$. Then, based on the new system cost function, the minimization problem has become a problem of finding the number of $x_{ij}$, which will result in the optimal minimization of system cost.

Even though the average cost function is nonlinear, it can be perceived as constant slope for a range of change in load levels, denoted by the limits of the range $\left(\overline{P} - P\right)$. Thus, the minimization of nonlinear system cost function is reduced into a linearized problem.

*7.2.4 Formulation*

The reduced cost of deferring the energy is as follows:

$$DF = \sum_{i=1}^{n} \sum_{j=1}^{m} \sum_{k=1}^{\overline{K_i}} \frac{x_{i,j,k}}{g_i} \sum_{v=1}^{k} P_{i,j+v-1}(pr_{j+v-1} + c) \tag{7.6}$$

The increased in cost for paying back the energy is as follows:

$$PB = \sum_{i=1}^{n} \sum_{j=1}^{m} \sum_{k=1}^{\overline{K_i}} \frac{x_{i,j,k}}{g_i} \sum_{v=1}^{k} P_{i,j+v-1} \sum_{s=1}^{q(i,j,k)} \alpha_{i,j,k,s} Pr_{j+v+s-1} \tag{7.7}$$

If the summation of payback ratio of any customer/load type at any period j, $\sum_{s=1}^{q(i,j,k)} \alpha_{i,j,k,s}$, dos not equal 1, there is a loss or increase in revenue. Thus, the possible revenue loss must be included as a penalty.

$$PT = \sum_{i=1}^{n} R_i \sum_{j=1}^{m} \sum_{k=1}^{\overline{K_i}} \frac{x_{i,j,k}}{g_i} \left( \sum_{v=1}^{k} P_{i,j+v-1} \right) \left( 1 - \sum_{s=1}^{q(i,j,k)} \alpha_{i,j,k,s} \right) \tag{7.8}$$

Then, the objective of controlling the load is to minimize the overall cost, which is

$$\text{minimize} \qquad -DF + PB + PT \tag{7.9}$$

subject to:

$$x_{i,j,k} \geq 0 \tag{7.10}$$

$$\sum_{k=1}^{\overline{K_i}} x_{i,j,k} + \sum_{\substack{a=j-1 \\ a \geq 1}}^{1} \sum_{k=1}^{\overline{K_i}} x_{i,a,k} u(a+k+q(i,a,k)-j- \leq g_i \quad \forall i, j \tag{7.11}$$

where at any period, for any customer/load type i, the controlled (deferred or paybacked) groups can never exceed the preset maximum value, $g_i$.

Since the nonlinear objective is approximated by piecewise linear segments, extra constraints, (7.12 and 7.13) should be included to ensure that the change in load at any period will not exceed the maximum allowable increase or decrease.

$$\sum_{i=1}^{n} \Bigg( \sum_{k=1}^{\overline{K_i}} \frac{x_{i,j,k} P_{i,j}}{g_i} + \sum_{\substack{a=j-1 \\ a \geq 1}}^{1} \sum_{k=1}^{\overline{K_i}} \frac{x_{i,a,k}}{g_i} P_{i,a} u(a+k-j- \\ - \sum_{\substack{a=j-1 \\ a \geq 1}}^{1} \sum_{k=1}^{\overline{K_i}} \frac{\alpha_{i,j,k,j-a-k+1}}{g_i} \sum_{v=1}^{k} P_{i,j+v-1} \Bigg) \leq -\underline{P} \qquad \forall j \tag{7.12}$$

and

$$\sum_{i=1}^{n}\Big(-\sum_{k=1}^{\overline{K_i}}\frac{x_{i,j,k}P_{i,j}}{g_i}-\sum_{\substack{a=j-1\\a\geq 1}}^{1}\sum_{k=1}^{\overline{K_i}}\frac{x_{i,a,k}}{g_i}P_{i,a}u(a+k-j-\\
+\sum_{\substack{a=j-1\\a\geq 1}}^{1}\sum_{k=1}^{\overline{K_i}}\frac{\alpha_{i,j,k,j-a-k+1}}{g_i}\sum_{v=1}^{k}P_{i,j+v-1}\Big)\leq \overline{P} \qquad \forall j \tag{7.13}$$

A cautious step should be noted, where, for any period, the largest deferred and paybacked load, denoted by $\frac{P_{i,j}}{g_i}$, and $\alpha_{i,j,k,s}\sum_{v=1}^{k}P_{i,j+v-1}$ respectively, should at least smaller than or equal to $\overline{P}$ or $\boxed{P}$.

Graphical Representation:

To visualize (7.6) through (7.13), a graphical representation is shown in Figure 7-4. There are n pieces lying on top of each other to represent the individual customer/load type. For each individual category, the studied m periods are listed from left to right. From top to bottom of each layer are the load-shifting choices and how they interact throughout the time frame. The boxes (undashed) on each layer represent the periods when the load is deferred while the dashed boxes represent the period of paybacked load. For simplicity, the load-shifting choices of each customer/load type is set to be three, i.e., $\overline{k}_1 = \cdots = \overline{k}_n = 3$, and $q(i,j,k)$s of all customer/load types are equal.

In accordance with (7.6), the DF coefficient of each control choice, $x_{i,j,k}$, is represented by the undashed boxes. For instance, when k = 3, the DP coefficient is found as follow:

$$DF|_{x_{i,j,3}} = \frac{P_{i,j}(pr_j + c) + P_{i,j+1}(pr_{j+1} + c) + P_{i,j+2}(pr_{j+2} + c)}{g_i} \qquad \forall\, i, j$$

From (7.7), the PB coefficient of each control choice, $x_{i,j,k}$, is represented by the dashed boxes. The PB coefficient of k = 3, for instance, is determined as follows:

$$PB|_{x_{i,j,3}} = \frac{(\alpha_{i,j,3,1}\, pr_{j+3} + \alpha_{i,j,3,2}\, pr_{j+4})(P_{i,j} + P_{i,j+1} + P_{i,j+2})}{g_i} \qquad \forall\, i, j$$

where $\alpha_{i,j,k,s}$ is determined by statistical measurement of past data and present control load pattern.

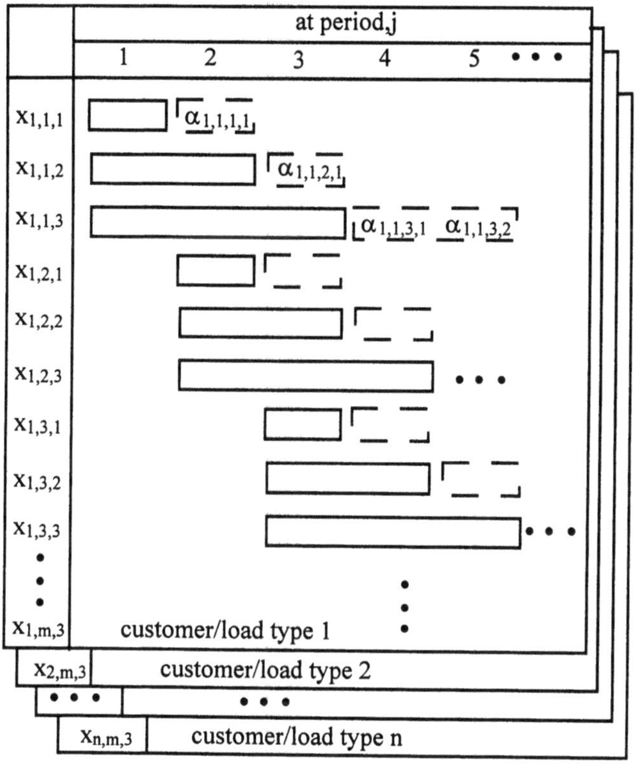


Figure 7-4. Graphical Representation of Control Pattern.

The PT function, (7.8), takes action only when the payback ratio of $x_{i,j,k}$ does not sum up to 1. For instance, The PT coefficient of k = 3 is calculated as follow:

$$PT|_{x_{i,j,3}} = \frac{R_{i,j}(P_{i,j} + P_{i,j+1} + P_{i,j+2})(1 - \alpha_{i,j,k,1} - \alpha_{i,j,k,2})}{g_i} \qquad \forall\, i, j$$

If the payback back ratio sum up to 1, then the expression, $(1 - \alpha_{i,j,k,1} - \alpha_{i,j,k,2})$, will become zero, and the penalty will not take any action. If the sum of payback ratio is greater than 1, the PT coefficient becomes negative and is considered as bonus rather than the penalty.

By (7.10), each control choice, $x_{i,j,k}$, has to be greater or equal to zero, i.e., $x_{1,1,1} \geq 0$, $x_{1,1,2} \geq 0$, $x_{1,1,3} \geq 0$ and etc.

From (7.11), the maximum control choices of any period, $x_{i,j,k}$, of individual customer/load type should be smaller than or equal to the maximum allowable divisible groups. At period 3, for instance, the limitation on control choices are found to be:

$$x_{i,1,2} + x_{i,1,3} + x_{i,2,1} + x_{i,2,2} + x_{i,2,3}$$
$$+ x_{i,3,1} + x_{i,3,2} + x_{i,3,3} \leq g_i \qquad \forall\, i$$

In short, at any period, for any customer/load type, if the box (dashed or undashed) corresponding to $x_{i,j,k}$ is extended to the observed period, $x_{i,j,k}$ is then considered as one of the elements that will confine the maximum controllable groups during the period.

According to (7.12) and (7.13), at any period, the difference between the energy deferred and paybacked should be less than the predetermined values $\overline{P}$ and $\underline{P}$. At period 3, for instance, (7.12) and (7.13) are expressed as follows:

$$\sum_{i=1}^{n} [\alpha_{i,1,2,1} \frac{(P_{i,1} + P_{i,2})}{g_i} x_{i,1,2} + \alpha_{i,2,1,1} \frac{P_{i,2}}{g_i} x_{i,2,1}$$
$$- \frac{P_{i,3}}{g_i} (x_{i,1,3} + x_{i,2,2} + x_{i,2,3} + x_{i,3,1} + x_{i,3,2} + x_{i,3,3})] \leq \overline{P}$$

and

$$\sum_{i=1}^{n} [-\alpha_{i,1,2,1} \frac{(P_{i,1} + P_{i,2})}{g_i} x_{i,1,2} - \alpha_{i,2,1,1} \frac{P_{i,2}}{g_i} x_{i,2,1}$$
$$+ \frac{P_{i,3}}{g_i} (x_{i,1,3} + x_{i,2,2} + x_{i,2,3} + x_{i,3,1} + x_{i,3,2} + x_{i,3,3})] \leq -\underline{P}$$

Iterative Procedure:

To obtain optimal integer solutions, the objective function and constraints have to be updated to reflect the change in the average cost and in available controlling load group. To achieve the mentioned task, an iterative process is employed.

To facilitate the explanation of the iterative process, the following representations are defined:

From (7.10):

$$\overline{A}(x_{i,j,k}) = x_{i,j,k}, \qquad \forall\, i, j, k \tag{7.14}$$

$$\overline{B}(x_{i,j,k}) = 0, \qquad \forall\, i, j, k \tag{7.15}$$

From (7.11):

$$\overline{C}(x_{i,j,k}) = \sum_{k=1}^{\overline{k_i}} x_{i,j,k} + \sum_{\substack{a=j-1 \\ a \geq 1}}^{1} \sum_{k=1}^{\overline{k_i}} x_{i,a,k} u(a+k+q(i,a,k)-j- \qquad \forall\, i, j \tag{7.16}$$

$$\overline{D}(x_{i,j,k}) = g_i \qquad \forall\, i, j \tag{7.17}$$

With these defined representation, an iterative process searches for the optimal solution as shown in Figure 7-5.

Evaluation:

Upon completion of iterative procedure where $x_{i,j,k}^{sol}$ and $pr_j^{sol}$ are found, the final step is needed to exclude the operating and maintenance cost, c, when evaluating the total savings of achieved from shifting the load.

### *7.2.5 Results*

To show how the algorithm achieves cost saving, a program written in MATLAB code is used to examine a system consisting of two-customer/load type for a 12-hour duration. Five percent of each customer/load type is assumed directly controllable. At individual hour, three control choices are available, i.e., $\overline{k_1} = \overline{k_2} = \overline{k} = 3$. Also, for simplicity, the payback pattern was assumed equal for both customer/load type at any hour. For k = 1, a 100% payback occurs at the next hour. For k = 2, a 100% payback occurs at the third hour. For k = 3, a 60% payback occurs at the fourth hour, and a 40% payback occurs at the fifth hour. For simplicity, the effect of penalty function is not examined.

Before any load shifting, the system cost is $ 201,760. Upon completion of the load shifting, the system cost is $199,860, while the operating and maintenance cost of DLC is $28.82. This is a total savings of $1871.18.

### *7.2.6 Summary*

A cost-based load management solution using LP is presented. This new approach includes the benefits of other approaches [Chen, 1995; Cohen, 1987, 1988; Hsu, 1991; Lee, 1984; Kurucz, 1996]. The algorithm relates the control devices to the DLC. Furthermore, the integer solutions achieved by the approach are more practical representation of the control system.

## 7.3 Profit-based Load Management Using LP

### *7.3.1 Introduction*

Despite various approaches [Chen, 1995; Cohen, 1987; 1988; Hsu, 1991; Lee, 1984; Kurucz, 1996] that have been successfully developed to minimize the cost of producing, both utilities and customers doubt the benefit of load management. From the utilities' perspective, customers participating in load management programs may cause a revenue loss. Also, customers have doubted load management may actually reduces their cost. Additionally, the introduction of time of use (t.o.u.) and time varying (t.v.) rate structure have created more controversies over the effectiveness of load management. For instance, despite cost savings, a load shifting from a high-cost, high-rate period to a low-cost, low-rate period may result in even more revenue loss. Finally, the upcoming deregulation in the electric power industry and the corresponding competition in the marketplace have made load management more controversial. The open market environment forces utilities to absorb the ultimate consequence of their decision making. Beside the need for a more efficient algorithm to optimize load management, an explanation on how load management may benefit both utilities and customers has yet to be formulated.

This work introduces the scheduling of profit-based load management, under a t.o.u. rate structure, to clarify how utilities may benefit from load management. Instead of reducing the overall system cost in a cost-based operation, a profit-based load management refers to the profit margin to determine how the load should be shifted.

Moreover, the developed algorithm includes various benefits of others [Chen, 1995; Cohen, 1987, 1988; Hsu, 1991; Lee, 1994; Kurucz, 1996]. First, rather than determining the amount of energy to be controlled during a certain period, which is a rather indirect approach, the new algorithm tries to decide the number of groups to be controlled during a certain period. Second, instead of a "two-phase" strategy [Kurucz, 1996], the developed algorithm is capable of including all possible control duration. Third, including the nonfixed maximum controllable load and payback pattern over time, the algorithm resolves both problems through determining groups to be controlled. Introducing paybacked energy that depends on the load during the controlled period resolves the varied payback pattern. Finally, by changing the constant parameters of the objective function and constraints from period to period, the LP algorithm considers a nonlinear cost function and obtains integer solutions that are optimal.

### *7.3.2 Nomenclature*

$Pt_j$: distribution load level at period j
$Pn_j$: uncontrollable distribution load level at period j
$Pc_j$: controllable distribution load level at period j
$pf_{i,j}$: profit experienced by the utilities at period j for selling energy to customer/load type, c/l, i
$P_{i,j}$: load available for DLC at period j by c/l i
$r_{i,j}$: rate charged on noncontrollable load for c/l i (\$/MW) at period j
$R_{i,j}$: rate charged on controllable load of c/l i (\$/MW) at period j
$pr_j(*)$: c/m at period j in \$/MW
$pm_{i,j}$: the profit margin of selling power to c/l i at period j (\$/MW)
$x_{i,j}$: number of groups of c/l i that are controllable at period j
$x_{i,j,k}$: number of groups of c/l i that are undergoing load shedding at period j of length k
n: maximum number of c/l
m: maximum period under study
$\bar{k}_i$: maximum length of period for load shedding for c/l i
$\alpha_{i,j,k,s}$: payback ratio at period (j+k+s-1) for load shedding at period j of length k on c/l i
q(i,j,k):the maximum payback period for load shedding at period j of length k on c/l i

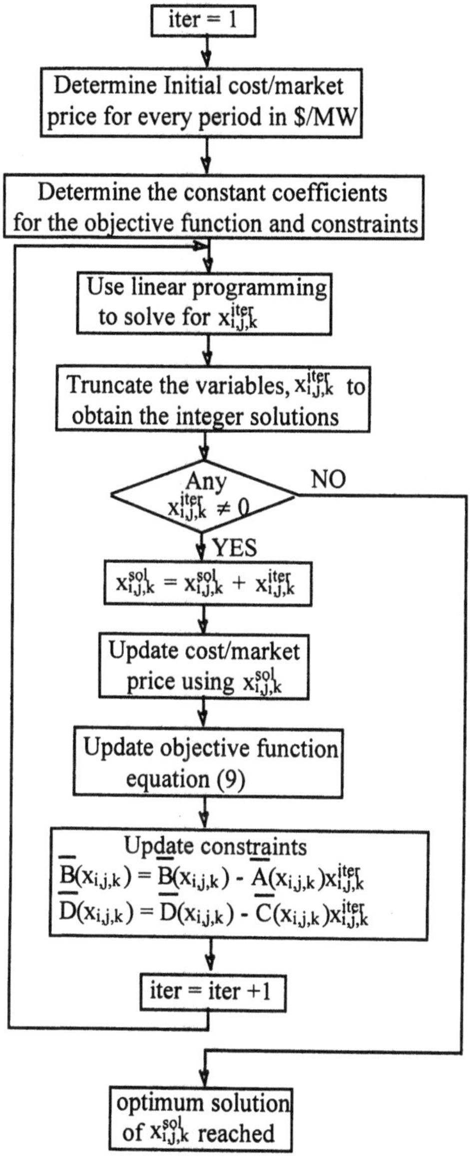


Figure 7-5. LP Iterative Procedure.

u(*): unit step function where u(*) = 1 if * ≥ 0,
and u(*) = 0 if * < 0

c: cost of control device in load shedding

$g_i$: maximum divisible groups of c/l i

$\overline{\Delta P}$: maximum load that can be increased at any period for the price to remain in the feasible range
$\underline{\Delta P}$: minimum load that can be decreased at any period for the price to remain in the feasible range
$incpft^{sol}$: optimal increased in profit experienced by the utilities after optimal scheduling is reached
$x^{sol}_{i,j,k}$: optimal scheduling of DLC

*7.3.3 Model*

Dividing the load forecast into two sections, the amount of power available for control from one to another period is obtained. Distinguishing the controllable load of one customer/load type (c/l) to another is then categorized for more detail. A lower t.o.u. rate structure, denoted $R_{i,j}$, (original t.o.u. rate structure is denoted by $r_{i,j}$) is offered to the customers for controllability over their load.

The profit experienced by the utilities at any period depends on the cost/market (c/m) price function, $pr_j(*)$, and the rate structure offered to the customers.

$$pf_{i,j} = [R_{i,j} - pr_j(Pt_{i,j})]Pc_{i,j} + [r_{i,j} - pr_j(Pt_{i,j})]Pn_{i,j}$$

where

$$Pc_{i,j} + Pn_{i,j} = Pt_{i,j} \tag{7.18}$$

A further restructuring of (7.18) reveals that the profit may change if the controllable load, $Pc_{i,j}$, is allowed to be shifted from a low-profit to a high-profit period.

$$\begin{aligned} pf_{i,j} &= R_{i,j}Pc_{i,j} + r_{i,j}Pn_{i,j} - pr_j(Pt_{i,j})[Pc_{i,j} + Pn_{i,j}] \\ &= R_{i,j}Pc_{i,j} + r_{i,j}Pn_{i,j} - Pc_{i,j}pr'_j(Pc_{i,j}) \\ &= [R_{i,j} - pr'_j(Pc_{i,j})]Pc_{i,j} + r_{i,j}Pn_{i,j} \end{aligned} \tag{7.19}$$

Divide the control choice of each c/m by $g_i$, at any period:

$$\begin{aligned} pf_{i,j} &= [R_{i,j} - pr'_j(\frac{x_{i,j}P_{i,j}}{g_i})]\frac{x_{i,j}P_{i,j}}{g_i} + r_{i,j}Pn_{i,j} \\ &= [R_{i,j} - pr''_j(x_{i,j})]\frac{x_{i,j}P_{i,j}}{g_i} + r_{i,j}Pn_{i,j} \end{aligned}$$

where

$$Pc_{i,j} = P_{i,j} \tag{7.20}$$

Then, redefine:

$$pr_j = pr''_j(x_{i,j}) \tag{7.21}$$

the c/m is transformed into a function of the number of controllable groups, $x_{i,j}$. Then, based on the new c/m function, the profit maximization problem has become a problem of finding the number of $x_{i,j}$, which will result in the optimal profit maximization. Furthermore, since $r_{i,j}Pn_{i,j}$ does not change for any period, the profit maximization problem solely depends on the shifting of $x_{i,j}$ of the controllable load, $Pc_{i,j}$.

Even though the c/m is nonlinear, it can be perceived as constant slope for a range of change in load levels, denoted by $(\overline{P} - P)$. Thus, the profit maximization of nonlinear c/m is reduced into a linearized problem.

### *7.3.4 Formulation*

Defining the profit margin for each c/l i, at any period, j, as

$$pm_{i,j} = R_{i,j} - pr_j \tag{7.22}$$

The decreased profit for deferring the energy is as follows:

$$DP = \sum_{i=1}^{n} \sum_{j=1}^{m} \sum_{k=1}^{\overline{k_i}} x_{i,j,k} \sum_{v=1}^{k} \left(\frac{P_{i,j+v-1}\, pm_{i,j+v-1}}{g_i} + c\right) \tag{7.23}$$

where c denotes the average fixed, operating, and maintaining cost of the control devices to switch off the customers' load during the period.

The increased profit for paying back the energy is as follows:

$$IP = \sum_{i=1}^{n} \sum_{j=1}^{m} \sum_{k=1}^{\overline{k_i}} \frac{x_{i,j,k}}{g_i} \sum_{v=1}^{k} P_{i,j+v-1} \sum_{s=1}^{q(i,j,k)} \alpha_{i,j,k,s}\, pm_{i,j+v+s-1} \tag{7.24}$$

Then, the objective of controlling the load is to increase the profit, which is

$$\text{maximize} \qquad IP - DP \tag{7.25}$$

subject to:

$$x_{i,j,k} \geq 0 \tag{7.26}$$

$$\sum_{k=1}^{\overline{k_i}} x_{i,j,k} + \sum_{\substack{a=j-1 \\ a \geq 1}}^{1} \sum_{k=1}^{\overline{k_i}} x_{i,a,k}\, u(a+k+q(i,a,k)-j- \leq g_i \qquad \forall\, i, j \tag{7.27}$$

where at any period, for any c/l i, the controlled (deferred or paybacked) groups can never exceed the preset maximum value, $g_i$.

Since the nonlinear objective is approximated by piecewise linear segments, extra constraints, (7.28) and (7.29), should be included to ensure that the change in load at any period will not exceed the maximum allowable increase or decrease. A cautious step should be noted, where, for any period, the largest deferrable load, denoted by $P_{i,j}/g_j$, or payback load, denoted by $\alpha_{i,j,s}\sum_{v=1}^{k} P_{i,j+v-1}$, should at least smaller than or equal to $\overline{\Delta P}$ or $\Delta\underline{P}$.

$$\sum_{i=1}^{n}\Big(\sum_{k=1}^{\overline{K_j}} \frac{x_{i,j,k}P_{i,j}}{g_i} + \sum_{\substack{a=j-1\\ a\geq 1}}^{1}\ \sum_{k=1}^{\overline{K_j}} \frac{x_{i,a,k}}{g_i}P_{i,a}u(a+k-j-$$
$$-\sum_{\substack{a=j-1\\ a\geq 1}}^{1}\ \sum_{k=1}^{\overline{K_j}} \frac{\alpha_{i,j,k,j-a-k+1}}{g_i}\sum_{v=1}^{k} P_{i,j+v-1}\Big) \leq \Delta\underline{P} \qquad \forall j$$

(7.28)

$$\sum_{i=1}^{n}\Big(-\sum_{k=1}^{\overline{K_j}} \frac{x_{i,j,k}P_{i,j}}{g_i} - \sum_{\substack{a=j-1\\ a\geq 1}}^{1}\ \sum_{k=1}^{\overline{K_j}} \frac{x_{i,a,k}}{g_i}P_{i,a}u(a+k-j-$$
$$+\sum_{\substack{a=j-1\\ a\geq 1}}^{1}\ \sum_{k=1}^{\overline{K_j}} \frac{\alpha_{i,j,k,j-a-k+1}}{g_i}\sum_{v=1}^{k} P_{i,j+v-1}\Big) \leq \overline{\Delta P} \qquad \forall j$$

(7.29)

Graphical Representation:

To visualize (7.23) through (7.29), a graphical representation is created in Figure 7-5. There are n pieces lying on top of each other to represent the individual customer/load type. For each individual piece, monitored m periods are listed from left to right. Listing from top to bottom of each piece are the load shifting choices and how they acted throughout the time frame. The boxes (undashed) on each piece represent the periods when the load is deferred while the dashed boxes represent the period of paybacked load.

In accordance with (7.6), the DP coefficient of each control choice, $x_{i,j,k}$, is represented by the undashed boxes. On the other hand, the IP coefficient, from (7.7), of $x_{i,j,k}$, is represented by the dashed boxes. The notation, $\alpha_{i,j,k,s}$, determines how the total deferred energy is paid back. It may be retrieved from statistical measurement of past data and present control load pattern.

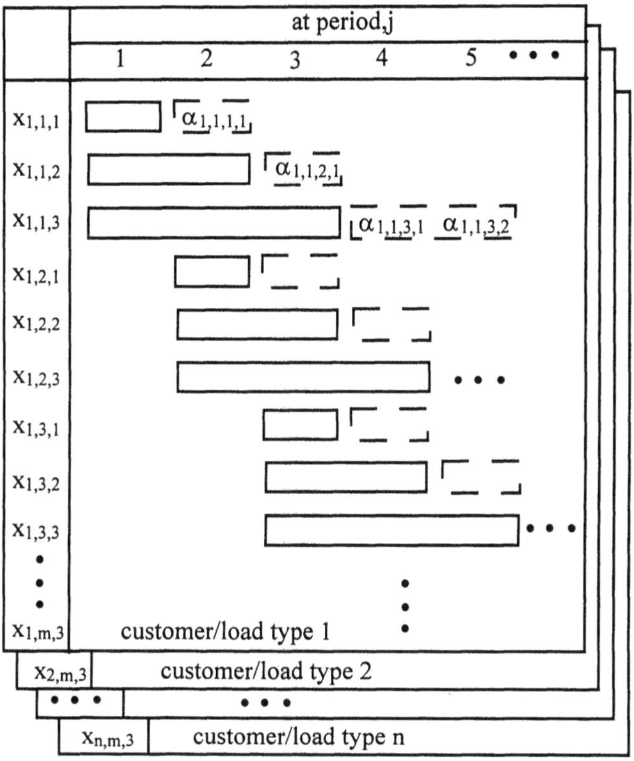


Figure 7-5. Graphical Representation of Control Pattern.

By (7.25), each control choice, $x_{i,j,k}$, has to be greater or equal to zero, i.e., $x_{1,1,1} \geq 0$, $x_{1,1,2} \geq 0$, $x_{1,1,3} \geq 0$ and etc. From (7.27), the maximum control choices of any period, $x_{i,j,k}$, of individual customer/load type should be smaller than or equal to the maximum allowable divisible groups. At any period, for any c/l, if the box (dashed or undashed) corresponding to $x_{i,j,k}$ is extended to the observed period, $x_{i,j,k}$ is then considered as one of the elements that will confine the maximum controllable groups during the period.

According to (7.28), at any period, the difference between the energy deferred and paybacked should be less than the predetermined values $\overline{P}$ and $\underline{P}$. For any period j, adding up the power deferred and paybacked by all c/l, the predetermined $\overline{P}$ and $\underline{P}$ should not be violated.

Iterative Procedure:

To obtain optimal, integer solutions, the objective function and constraints have to be updated to reflect the change in the average cost and in available controlling load group. To achieve the mentioned task, an iterative process is employed.

To facilitate the explanation of the iterative process, the following representations are defined as from (7.26):

$$\overline{A}(x_{i,j,k}) = x_{i,j,k}, \qquad \forall\, i, j, k \tag{7.30}$$

$$\overline{B}(x_{i,j,k}) = 0, \qquad \forall\, i, j, k \tag{7.31}$$

From (7.26):

$$\overline{C}(x_{i,j,k}) = \sum_{k=1}^{\overline{K_j}} x_{i,j,k} + \sum_{\substack{a=j-1 \\ a \geq 1}}^{1} \sum_{k=1}^{\overline{K_j}} x_{i,a,k}\, u(a+k+q(i,a,k)-j- \qquad \forall\, i, j \tag{7.32}$$

$$\overline{D}(x_{i,j,k}) = g_i \qquad \forall\, i, j \tag{7.33}$$

With these defined representation, an iterative process searches for the optimal solution as shown in Figure 7.7.

Evaluation:

Before load shifting, the revenue loss experienced by utilities, (equivalent to cost reduction experienced by customers) for controllability over the load are as follows:

$$RL_{util}^{pre\text{-}DLC} = \sum_{i=1}^{n} \sum_{j=1}^{m} (r_{i,j} - R_{i,j})P_{i,j} \tag{7.34}$$

During the observed duration, as long as the increased in profit experienced from the optimal scheduling, $incpft^{sol}$, is greater than the reduced revenue experienced from lower rate structure, the utilities will be benefited from load management.

### *7.3.5 Results*

Two cases, flat rate and t.o.u. rate structure, are used to study the difference between cost-based and profit-based DLC. For detailed construction of cost-based DLC, please refer to Section 7.3.7.

Case 1: Flat Rate Structure

Under the flat rate structure, both profit-based DLC and cost-based DLC operate in the same way. The resulting scheduling pattern, cost savings, and increased profit from both approaches are exactly the same. Table 7-1 summarizes the resulting cost savings and increased profit of both approaches.

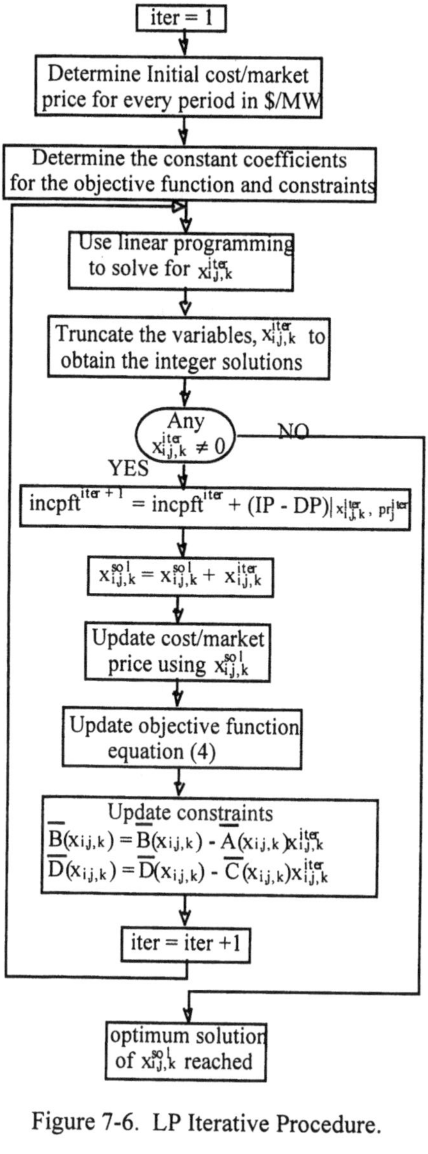


Figure 7-6. LP Iterative Procedure.

Case 2: t.o.u. Rate Structure

Under a t.o.u. rate structure, cost-based and profit-based DLC result in different scheduling. Profit-based DLC tries to increase the profit by scheduling the load from a low-profit to a high-profit period. However, cost-based DLC tries to reduce the system operating cost by shifting the load from a high-cost to a low-cost period. Thus, there is the possibility that, under cost-based DLC, the load be shifted from a high-profit to a low-profit period, to achieve the mentioned task. As may be seen from Table 7-2, despite cost-based DLC results in bigger cost savings, \$850.82, the increased profit from the approach is relatively smaller than what is achieved by profit-based DLC.

Table 7-1. Results Under Flat Rate Structure.

| | | Cost-based Approach | Profit-based Approach |
|---|---|---|---|
| Without DLC Program | Revenue | \$ 309300.00 | \$ 309300.00 |
| | Cost | \$ 183317.17 | \$ 183317.17 |
| With DLC Program | Pre-S revenue | \$ 308887.60 | \$ 308887.60 |
| | Post-S revenue | \$ 308887.60 | \$ 308887.60 |
| | Post-S cost | \$ 182466.35 | \$ 182466.35 |
| | DLC O&M cost | \$ 74.50 | \$ 74.50 |
| | Cost saving | \$ 850.82 | \$ 850.82 |
| | Increased profit | \$ 363.02 | \$ 363.02 |

note: pre-S - pre-scheduling post-S - post-scheduling
O & M - fixed, operating and maintenance cost

Table 7-2. Results Under t.o.u. Rate Structure.

| | | Cost-based Approach | Profit-based Approach |
|---|---|---|---|
| Without DLC Program | Revenue | \$ 309300.00 | \$ 309300.00 |
| | Cost | \$ 183317.17 | \$ 183317.17 |
| With DLC Program | Pre-S revenue | \$ 308935.95 | \$ 308935.95 |
| | Post-S revenue | \$ 308856.80 | \$ 308910.54 |
| | Post-S cost | \$ 182466.35 | \$ 182499.59 |
| | DLC O&M cost | \$ 75.40 | \$ 79.40 |
| | Cost saving | \$ 850.82 | \$ 787.82 |
| | Increased profit | \$ 332.22 | \$ 398.35 |

Note: pre-S - pre-scheduling post-S - post-scheduling
O & M - fixed, operating and maintenance cost

### *7.3.6 Summary*

A profit-based load management scheduling program using LP is presented. This new approach prepares the utilities to act more decisively in the upcoming deregulated environment. A comparison with the conventional cost-based approach clearly shows that profit-based DLC achieves a better solution when the rate structure varies each time. Furthermore, the algorithm schedules each DLC control devices. The integer solutions achieved by the approach are the practical representation of the control system.

### *7.3.7 Comparison*

A cost-based DLC using the same approach as depicted in Section 7.3.4 will have the same constraints as for the profit-based DLC. However, the objective function will be changed as follows:

$$\text{Minimize } - DF + PB + PT \tag{7.35}$$

where:

DF reflects the reduced cost of deferring the energy:

$$DF = \sum_{i=1}^{n} \sum_{j=1}^{m} \sum_{k=1}^{k_i} x_{i,j,k} \sum_{v=1}^{k} \left( \frac{P_{i,j+v-1}\, pr_{j+v-1}}{g_i} - c \right) \tag{7.36}$$

PB reflects the increased in cost for paying back the energy:

$$PB = \sum_{i=1}^{n} \sum_{j=1}^{m} \sum_{k=1}^{k_i} \frac{x_{i,j,k}}{g_i} \sum_{v=1}^{k} P_{i,j+v-1} \sum_{s=1}^{q(i,j,k)} \alpha_{i,j,k,s}\, Pr_{j+v+s-1} \tag{7.37}$$

Table 7-3. Load Data and Rate Structure.

| eriod, j | $P_{1,j}$, MW | $P_{2,j}$, MW | $Pt_j$, MW | $_{1,j} = R_{2,j}$ case 1 | $_{1,j} = R_{2,j}$ case 2 |
|---|---|---|---|---|---|
| 1 | 24 | 18 | 840 | $ 29.30/MW | $ 29.00/MW |
| 2 | 18 | 15 | 660 | $ 29.30/MW | $29.00/MW |
| 3 | 20 | 22.5 | 850 | $ 29.30/MW | $29.00/MW |
| 4 | 26 | 18 | 880 | $ 29.30/MW | $29.80/MW |
| 5 | 38 | 22.5 | 1210 | $ 29.30/MW | $29.80/MW |
| 6 | 34 | 21 | 1100 | $ 29.30/MW | $29.80/MW |
| 7 | 30 | 13.5 | 870 | $ 29.30/MW | $29.80/MW |
| 8 | 24 | 18 | 840 | $ 29.30/MW | $ 29.00/MW |
| 9 | 24 | 16.5 | 810 | $ 29.30/MW | $29.00/MW |
| 10 | 20 | 15 | 700 | $ 29.30/MW | $29.00/MW |
| 11 | 18 | 19.5 | 750 | $ 29.30/MW | $29.00/MW |
| 12 | 19 | 21 | 800 | $ 29.30/MW | $29.00/MW |

PT reflects the revenue loss when payback ratio of the control choice, $x_{i,j,k}$, does not sum up to 1:

$$PT = \sum_{i=1}^{n} R_i \sum_{j=1}^{m} \sum_{k=1}^{k_i} \frac{x_{i,j,k}}{g_i} \left( \sum_{v=1}^{k} P_{i,j+v-1} \right) \left( 1 - \sum_{s=1}^{q(i,j,k)} \alpha_{i,j,k,s} \right) \tag{7.38}$$

If the summation of payback ratio of any c/l at any period j, $\sum_{s=1}^{q(i,j,k)} \alpha_{i,j,k,s}$ is greater than 1, the revenue increases.

## 7.4 Battery Stored Energy Control for Price-based Operation

### *7.4.1 Introduction*

The energy storage system (ESS) is one of the load management strategies introduced in the 1970s. It allows both utilities and customers to store and consume electric energy during scheduled periods. Within energy storage system, BESS has seen significant technological advancement during the past few years. This work examines a price-based, utility-owned BESS using LP.

Table 7.4. Miscellaneous Data.

| Parameters | Values |
|---|---|
| N | 2 |
| M | 12 |
| $\overline{k}_i$ | 3 $\forall$ i |
| q(i,j,k) | 1 for q(i,j,1) $\forall$ i, j<br>1 for q(i,j,2) $\forall$ i, j<br>2 for q(i,j,3) $\forall$ i, j |
| $\alpha_{i,j,k,s}$ | $\alpha_{i,j,1,1} = 1 \quad \forall\ i, j \qquad \alpha_{i,j,2,1} = 1 \quad \forall\ i, j$<br>$_{i,j,3,1} = 0.6 \quad \forall\ i, j \qquad \alpha_{i,j,3,2} = 0.4 \quad \forall\ i, j$ |
| $g_i$ | 80 $\forall$ i |
| C | \$0.20/period/device |
| $pr_j$ | $7 + 0.012\ Pt_j + 0.0000002\ Pt_j^2 \quad \forall\ j$ |
| $r_{i,j}$ | \$30.00/MW $\forall$ i, j, cases |

There are various ESSs existing in the literature and practice. Investigated algorithms for ESSs are commonly found on pumped hydroelectric storage [Aoki, 1987; Guan, 1994; Jeng, 1996; Lee, 1992; Lidgate, 1984; Prasannanm, 1996; Rakic, 1994; Walsh, 1997; Wood, 1996], batteries [Alt, 1996; Maly, 1995; Lee, 1992, 1993, 1994], and cool storage [Baughman, 1993; Chen, 1993; Rupanagunta, 1995]. Pumped storage is the ESS that receives most attention. Different algorithms were developed to intergrate its operation with hydrothermal system [Aoki, 1987; Guan, 1994; Jeng, 1996; Lidgate, 1984; Prasannanm, 1996; Rakic, 1994; Walsh, 1997; Wood, 1996] or batteries [Lee, 1992]. Reported algorithms include dynamic programming [Rakic, 1994; Lee, 1992], gradient method [Wood, 1996], LaGrangian relaxation [Aoki, 1987; Guan, 1994; Prasannanm, 1996], linear programming [Jeng, 1996; Lidgate, 1984], etc. Within the ESS field, battery energy storage system (BESS) has seen significant technological advancement during the past few years. Algorithms for scheduling BESS are mostly dynamic programming [Lee, 1992, 1993, 1994; Maly, 1995] and linear programming [Daryanian, 1989; Ng, 1997]. Coordinately of battery operation with others are rare, with the exception of [Lee, 1992], coordinating batteries with pumped storage, and [Ng, 1997], coordinating batteries with DLC. Research work conducted on cool storage increases each year. Reported scheduling algorithms, however, are few and solved mainly using nonlinear programming technique.

This work introduces price-based BESS using LP. The developed algorithm decides the number of cells to be charged/discharged at each period. Full charge/discharge is assumed, *partial* charge/discharge is prohibited because it is not expected to occur. This technique is particularly useful for ESS composing of large numbers of small capacity units that operate independent of each other (e.g., batteries). Finally, the price-based algorithm tries to increase utility profit rather than reducing system production cost.

### 7.4.2 Nomenclature

$y_{i,j,k}$: variable that determines the number of cells type $i$ to be fully charged for $k$ periods beginning at period $j$

$z_{i,j,k}$: variable that determines the number of cells type $i$ to be fully discharged for $k$ periods beginning at period $j$

$\Delta P_j^+$: variable that determines the increase in load at period $j$

$\Delta P_j^-$: variable that determines the decrease in load at period $j$

$pr_j^+$: per MW average marginal cost/market price at period $j$

$pr_j^-$: per MW average marginal cost/market price at period $j$

$Ps_{i,k,v}$: required storage energy for cell type $i$ at $v$ charge stage for a $k$ periods of charging duration

$P_{i,r,w}$: required discharge energy for cell type $i$ at $w$ discharge stage for a $r$ periods of discharging duration

$\underline{k}_i$: minimum charge duration for cell type $i$

$\overline{k}_i$: maximum charge duration for cell type $i$

$\underline{r}_i$: minimum discharge duration for cell type $i$

$\overline{r}_i$: maximum discharge duration for cell type $i$

$n$: maximum number of cell types

$m$: maximum period under study

$u(*)$: unit step function, where

$u(*) = 1$ if $* \geq 0$

$u(*) = 0$ if $* < 0$

Additional parameters are explained when introduced.

### 7.4.3 Model

Utilities usually adopt a *market-based* pricing mechanism, i.e., customers are categorized and charged differently. When energy is regenerated, there is no certainty of which customer will use the energy. In fact, the regenerated energy will be shared by the system load during the discharge period. Since a reference to the rate structure is needed when a price-based approach is discussed, an averaged rate structure, $\overline{R}_j$, is introduced. It can be derived from (7.39). $P_{i,j}$ represents the load of customer $i$ at period $j$. $R_{i,j}$ represents the rate charged on customer $I$iat period $j$.

$$\overline{R}_j\left(\sum_{i=1}^{n} P_{i,j}\right) = \sum_{i=1}^{n} P_{i,j} R_{i,j} \tag{7-39}$$

Depending on the charge/discharge characteristics, units within ESS are categorized into different groups. The following assumptions were made in modeling ESS I:

- At each period, load dispatcher determines (1) the number of units to be used for charge/discharge, and (2) the duration of each ESS unit should be charged/released.

- *Full* charge/discharge of units is assumed. *Partial* charge/discharge can be achieved by assuming the next charge/discharge state as a new set of variables.

### *7.4.4 Formulation*

During the charge and discharge phases, the energy required during each period is fixed by the current and the length of charging. Denoting $Ps_{i,k,v}$ to be the required storage energy, $i$ represents the cell type, $k$ represents the total length of charging, and $v$ represents the current charge stage. Similarly, $Pu_{i,r,w}$ is the required discharge energy, $i$ represents the cell type, $r$ represents the total length of discharging, and $w$ represents the current discharge stage. Detailed information on how to determine $Ps_{i,k,v}$ and $Pu_{i,r,w}$ could be found in [1-5].

The profit for charging the energy at earlier periods, BDP, is shown in (7.40):

$$BDP = -\sum_{i=1}^{n}\sum_{j=1}^{n}\sum_{k=\underline{k}_i}^{\overline{k}_i}\left(y_{i,j,k}\sum_{v=1}^{k} Ps_{i,k,v}\overline{R}_{i+v-1}\right) - pr_j^{+}\Delta P_j^{+} \tag{7.40}$$

where the first term of the right side of (7.40) represents the revenue of stored energy and the second term represents the cost of increasing the load. For each control choice $y_{i,j,k}$, $i$ denotes the differences in group characteristics, $j$ denotes the period when the full charge phase begins, and $k$ denotes the number of periods needed for a full charge.

The profit for discharging the energy at earlier periods, BIP, is shown in (7.41):

$$BIP = \sum_{i=1}^{n}\sum_{j=1}^{m}\sum_{r=\underline{r}_i}^{\overline{r}_i}\left(z_{i,j,r}\sum_{w=1}^{r} Pu_{i,r,w}\overline{R}_{j+w-1}\right) + pr_j^{-}\Delta P_j^{-} \tag{7.41}$$

where the first term of the right side of (7.41) represents the revenue of charged energy and the second term represents the cost of decreasing the load. For each control choice $z_{i,j,k}$, $i$ denotes the differences in group characteristics, $j$ denotes the period when the full discharge phase begins, and $k$ denotes the number of periods needed for a full discharge.

Then, the objective of BESS is to store the energy at a high-profit period and discharge it at low-profit period for the utility:

$$\max \quad BDP + BIP \tag{7.42}$$

subject to:

$$y_{i,j,k} \geq 0 \quad \forall i,j,k \tag{7.43}$$

which requires non-negativity of the number of charging choices, $y_{i,j,k}$,

$$z_{i,j,r} \geq 0 \qquad \forall i,j,r \tag{7.44}$$

which requires non-negativity of the number of discharging choices, $z_{i,j,r} \geq 0$,

$$\sum_{a=1}^{j}\sum_{k=\underline{k}_i}^{\overline{k}_i} y_{i,a,k} - \sum_{a=1}^{j}\sum_{r=\underline{r}_i}^{\overline{r}_i}\left[z_{i,a,r}u(j+1-a-r) - z_{i,a,r}u(a+r-j-1)\right] \leq G_i \quad \forall i,j \tag{7.45}$$

which indicated that at any period, the total number of charged cells, $y_{i,j,k}$, and discharged cells, $z_{i,j,k}$, is limited by the available number of cells, $G_i$; the unit step functions act as ON/OFF switches to determine if the battery cells under control choice, $z_{i,a,r}$, are at fully or partially discharge stage,

$$\sum_{a=1}^{j}\left[\sum_{k=\underline{k}_i}^{\overline{k}_i} y_{i,a,k}u(j-a-k) - \sum_{r=\underline{r}_i}^{\overline{r}_i} z_{i,a,r}\right] \geq 0 \qquad \forall i,j \tag{7.46}$$

which indicated that at any period, the total number of dischargeable cells should not be more than the total number of available fully charged cells. The unit step functions act as ON/OFF switches to decide if the cells under control choice $y_{i,a,k}$ are fully charged at period $j$.

Also, (7.47) is included to relate the number of charging cells, $y_{i,j,k}$, and discharging cells, $z_{i,j,k}$, with the increase in energy, $\Delta P_j^+$, or decrease in energy, $\Delta P_j^-$, at period $j$:

$$\sum_{i=1}^{n}\sum_{a=j}^{a}\sum_{k=\underline{k}_i}^{\overline{k}_i} y_{i,a,k}Ps_{i,k,j-a+1}u(j-a-k) - \sum_{i=1}^{n}\sum_{a=j}^{a}\sum_{r=\underline{r}_i}^{\overline{r}_i} z_{i,a,r}Pu_{i,a,j-a+1}u(j-a-k) = \Delta P_j^+ - \Delta P_j^- \quad \forall j \tag{7.47}$$

where all energy charged/discharged by all control choices, $y_{i,j,k}$ and $z_{i,j,k}$, at period $j$ should be equivalent to $\Delta P_j^+$ or $\Delta P_j^-$; the unit step functions decide if the control choices, $y_{i,a,r}$ and $z_{i,a,r}$, have extended to period $j$.

*Graphical Representation:*

To visualize (7.45), (7.46), and (7.49), a graphical representation is presented in Figure 7.7. The $n$ pieces lying on top of each other represent cells with different charge/discharge characteristics. The charging phase, with variables $y_{i,j,k}$, is shown on the left. The discharging phase, with variables $z_{i,j,k}$, is shown on the right. Governing equations connect the charge and discharge phases for differently grouped cells.

Equations (7.45) and (7.46) are *individualized* constraints while (7.47) is a *group* constraint. According to (7.45), the maximum number of charge and discharge cells of any group is restricted by the existing cells, $G_i$. Equation (7.46) further restricts the available discharge cells to available fully charged cells. Finally, (7.47) is included to relate the total increase/decrease in energy at each period to the number of charge and discharge cells, $y_{i,j,k}$ and $z_{i,j,k}$.

*Iterative Procedure:*

Even though the average marginal cost/market price function is nonlinear, it can be perceived as a constant slope for a range of change in load levels, as shown in Figure 7.8. Thus, the nonlinear average marginal cost/market price function is reduced into a piecewise linear function. To obtain the final optimal solution, an iterative procedure is employed.

Defining $\Delta P_j^{iter}$ to be the increased load at period $j$, then, the relationship between the value of $i$-th iteration increased load to previous iteration increased load is described in (7.48):

$$\Delta P_j^{iter} = \Delta P_j^{iter-1} + \Delta P_j^+ - \Delta P_j^- \qquad \forall j \qquad (7.48)$$

Define one set of variables, $Q_j$, to represent the integer part of $\Delta P_j^{iter}$ divided by $\overline{\Delta P}$ ($\overline{\Delta P}$ is the maximum decreaseable or increaseable load for period $j$ at each iteration,)

$$Q_j = quotient\left(\frac{\Delta P_j^{iter}}{\overline{\Delta P}}\right) \qquad \forall j \qquad (7.49)$$

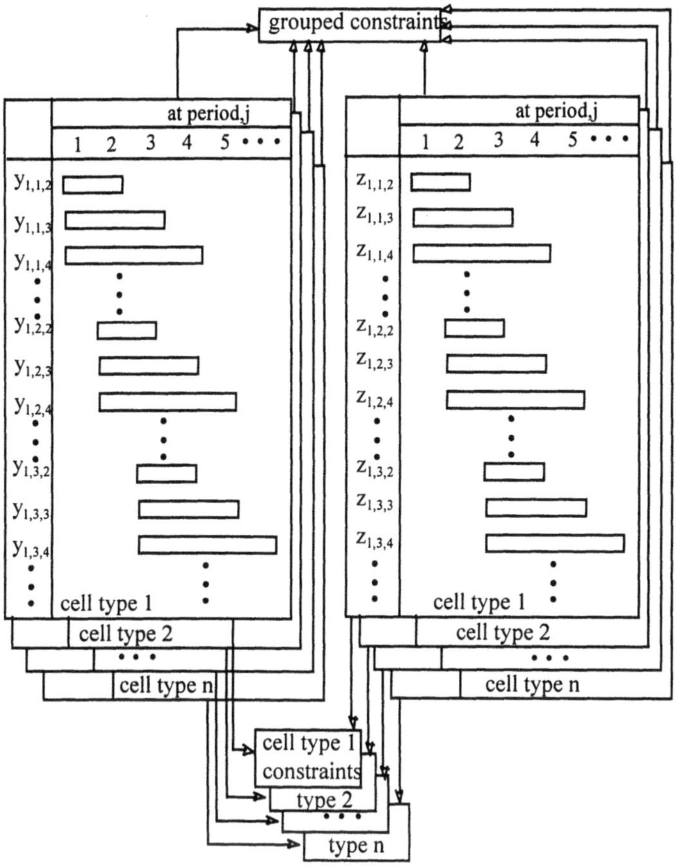


Fig. 7.7. BESS Graphical Representation.

and another set of variables, $r_j$, to represent the remainder of $\Delta P_j^{iter}$ divided by $\overline{\Delta P}$,

$$r_j = remainder\left(\frac{\Delta P_j^{iter}}{\overline{\Delta P}}\right) \tag{7.50}$$

With known $Q_j$, the average linearized average marginal cost/market price for $Q_j$-th increase in load, $pr_j(Q_j)$, is shown in (7.51):

$$pr_j(Q_j) = \frac{S_j\left(P_j + (Q_j + 1)\overline{\Delta P}\right) - S_j\left(P_j + Q_j\overline{\Delta P}\right)}{\overline{\Delta P}} \quad \forall j \tag{7.51}$$

where $S_j(*)$ is the system production cost at * load level; $P_j$ is the initial system load level at period $j$.

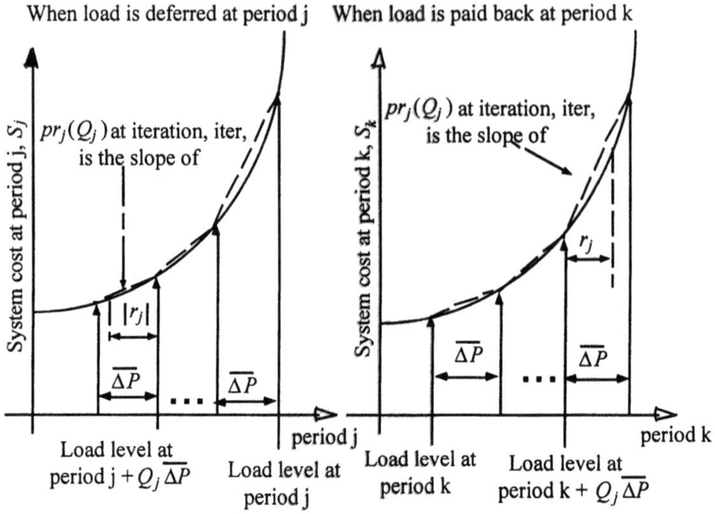


Figure 7-8. Piecewise Linearization of System Cost Function.

To conduct iterative procedure, (7.47) needs to be modified as in (7.52):

$$\sum_{i=1}^{n}\sum_{a=j}^{a}\sum_{k=\underline{k}_i}^{\bar{k}_i} y_{i,a,k} Ps_{i,k,j-a+1} u(j-a-k)$$
$$-\sum_{i=1}^{n}\sum_{a=j}^{a}\sum_{r=\underline{r}_i}^{\bar{r}_i} z_{i,a,r} Pu_{i,a,j-a+1} u(j-a-k) = \Delta P_j^{iter} + \Delta P_j^{+} - \Delta P_j^{-}$$
$$\forall j \tag{7.52}$$

which is formed by adding $\Delta P_j^{iter}$ at the end of (7.47)

With the defined parameters, a six-step iterative procedure shown below could then be carried out to achieve the final optimal objective function.

**STEP 1**
Initializing all variables, where

iter = 1

$\Delta P_j^0 = 0$

**STEP 2**
Determine the values of $Q_j$ and $r_j$ as defined in (7.49) and (7.50).

**STEP 3**
Finding/updating values for $pr_j^+$ and $pr_j^-$ :

*if* $r_j = 0$

$pr_j^+ = pr_j(Q_j)$

$$pr_j^- = pr_j(Q_j - 1)$$
$$0 \le \Delta P_j^+ \le \overline{\Delta P}$$
$$0 \le \Delta P_j^+ \le \overline{\Delta P}$$

*else if* $r_j > 0$

$$pr_j^+ = pr_j^- = pr_j(Q_j)$$
$$0 \le \Delta P_j^+ \le \overline{\Delta P} - r_j$$
$$0 \le \Delta P_j^- \le r_j$$

*else*

$$pr_j^+ = pr_j^- = pr_j(Q_j - 1)$$
$$0 \le \Delta P_j^+ \le -r_j$$
$$0 \le \Delta P_j^- \le \overline{\Delta P} + r_j$$

**STEP 4**

Solve for linear program as described in (7.40) through (7.46), and (7.52), in addition to the constraints set in Step 3.

**STEP 5**

Updating increased and decreased load as defined in (7.48).

**STEP 6**

Checking for optimality:

*if any* $\Delta P_j^+$ *or* $\Delta P_j^-$ *value is greater than 0*

*iter = iter +1*

*GO TO STEP 1*

*else*

*Optimal solutions are found*

### *7.4.5 Results*

Successful clarification of the concept presented in Section 7.4.3 through 7.4.7 is accomplished by an example consisting of two ESS cell types and six periods (in this case, 1 period = 1 hour) test system. $\overline{\Delta P}$ was set to 50 kW for all periods. However, it should be noted that $\overline{\Delta P}$ can be set differently for each period under consideration. Additional details of the model and relevant data used are listed in the Section 7.4.6.

Using the developed algorithm to test the system, Table 7-6 shows the resulting dispatching sequences and Table 7-7 shows all relevant results.

Table 7-6. Dispatching Sequences for Price-based BESS.

| Control Choice | Battery type 1 | | Battery type 2 | |
|---|---|---|---|---|
| | Chargi ng | Dischargi ng | Chargin g | Dischargi ng |
| i,1,2 | 30 | 0 | 15 | 0 |
| i,3,1 | 0 | 10 | 0 | 5 |
| i,4,1 | 0 | 20 | 0 | 10 |

The discrepancy between increased utility profit and system cost savings, \$46.91, is due to the energy being discharged at higher rate periods, periods 3 and 4. However, when energy is stored, it is stored within the equipment of the utilities and when stored energy is discharged, customers are the consumers. Therefore, the customers are not hurt by paying a higher bill. In fact, customers are paying the same amount of money for the electric usage. Using a price-based algorithm, utilities base on the profit margin of observed duration to decide the dispatching sequences of the battery cells.

Also, the results show that maximum energy efficiency, 90%, may not be desirable when system conditions are considered. The actual efficiency level achieved by the batteries is 86.49% and 86.59%, respectively.

Table 7-7. Results from Price-based BESS.

| Description | Results |
|---|---|
| Energy efficiency of battery | |
| Type 1 | 86.49% |
| Type 2 | 86.59% |
| Increased utilities' profit | \$123.03 |
| Increased system cost savings | \$76.12 |

Table 7-8 shows the system load before and after dispatching is conducted.

Table 7-8. Change in System Load.

| Period, | System load, kW | |
|---|---|---|
| Hour | Before scheduling | After scheduling |
| 1 | 7300 | 7472.5 |
| 2 | 7600 | 7792.5 |
| 3 | 12000 | 11900.5 |
| 4 | 12900 | 12701 |
| 5 | 7900 | 7900 |
| 6 | 8600 | 8600 |

The sum of system load after scheduling is 46.50 kW higher than the sum of system load before scheduling. The difference is the storage losses during energy conversion.

Table 7-9 shows the average marginal cost/market price before and after the dispatching is conducted. Table 7-10 shows the profit margin of utilities for additional increased load before and after the dispatch is conducted.

From Table 7-9, the cost differences between periods 1 and 2 and periods 3 and 4 are large, as high as 27.625 cents/kW (34.02 cents/kW - 6.395 cents/kW). However, from Table 7-10, the profit margins between periods 1 and 2 and periods 3 and 4 are smaller, with the largest difference of 7.625 cents/kW (21.205 cents/kW - 13.58 cents/kW.) The possible increased profit margin of 7.625 cents/kW is not large enough to compensate for the lost revenue of storage losses. Therefore, five battery cells of both cell types are not excited for load shifting.

Table 7-9. Change in Average Marginal Cost/market Price.

| Periods, | Average marginal cost/market price, cents/kW | |
|---|---|---|
| Hour | Before dispatching | After dispatching |
| 1 | 6.395 | 6.485 |
| 2 | 6.575 | 6.665 |
| 3 | 31.86 | 31.62 |
| 4 | 34.02 | 33.54 |
| 5 | 6.755 | 6.755 |
| 6 | 7.175 | 7.175 |

Table 7-10. Change in Profit Margin for Increased Load.

| Periods, | Profit margin for increased load, cents/kW | |
|---|---|---|
| Hour | Before dispatching | After dispatching |
| 1 | 21.205 | 21.115 |
| 2 | 21.055 | 20.965 |
| 3 | 15.85 | 16.09 |
| 4 | 13.58 | 14.06 |
| 5 | 20.775 | 20.775 |
| 6 | 20.265 | 20.265 |

### *7.4.6 Summary*

A BESS dispatching algorithm was successfully formulated for a price-based operation. Significant improvements have been made over previous work to include storage losses and full charge/discharge characteristics of batteries. Since the algorithm decides the number of battery cells to be charged or discharged during observed periods, physical feel of actual system is enhanced. Besides providing a physical feel to actual systems developed in LP, the BESS algorithm considers nonlinear cost/market price function through successive approximation. Finally, results show that maximum energy efficiency of BESS may not be desirable when system cost is considered.

Table 7-11. Initial System Load, $P_j$.

| Period, $j$ | Load level, kW | | |
|---|---|---|---|
| Hour | $P_j$ | $P_{1,j}$ | $P_{2,j}$ |
| 1 | 7300 | 3500 | 3800 |
| 2 | 7600 | 3600 | 4000 |
| 3 | 12000 | 5500 | 6500 |
| 4 | 12900 | 6200 | 6700 |
| 5 | 7900 | 3900 | 4000 |
| 6 | 8600 | 4400 | 4200 |

Table 7-12. Rate Structure.

| Period, $j$ Hour | Rate, cents/kW | | |
|---|---|---|---|
| | $\bar{R}_j$ | $R_{1,j}$ | $R_{2,j}$ |
| 1 | 27.60 | 25 | 30 |
| 2 | 27.63 | 25 | 30 |
| 3 | 48.79 | 45 | 50 |
| 4 | 47.60 | 45 | 50 |
| 5 | 27.53 | 25 | 30 |
| 6 | 27.44 | 25 | 30 |

Table 7-13. System Cost Function, $S_j$.

| Period, $j$ | System cost function, $S_j$ |
|---|---|
| 1,2,5,6 | $0.02\,P_j + 0.000003\,P_j^2$ |
| 3,4 | $0.03\,P_j + 0.000012\,P_j^2$ |

Table 7-14. Data Pertinent to BESS.

| Parameter | Description |
|---|---|
| $G_1$ | 35 |
| $G_2$ | 20 |
| Maximum energy efficiency of battery | |
| type 1 | 90.00% |
| type 2 | 90.00% |
| Minimum energy efficiency of battery | |
| type 1 | 84.21% |
| type 2 | 84.52% |

Table 7-15. Charging and Discharging Characteristic of BESS.

| Energy required during each period | Duration of charging/discharging Period 1 | 2 | 3 |
|---|---|---|---|
| Battery type 1, kW | | | |
| Charging | 7.60 | 3.70 | 2.40 |
| Discharging | 6.40 | 3.22 | 2.16 |
| Battery type 2, kW | | | |
| Charging | 8.40 | 4.10 | 2.70 |
| Discharging | 7.10 | 3.60 | 2.43 |

# Bibliography


[Abel, 1992] A. Abel, "Public Utility Regulatory Policies Act of 1978: A Fact Sheet," *Congressional Research Service*, The Library of Congress, July 30, 1992.

[Ahuja, 1993] R. K. Ahuja, T. L. Magnanti, and J. B. Orlin. *Network Flows: Theory, Algorithms and Applications,* Prentice Hall, Englewood Cliffs, NJ 1993.

[Alsac, 1990] O. Alsac, J. Bright, M. Paris, and B. Stott, "Further Developments in LP-based Optimal Power Flow," *IEEE Transactions on Power Systems,* Vol. PWRS-5, no. 3, pp. 697-711, August 1990.

[Alt, 1996] J. T. Alt, M. D. Anderson and R. G. Jungst, "Assessment of Utility Side Cost Savings from Battery Storage-type Customers," 96 SM 474-7 PWRS, *IEEE Summer Meeting,* 1996.

[Alvarado, 1991] F. Alvarado, Y. Hu, D. Ray, R. Stevension and E Cashmam, "Engineering Foundation for the Determination of Security Costs," *IEEE Transactuions on PWS*, pp.1175-1182, August 1991.

[Alvarado, 1995] F. Alvarado, M. Baughman, A. Bose, A. Breipohl, G. Gross, M. Ilic, G. Sheblé, B. Wollenberg, and F. Wu, "Opinions on Ancillary Services." Presented to the Federal Energy Regulatory Commission, Washington D. C., August 1995.

[Alvarado, 1998] F. L. Alvarado, "Market power: a dynamic definition," International Symposium on Bulk Power Systems Dynamics and Control—IV, Restructuring, Santorini, Greece, 1998, proceedings in press.

[Alvey, 1997] T. Alvey, D. Goodwin, X. Ma, D. Streiffert, and D. Sun, "A Security-Constrained Bid-Clearing System for the New Zealand Wholesale Electricity Market," presented at the 1997 IEEE PES Summer Power Meeting, Berlin, Germany, in press.

[Anderson, 1996] E. Anderson, and Y. Ye, "Combining Interior-point and Pivoting Algorithms for Linear Programming," *Management Science,* vol. 42, no. 12, December 1996, pp. 1719-1731.

[Andrew, 1983] Andrew D. Seidel and Philip M. Ginsberg, Commodities Trading: Foundations, Analysis, and Operations, Prentice Hall, Inc., Englewood Cliffs, New Jersey, 1983.

[Andrews, 1994] M. Andrews and R. Prager, "Genetic Programming for the Acquisition of Double Auction Market Strategies," in *Advances in Genetic Programming*, K. Kinnear, Jr., Cambridge, Massachusetts: The MIT Press, 1994.

[Angelidis, 1993] G. Angelidis, and A. Semlyen, "Optimal Power Flow Using a Generalized Power Balance Constraint," Canadian *Journal of Electrical and Computer Engineering*, Vol. 18, pp. 191-198, October 1993.

[Anwar, 1994] D. Anwar and G. B. Sheblé. "Application of Optimal Power Flow to Interchange Brokerage Transaction," *Electric Power Systems Research Journal*, Vol. 30, no. 1, pp. 83-90, 1994.

[Aoki, 1989] K. Aoki, M. Itoh, T. Satoh, K. Nara, and M. Kanezashi, "Optimal Long-Term Unit Commitment in Large Scale Systems Including Fuel

Constrained Thermal and Pumped-Storage Hydro," *IEEE Transactions on Power Systems,* Vol. 4, pp. 1065-1073, August 1989.

[Aoki, 1987] K. Aoki, T. Satoh, M. Itoh, T. Ichimori, and K. Masegi, "Unit Commitment in a Large Scale Power System Including Fuel Constrined Thermal and Pumped-Storage Hydro," *IEEE Transactions on Power Systems*, Vol. PWRS-2, no. 4, 1987.

[Arbel, 1993] A. Arbel, *Exploring Interior-Point Linear Programming.* The MIT Press, London, 1993.

[Asgarpoor, 1995] S. Asgarpoor, and S. K. Panarelli, "Expected Cost Penalty Due to Deviation From Economic Dispatch for Interconnected Power Systems," *IEEE Transactions on Power Systems,* Vol. 10, no. 1, p. 441-447, February 1995

[Ashlock, 1997] D. Ashlock and C. Richter, "The Effect of Splitting Populations on Bidding Strategies." *GP-97 Proceedings of the Second Annual Conference,* pp. 27-34. San Francisco, CA: Morgan Kaufmann Publishers, 1997.

[Ashlock, 1995] D. Ashlock, "GP-automata for Dividing the Dollar." Mathematics Department, Iowa State University, Ames, IA, 1995.

[Au, 1972] Tung Au, Richard M. Shane, Lester A. Hoel, *Fundamentals of Systems Engineering: Probabilistic Models*, Addison Wesley, Reading, MA, 1972.

[Baldick, 1995] R. Baldick, "Generalized Unit Commitment Problem," *IEEE Transactions on Power Systems,* Vol. 10, pp. 465-475, February 1995.

[Ballance, 1996] J. W. Ballance "Setting Up the Western Power Exchange," presented at Potential Impacts of FERC Mega NOPR on System Operations - Panel Discussion, *IEEE 96 Winter Power Meeting*, Baltimore, Maryland, January 22, 1996.

[Bard, 1988] J. Bard, "Short-term Scheduling of Thermal-electric Generators Using LaGrangian Relaxation," *Operations Research,* Vol. 36, pp. 756-766, September-October 1988.

[Barish, 1978] Barish, Kaplan, Economic Analysis: For Engineering And Managerial Decision-Making, McGraw-Hill, New York, 1978.

[Barkovich, 1996] B. Barkovich and D. Hawk, "Charting A New Course in California," IEEE *Spectrum*, Vol. 33, p. 26, July 1996.

[Barry, 1985] D. Barry, "Socket to Them," in *Dave Barry's Bad Habits,* New York, NY: Henry Holt and Company, Inc., 1985.

[Batut, 1991] J. Batut and A. Renaud, "Daily Generation Scheduling Optimization with Transmission Constraints: A New Class of Algorithms," Electricité de France, Clamart, France 1991.

[Baughman, 1997b] M. L. Baughman, S. N. Siddiqi, and J. W. Zarnikan, "Advanced Pricing in Electrical Systems, Part I: Theory," *IEEE Transactions on Power Systems*, Vol. 12, no. 1, February 1997.

[Baughman, 1997a] M. L. Baughman, S. N. Siddiqi, and J. W. Zarnikan, "Advanced Pricing in Electrical Systems, Part II: Implications," *IEEE Transactions on Power Systems*, Vol. 12, no. 1, February 1997.

[Baughman, 1993] M. L. Baughman, J. W. Jones, and A. Jacob, "Model for Evaluating the Economics of Cool Storage Systems," *IEEE Transactions on Power Systems*, Vol. 8, no. 2, May 1993.

[Bayes, 1763] Reverend Thomas Bayes, "An Essay toward solving a Problem in the Doctrine of Chance, "Philosophical Transactions of the Royal Society, 1763.

[Bazaraa, 1990] M. Bazaraa, J. Jarvis, and H. Sherali, *Linear Programming and Network Flows*. John Wiley and Sons, New York, 2nd Edition, 1990.

[Bazaraa, 1993] M. S. Bazaraa, H. 0. Sherali, and C. M. Shetty, *Nonlinear Programming: Theory and Algorithms*, John Wiley and Sons, New York, 2nd Edition, 1993, pp. 225 and following.

[Bently, 1987] W. G. Bently and J. C. Evelyn. "Customer Thermal Energy Storage: A Marketing Opportunity for Cooling Off Electric Peak Demand," *IEEE Trans. on Power Systems*, Vol. 1, no. 4, pp. 973-979, November 1987.

[Bernouli, 1738] Daniel Bernoulli, "Exposition of a New Theory of the Measurement of Risk, "Econometrica (1954), pp. 23-36, Translation of a paper "Specimen Theoriae Novae de Mensura Sortis," Papers of the Imperial Academy of Sciences in Petersburg, V, 1738.

[Billinton, 1991a] R. Billinton and G. Lian, "Monte Carlo Approach to Substation Reliability Evaluation," *IEE Proceedings-C*, Vol. 140, no. 2, p. 147-152, March 1991.

[Billinton, 1991b] R. Billinton and L. Wenyuan, "Hybrid Approach for Reliability Evaluation of Composite Generation and Transmission Systems Using Monte-Carlo Simulation and Enumeration Technique," *IEE Proceedings-C*, Vol. 138, no. 3, p. 233-241, May 1991

[Billinton, 1992] R. Billinton and W. Li, "A Monte Carlo Method for Multi-Area Generation System Reliability Assessment," *IEEE Transactions on Power Systems*, Vol. 7, no. 4, p. 1487-1492, November 1992

[Billinton, 1993] R. Billinton and L. Gan, "Monte Carlo Simulation Model for Multiarea Generation System Reliability Studies," *IEE Proceedings-C*, Vol. 140, no. 6, p. 532-538, November 1993

[Binger, 1988] B. R. Binger and E. Hoffman. *Microeconomics with Calculus*. Glenview: Scott, Foresman and Company, 1988.

[Bisat, 1997] M. Bisat, Probabilistic Production Costing: A comparison of Four Methods. Master's research project, Iowa State University, Ames, IA, 1997.

[Bixby, 1994] R. Bixby, and M. Saltzman, "Recovering an Optimal LP Basis from an Interior-point Solution," *Operations Research Letters,* vol. 15, no. 4, May 1994, pp. 169-178.

[Bos, 1996] M. Bos, R. Beune, and R. van Amerongen, "On the Incorporation of A Heat Storage Device in LaGrangian Relaxation Based Algorithms for Unit Commitment," International Journal *of Electrical Power and Energy* System, Vol. 18, pp. 207-214, May 1996.

[Bosch, 1985] P. P. J. V. D. Bosch, "Optimal Static Dispatch with Linear, Quadratic and Non-linear Functions of the Fuel Costs," *IEEE Transactions on Power Apparatus and Systems,* Vol. PAS-104, pp. 3402-3408, December 1985.

[Bui, 1982] R. Bui and S. Ghaderpanah, "Real Power Rescheduling and Security Assessment," *IEEE Transactions on Apparatus and Power Systems,* Vol. PAS-101, pp. 2906-2915, August 1982.

[Bussey, 1981] Lynn E. Bussey, The Economic Analysis of Industrial Projects, Prentice-Hall, Englewood Cliffs, New Jersey, 1981.

[Caramanis, 1982] M. C. Caramanis, R. E. Bohn, and F. C. Schweppe. "Optimal Spot Pricing: Practice and Theory," *IEEE Trans. on Power Apparatus and Systems*, Vol. PAS-101, no. 9, pp.3234-3244, 1982.

[Caramanis, 1986] M. C. Caramanis, R. E. Bohn, and F. C. Schweppe. 1986. The Costs of Wheeling and Optimal Wheeling Rates. *IEEE Transactions on Power Systems* 1 (1): 63-73.

[Carpentier, 1987] J. Carpentier, "Towards A Secure and Optimal Automatic Operation of Power Systems," in *Proceedings of 1987 Power Industry Computer Application Conference,* pp. 2-37, 1987.

[CBOT, 1996] CBOT, "Actions in the Marketplace," *Chicago Board of Trade*, Publication Department, Chicago, IL, 1996.

[Chamorel, 1982] P. Chamorel, and A. Germond, "Efficient Constrained Power Flow Technique Based on Active-reactive Decouplina, and the Use of Linear Programming," *IEEE Transactions on Power Apparatus* and *Systems,* Vol. PAS-101, pp. 158-167, January 1982.

[Chao, 1983] H. P. Chao, "Peak Load Pricing And Capacity Planning with Demand and Supply Uncertainty," *Bell Journal of Economics*, Vol. 14, no.1, November 1983, pp. 179-190.

[Chao, 1986] H. Chao, S. S. Oren, S. A. Smith and R. B. Wilson. 1986. Unbundling the Quality Attributes of Electric Power: Models of Alternative Market Structures. A UERG California Energy Studies Report, UER-165, (April).

[Chattopadhyay, 1995] D. Chattopadhyay, "Energy Brokerage System with Emission Trading and Allocation of Cost Savings," *IEEE Transactions on Power Systems,* Vol. 10, no. 4, pp. 1939-1945, November 1995.

[Chen, 1995] J. Chen, F. N. Lee, A. M. Breipohl, R. Adapa, "Scheduling Direct Load Control to Minimize System Operational Cost," *IEEE Transactions on Power Systems,* Vol. 10, November 1995.

[Chen, 1993] C. S. Chen, and J. N. Sheen, "Cost Benefit Analysis of a Cooling Energy Storage System," *IEEE Transactions on Power Systems*, Vol. 8, no. 4, November 1993.

[Christie, 1995] R. D. Christie and A. Bose. "Load Frequency Control Issues in Power System Operations After Deregulation," *Proceedings of the 1995 PICA Conference,* pp.18-23, 1995.

[Chu, 1993] W. Chu, B. Chen, and C. Fu, " Scheduling of Direct Load Control to Minimize Load Reduction for a Utility Suffering from Generation Shortage," *IEEE/PES Winter Meeting*, 1993, Columbus, Ohio.

[Chua, 1984] L. O. Chua and G. N. Lin. "Nonlinear Programming Without Computation," *IEEE Transactions on Circuit and Systems*, CAS - 31, pp. 182-188, February 1984.

[Clayton, 1990] J. S. Clayton, S. R. Erwin and C. A. Gibson. Interchange Costing and Wheeling Loss Evaluation by Means of Incrementals. *IEEE Transactions on Power Systems,* Vol. 5, no. 3, pp. 759-765, 1990.

[Clayton, 1996] R. E. Clayton and R. Mukerji, "System Planning Tools for the Competitive Market," *IEEE Computer Applications in Power*, ISSN 0895-0156/96, p. 50, 1996.

[Cohen, 1987] A. I. Cohen, J. W. Patmore, D. H. Oglevee, R. W. Berman, L. H. Ayers and J. F. Howard, "An Integrated System for Load Control," *IEEE Transactions on Power Systems*, Vol. PWRS-2, no. 3, August 1987.

[Cohen, 1988] A. I. Cohen and C. C. Wang, "An Optimization Method for Load Management Scheduling," *IEEE Transactions on Power Systems* PWRS, Vol. 3, no. 2, 1988.

[Conejo, 1998] A. Conejo, and J. Arroyo, "Lagrangian Relaxation Case Studies," *University Report*, January 1998.

[Contaxis, 1986] G. Contaxis, C. Delkis, and G. Korres, "Decoupled Optimal Load Flow Using Linear or Quadratic Programming," *IEEE Transactions on Power Systems,* Vol. PWRS-1, pp. 1-7, May 1986.

[Cooper, 1974] L. Cooper and D. Steinberg, *Methods and Applications of Linear Programming*, Philadelphia: W.B. Saunders Company, 1974.

[Coppinger, 1980] V. M. Coppinger, V. L. Smith, and J. A. Titus. Incentives and Behavior in English, Dutch, and Sealed-Bid Auctions. *Economic Inquiry,* Vol. 18, no. 1, pp. 1-22, 1980.

[Coppinger, 1995] S. S. Coppinger and G. B. Sheblé, "Interchange of Electric Power By Using A Double Auction Mechanism," *Proceedings of the 27th Annual North American Power Symposium,* pp. 299-303, Bozeman, MT, 1995.

[Coursey, 1983] D. L. Coursey and V. L. Smith. Price Controls in a Posted Offer Market. *American Economic Review,* Vol. 73, no. 1, pp. 218-221, 1983.

[CPLEX, 1993] CPLEX Optimization, Inc. Using the CPLEX Callable Library and CPLEX Mixed Integer Library, Incline Village Nevada, 1993.

[Dansby, 1979] R. E. Dansby and R. D. Willig, "Industry performance gradient indexes," *The American Economic Review*, vol. 69, no. 3, pp. 249-260, June 1979.

[Dantzig, 1960] G. B. Dantzig and P. Wolfe, "The Decomposition Algorithm for Linear Programs," *Operations Research,* Vol. 8, pp.101-111, 1960.

[Daryanian, 1989] B. Daryanian, R. E. Bohn, and R. D. Tabors, "Optimal Demand-side Response to Electricity Spot Prices for Storage-type Customers," *IEEE Transactions on Power Systems,* Vol. 4, no. 3, pp. 897-903, August 1989.

[David, 1993] A. K. David, and Y. Z. Li, "Effect of Inter-temporal Factors on the Real-time Pricing of Elasticity," *IEEE Transactions on Power Systems*, Vol. 8, no. 1, February 1993.

[Day, 1971] J. T. Day, "Forecasting Minimum Production Costs with Linear Programming," *IEEE Trans. on Power Apparatus and Systems,* Vol. PAS-90, no. 2, pp.814-823, March/April 1971.

[Dekrajangpetch, 1997] S. Dekrajangpetch, *Auction Implementations Using LaGrangian Relaxation, Interior-Point Linear Programming, and Upper-Bound Linear Programming*, Master's Thesis, Iowa State University, Ames, IA, 1997.

[Dekrajangpetch, 1998a] S. Dekrajangpetch and G. B. Sheblé, "Alternative implementations of electric power auctions," in *Proceedings of the 60th American Power Conference*, vol. 60-1, pp. 394-398, 1998.

[Dekrajangpetch, 1998b] S. Dekrajangpetch, and G. Sheblé, "Auction Implementation Problems Using LaGrangian Relaxation," *IEEE Trans.* on *Power Systems*, PE-279-PWRS-0-04-1998.

[Dekrajangpetch, 1998c] S. Dekrajangpetch, and G. Sheblé, "Interior-point Linear Programming Algorithm for Auction Methods," accepted for publishing in *IEEE Trans.* on *Power Systems*, 1998, in press.

[Dekrajangpetch, 1998d] Somgiat Dekrajangpetch and Gerald B. Sheblé, "Bidding Information to Generate Bidding Strategies for LaGrangian Relaxation-based Power Auction," submitted for review by the *Electric Power Systems Research Journal.*

[Delson, 1991] J. K. Delson, X. Feng, and W.C Smith, "A Validation Process for Probabilistic Production Costing Programs," *IEEE Transactions on Power Systems*, Vol. 6, no. 3, pp. 1326-1336, August 1991

[Department of Justice, 1997] Department of Justice and the Federal Trade Commission, *Horizontal Merger Guidelines*, April 8, 1997.

[Dorfman, 1987] *Linear Programming and Economic Analysis*, Dover Publications, Inc., New York, 1987.

[Doty, 1982] K. W. Doty and P. L. McEntire. "An Analysis of Electric Power Brokerage Systems," *IEEE Transactions on PAS,* Vol. PAS-101, no.2, pp. 389-396, 1982.

[EL&P, 1992] EL&P, "Focus on Wheeling & Access," *Electric Light And Power*, pp. 12-15, March 1992.

[El-Keib, 1994b] A. El-Keib, H. Ma, and J. Hart, "Environmentally Constrained Economic Dispatching the LaGrangian Relaxation Method," *IEEE Transactions on Power Systems,* Vol. 9, pp. 1723-1729, November 1994.

[El-Keib, 1994a] A. El-Keib and H. Ding, "Environmentally Constrained Economic Dispatch Using Linear Programming," *Electric Power Systems Research,* Vol. 29, pp. 155-159, May 1994.

[Engelbrecht, 1980] Richard Engelbrecht-Wiggans, Auctions and Bidding Models: A Survey. *Management Science,* Vol. 26, no. 2: 119-42, 1980.

[EPRI, 1983] Electric Power Research Institute, *Development of the Zinc Chloride Battery for Utility Applications*, EPRI EM-3136, Research Project 226-5 Interim Report. Palo Alto, California: EPRI, 1983.

[EPRI, 1987a] Electric Power Research Institute, *Development of the Zinc Chloride Battery for Utility Applications,* EPRI SP-5018, Projects 226-5, -9 Final Report. Palo Alto, California: EPRI, 1987.

[EPRI, 1987b] EPRI, "MIDAS, Multi Objective Integrated Decision Analysis System," *EPRI Journal*, Vol. 12:57-58, May 1987.

[EPRI, 1991] EPRI, "Simultaneous Transfer Capability: Direction for Software Development," EPRI EL-7351 Final Report, August 1991.

[Ethier, 1987] M. Ethier, G. Roy, P. Blondeau, D. Mukhedkar, "Discrete Algorithms for Residential Heating Strategies, an Application of Optimal Control Theory," *IEEE Transactions on Power Systems*, Vol-PWRS-2, no. 2, May 1987.

[Fahd 1992a] G. Fahd and G. Sheblé, "Optimal Power Flow of Interchange Brokerage System Using Linear Programming," *IEEE Transaction on Power Systems*, T-PWRS, Vol. 7, no. 2, pp. 497-504, May 1992.

[Fahd, 1992b] G. Fahd, Dan Richards, and Gerald B. Sheblé, "The Implementation of an Energy Brokerage System Using Linear Programming," *IEEE Transactions on Power Systems*, T-PWRS, Vol. 7, no. 1, pp. 90-96, February 1992.

[Fahd, 1992c] George Fahd, "Optimal Power Flow Emulation of Interchange Brokerage Systems Using Linear Programming." A Ph.D. Dissertation, Auburn University, 1992.

[Farghal, 1987] S. Farghal, A. Abou-Elela, and A. Abdel, "Efficient Technique for Real-time Control of System Voltages and Reactive Power," *Electric Power Systems Research,* Vol. 12, pp. 197-208, June 1987.

[Federal Energy Regulatory Commission, 1993] Federal Energy Regulatory Commission. "Notice Of Proposed Rulemaking: Docket No. RM93-3," *Congressional Record,* July 30, 1993.

[Federal Energy Regulatory Commission, 1995] Federal Energy Regulatory Commission. "Notice Of Proposed Rulemaking: Docket No. RM95-8-000," *Congressional Record,* March 29, 1995.

[Felak, 1992] R. P. Felak, "Reliability and Transmission Access," *Public Utilities Fortnightly,* July 15, 1992.

[Ferreira, 1993] L. Ferreira, "On the Duality Gap for Thermal Unit Commitment Problems," in *Proceedings of the 1993 IEEE International Symposium on Circuits and Systems,* Vol. 4, pp. 2204-2207, May 1993.

[Finlay, 1995] D. Finlay, Optimal Bidding Strategies in Competitive Electric Power Pools. Masters thesis, University of Illinois, Urbana-Champaign, IL, 1995.

[Fisher 1980] M. Fisher, "Applications Oriented Guide to LaGrangian Relaxation," Interfaces, Vol. 15, pp. 10-21, March-April 1985.

[Fisher, 1981] M. L. Fisher, "The LaGrangian Relaxation Method for Solving Integer Programming Problems," *Management Science*, vol. 27, no. 1, January 1981, pp. 1-18.

[Florida Electric Power Coordinating Group, Inc., 1980] Florida Electric Power Coordinating Group, Inc. *Peninsular Florida Central Dispatch Study, Part 1 Final Report*, May 14, 1980.

[Franklin, 1980] G. F. Franklin and J. D. Powell, *Digital Control of Dynamic Systems*, Additson Wesley, Reading, MA, 1980.

[Gass, 1958] S. I. Gass, *Linear Programming: Methods and Applications*, McGraw-Hill Inc., New York, 1958.

[Gedra, 1998] T. Gedra, *Power Economics and Regulation*, Gedra Publishing, Stillwater, 1998.

[Gellings, 1994] C. W. Gellings, and J. H. Chamberlin, *Demand-side Management: Planning.* Lilburn, Georgia: Fairmont Press, Inc., 1994.

[Gellings, 1993] C. W. Gellings, and J. H. Chamberlin, *Demand-side Management: Concepts and Methods.* Lilburn, Georgia: Fairmont Press, Inc., 1993.

[Gellings, 1988] C. W. Gellings, and J. H. Chamberlin, *Demand-side Management Planning.* Lilburn, Georgia: Fairmont Press, Inc., 1993.

[Gent, 1971] M. R. Gent and J. W. Lamont. "Minimum-Emission Dispatch," *IEEE Transactions,* Vol. PAS-90, pp. 2650-2660, 1971.

[Germond, 1992] A. J. Germond and D. Niebur, "Survey of Knowledge-Based Systems in Power Systems: Europe," *Proc. of IEEE,* Vol. 80, no. 5, pp. 732-744, May 1992.

[Gjengedal, 1996] T. Gjengedal, "Emission Constrained Unit-commitment (ECUC)," *IEEE Transactions on Energy Conversion,* Vol. 11, pp. 132-138, March 1996.

[Goldberg, 1989] D. Goldberg, *Genetic Algorithms in Search, Optimization & Machine Learning.* Reading, Massachusetts: Addison-Wesley Publishing Company, Inc., 1989.

[Gorenstin, 1993] B. G. Gorenstin, N. M. Campodónico, J. P. Costa, M. V. F. Pereira, "Power System Expansion Planning Under Uncertainty," 11th PSCC, Avignon, France, Set 1993.

[Granville, 1994] S. Granville, "Optimal Reactive Dispatch Through Interior Point Methods," *IEEE Transactions on Power Systems,* Vol. 9, pp. 134-146, February 1994.

[Grether, 1979] D. Grether, M. Isaac and C. Plott, "Alternative Methods of Allocating Airport Slots: Performance and Evaluation." CAB Report. Pasadena, California: Polynomics Research Laboratories, Inc., 1979.

[Grether, 1981] D. Grether, M. Isaac and C. Plott. "The Allocation of Landing Rights by Unanimity Among Competitors." *American Economic Review* 71 (May): 166-171, 1981.

[Guan, 1994] X. Guan, P. B. Luh, Yan Houzhong, and P. Rogan, "Optimization-Based Scheduling of Hydrothermal Power Systems with Pumped-Storage Units," *IEEE Transactions on Power Systems*, Vol. 9, no.2, May 1994, pp. 1023-1029.

[Guan, 1995] X. Guan, P. Luh, and L. Zhang, "Nonlinear Approximation Method in LaGrangian Relaxation-based Algorithms for Hydrothermal Scheduling," *IEEE Transactions on Power Systems,* Vol. 10, pp. 772-778, May 1995.

[Gustafson, 1987] M. W. Gustafson and J. S. Baylor, Transmission Loss Evaluations for Electric Systems. IEEE/PES 1987 Summer Meeting, San Francisco, California, July 12-17, 1987.

[Hao, 1997] S. Hao, G. Angelidis, H. Singh, and A. Papalexopoulos, "Consumer Payment Minimization in Power Pool Auctions," *Proceedings of the 20th International Conference on Power Industry Computer Applications,* Columbus, Ohio, May 1997, pp. 368-373.

[Happ, 1990] H. H. Happ, Report on Wheeling Costs. Case 88-E-238, The New York Public Service Commission, Feb. 1990.

[Harsyani, 1977] J. C. Harsyani, *Rational Behavior and Bargaining Equilibrium in Games and Social Sciences*, Cambridge University Press, New York, NY, 1977.

[Hazelrigg, 1996] G. A. Hazelrigg, Systems Engineering: An Approach to Information-Based Design, Prentice Hall, Upper Saddle River, NJ, 1996.

[Henderson, 1980] J. M. Henderson and R. E. Quandt, *Microeconomic Theory: A Mathematical Approach*, Third Edition, NewYork: McGraw Hill, 1980.

[Hertog, 1994] D. Hertog, *Interior Point Approach to Linear, Quadratic, and Convex Programming.* Kluwer Academic Publishers, Boston, 1994.

[Hillier, 1995] F. Hillier and G. Lieberman, *Introduction to Operations Research*, McGraw-Hill, New York, 6th edition, 1995.

[Hobbs, 1985] B. F. Hobbs and R. E. Schuler, "An Assessment of the Deregulation of Electric Power Generation Using Network Models of Imperfect Spatial Markets," *Papers of the Regional Science Association*, vol. 57, pp. 75-89, 1985.

[Hobson, 1980] E. Hobson, "Network Constrained Reactive Power Control Using Linear Programming," *IEEE Transactions on Apparatus and Power Systems,* Vol. PAS-99, pp. 868-877, May-June 1980.

[Hogan, 1997] W. W. Hogan, "A Market Power Model with Strategic Interaction in Electricity Networks," *The Energy Journal,* vol. 18, no. 4, pp. 107-141, 1997.

[Holland, 1975] J. H. Holland, *Adaptation in Natural and Artificial Systems*, MIT Press, Cambridge, MA, 1975.

[Howard, 1988] R. A. Howard, "Decision Analysis: Practice and Promise," Management Science, pp.675-679, 1988.

[Hsu, 1991] Y. Hsu, C. Su, "Dispatch of Direct Load Control Using Dynamic Programming," *IEEE Transactions* PWRS, Vol. 6, no. 3, 1991.

[Huang, 1993] S. R. Huang and S. L. Chen, "Evaluation and Improvement of Variance Reduction in Monte-Carlo Production Simulation," *IEEE Transactions on Energy Conversion*, Vol. 8, no. 4, p. 610-619, December 1993

[Huang, 1994] G. Huang, and K. Song, "Simple Two Stage Optimization Algorithm for Constrained Power Economic Dispatch," *IEEE Transactions on Power Systems,* Vol. 9, pp. 1818-1824, November 1994.

[Huneault, 1994] M. Huneault, C. Rosu, R. Manoliu, and F. D. Galiana, "A Study of Knowledge Engineering Tools in Power Engineering Applications," *IEEE Transactions on Power Systems*, Vol. 9, no. 4, November 1994.

[IEEE Committee Report, 1915] IEEE Committee Report. "Dynamic Models for Steam and Hydro Turbines in Power System Studies," Vol. PAS-92, November/December 1973, pp. 1904-1915.

[IEEE PSRC, 1996] IEEE Power System Reliability Subcommittee. "Reliability Test System," presented at IEEE PES Winter Meeting, 96 WM 183-7 PWRS, 1996.

[Ignizio, 1982] James P. Ignizio, *Linear Programming in Single and Multiple Objective Systems*, Englewood Cliffs, New Jersey: Prentice Hall Inc., 1982.

[Ilic, 1998] M. Ilic, F. Galiana, and L. Fink, *Power Systems Restructuring: Engineering and Economics*. Norwell, MA: Kluwer Academic Publishers. 1998.

[Ioannides, 1995] Y. Ioannides, "Evolution of Trading Structures," Department of Economics, Tufts University, Medford, MA 02155, 1995.

[Iwamoto, 1986] S. Iwamoto, S. Tamura, and Y. Tamura, "Fast VAR and Voltage Control Scheme Combining Fast Second-order Load-flow and Recursive Linear Programming," in *Proceedings of the IFAC, Planning and Operation of Electric Energy Systems,* pp. 399-404, 1986.

[Jacobs, 1997] J. Jacobs, "Artificial Power Markets and Unintended Consequences," *IEEE Trans.* on *Power Systems,* vol. 12, no. 2, May 1997, pp. 968-972.

[Jaleeli, 1992] N. Jaleeli L. VanSlyck, D. Ewart, L. Fink, and A. Hoffmann. "Understanding Automatic Generation Control," *IEEE Transactions on Power Systems,* Vol. 7, no. 3, pp. 1106-1122, August 1992.

[Jaleeli, 1995] N. Jaleeli and L. VanSlyck. "Tie-Line Bias Prioritized Energy Control," *IEEE Transactions on Power Systems,* Vol. 10, no. 1, pp. 51-59, February 1995.

[Jeloka, 1994] R. Jeloka, "Implementation LaGrangian Relaxation Method for Unit Commitment," Master's thesis, Iowa State University, 1994.

[Jeng, 1996] L.-H. Jeng, Y.-Y. Hsu, B. S. Chang, and K. K. Chen, "Linear Programming Method for the Scheduling of Pumped-Storage Units with Oscillatory Stability Constraints," *IEEE Transactions on Power Systems*, Vol. 11, no. 4, November 1996.

[Johnson, 1997] R. Johnson, S. Oren, and A. Svoboda, "Equity and Efficiency of Unit Commitment in Competitive Electricity Markets," *Utilities Policy: Strategy, Performance, Regulation,* Vol. 6, no. 1, 1997, pp. 9-19.

[Karmarkar, 1984] N. Karmarkar, "A New Polynomial-Time Algorithm for Linear Programming," *Combinatorica* 4, pp. 373-395, 1984.

[Kaufmann, 1968] Arnold Kaufmann, The Science of Decision-Making, McGraw-Hill Book Co., New York, New York, 1968.

[Kelley, 1987] K. Kelley, S. Henderson, P. Nagler and M. Eifert, Some Economic Principles for Pricing Wheeled Power. National Regulatory Research Institute, August 1987.

[Kennedy, 1987] M. P. Kennedy and L. O. Chua. "Unifying the Tank and Hopfield Linear Programming Circuit and the Canonical Nonlinear Circuit of Chua and Lin," *IEEE Transactions on Circuit and Systems,* CAS-34, pp. 210-221, February 1987.

[Kennedy, 1988] M. P. Kennedy and L. O. Chua. "Neural Networks for Nonlinear Programming," *IEEE Transactions on Circuit and Systems,* CAS - 35, pp. 554-562, May 1988.

[Ketcham, 1984] J. Ketcham, V. L. Smith and A. W. Williams. A Comparison of Posted-Offer and Double-Auction Pricing Institutions. *Review of Economic Studies* 51 (167): 595-614, 1984.

[Kirschen, 1988] D. Kirschen and H. V. Meeteren, "MW/voltage Control in a Linear Programming Based Optimal Power Flow," *IEEE Transactions on Power Systems,* Vol. PWRS-3, pp. 481-489, May 1988.

[Kojima, 1986] M. Kojima, "Determining Basic Variables of Optimal Solutions in Karmarkar's New LP Algorithm," *Algorithmica,* vol. 1, no. 4, 1986, pp. 499-515.

[Kondragunta, 1997] S. Kondragunta, "Genetic Algorithm Unit Commitment Program," M.S. Thesis, Iowa State University, Ames, IA, 1997.

[Koza, 1992] John Koza, *Genetic Programming*, MIT Press, Cambridge, Massachusetts, 1992.

[Krause, 1994] B. A. Krause and J. McCalley "Bulk Power Transaction Selection in a Competitive Electric Energy System with Provision of Security Incentives," *Proceedings of the 26th Annual North American Power Symposium,* Manhattan, Kansas, pp. 126-136, September 1994.

[Kumar, 1997] J. Kumar and G. B. Sheblé. "A Decision Analysis Approach to Transaction Selection Problem in a Competitive Electric Market," accepted in *Electric Power Systems Research Journal,* 1997.

[Kumar, 1993] J. Kumar, "Application of Artificial Neural Netowrks to Optimization Problem," Master's Thesis, Iowa State University, Ames, Iowa, December 1993.

[Kumar, 1994] J. Kumar and G. B. Sheblé. "A Framework for Transaction Selection Using Decision Analysis Based Upon Risk and Cost of Insurance," *Proceedings of the 29th North American Power Symposium*, Kansas State University, KS, pp.548-557, 1994.

[Kumar, 1995a] J. Kumar and G. B. Sheblé. "Clamped State Solution Of Artificial Neural Network For Real Time Economic Dispatch," *IEEE Transactions on PWRS,* Vol. 10, no. 2, pp. 925-931, 1995.

[Kumar, 1995b] J. Kumar and G. B. Sheblé. "Framework for Energy Brokerage Systems with Reserve Margin Constraints," *Proceedings of the 27th Annual North American Power Symposium*, pp. 41-46, Bozeman, MT, 1995.

[Kumar, 1996a] J. Kumar and G. B. Sheblé, "Auction Game in Electric Market Place," *Proceedings of the 1996 58th American Power Conference*, Vol. 58, Part 2, pp. 1272-1277, 1996.

[Kumar, 1996b] J. Kumar and G. B. Sheblé, "Framework for Energy Brokerage System with Reserve Margins and Transmission Losses," *IEEE Transactions on Power Systems*, Vol. 11, no. 4, pp. 1763-1769, November 1996.

[Kumar, 1996c] J. Kumar and G. B. Sheblé, "Transaction Selection Using Decision Analysis Based Upon Risk and Cost of Insurance," IEEE Winter Power Meeting 1996.

[Kumar, 1996d] J. Kumar "Electric Power Auction Market Implementation and Simulation," Ph.D. Dissertation, Iowa State University, 1996.

[Kumar, 1996e] J. Kumar, K. H. Ng, and G. B. Sheblé. "AGC Simulator For Price Based Operation, Part 1: A Model," presented at IEEE PES Summer Meeting, 96 SM 588-4 PWRS, 1996.

[Kumar, 1996f] J. Kumar, K. H. Ng, and G. B. Sheblé. "AGC Simulator For Price Based Operation, Part II: Case Study Results," presented at IEEE PES Summer Meeting, 96 SM 373-1 PWRS, 1996.

[Kumar, 1997] J. Kumar and G. B. Sheblé, "Auction Market Simulator for Price Based Operation," presented at the 1997 IEEE PES Summer Power Meeting, Berlin, Germany, in press.

[Kurucz, 1996] C. N. Kurucz, D. Brandt, S. Sim, "A Linear Programming Model for Reducing System Peak Through Customer Load Control Programs," presented at the IEEE PES Winter Meeting, 96 WM 239-9 PWRS, Baltimore, Maryland, 1996.

[Lamont, 1973] J. W. Lamont and M. R. Gent. "Environmentally-Oriented Dispatching Techniques," *Proceedings of the 8th PICA Conference*, Minneapolis, Minnesota, 1973.

[Landgren, 1971] G. L. Landgren, H. L. Terhune and R. K. Angel, "Transmission Interchange Capability-Analysis by Computer," IEEE PAS, pp. 2405-2414, 1971.

[Landgren, 1972] G. L. Landgren and S. W. Anderson, "Simultaneous Power Interchange Capability Analysis," IEEE PAS, pp. 1973-1986, 1972.

[Lasdon, 1970] L. S. Lasdon, *Optimization Theory for Large Systems*, Macmillan Publishing Co., Inc, New York, NY, 1970.

[Lazarus, 1992] E. Lazarus, "The Public Utility Holding Act of 1935: Legislative History and Background," *Congressional Research Service Report* 92-226 A, The Library of Congress, February 26,1992.

[Le, 1983] K. D. Le, R. F. Boyle, M. D. Hunter, K. D. Jones, "A Procedure for Coordinating Direct-Load-Control Strategies to Minimize System Production Cost," *IEEE Transactions Power Apparatus and Systems*, Vol. PAS-102, no. 6, June 1983.

[Lee, 1983] S. H. Lee, and C. L. Wilkins, "A Practical Approach to Appliance Load Control Analysis: A Water Heater Case Study," *IEEE Transactions on Power Apparatus and Systems*, Vol. 7, no. 4, December 1992.

[Lee, 1984] F. N. Lee and A. M. Breipohl, "Operational Cost Savings of Direct Load Control," *IEEE Trans. Power Apparatus and Systems,* Vol. PAS-103, no. 5, May 1984.

[Lee, 1988] F. N. Lee, Three-Area Joint Dispatch Production Costing. *IEEE Transactions on Power Systems* 3 (1): 294-300, 1988.

[Lee, 1992] T. Y. Lee and N. Chen, "The Effect of Pumped Storage and Battery Energy Storage Systems on Hydrothermal Generation Coordination," *IEEE Transactions on Energy Conversion*, Vol. 7, no. 4, pp. 631-637, December 1992.

[Lee, 1993a] F. Lee and A. Breipohl, "Reserve Constrained Economic Dispatch with Prohibited Operating Zones," *IEEE Transactions on Power Systems,* Vol. 8, pp. 246-254, February 1993.

[Lee, 1993b] T. Y. Lee and N. Chen, "Optimal Capacity of the Battery Storage System in a Power System," *IEEE Transactions on Energy Conversion*, Vol. 8, no. 4, pp. 667-673, December 1993.

[Lee, 1994] T. Y. Lee and N. Chen, "Effect of Battery Energy Storage System on the Time-of-Use Rates Industrial Customers," *IEE Proceedings: Generator Transmission Distribution*, Vol. 141, no. 5, pp. 5521-528, September 1994.

[Lerner, 1934] A. P. Lerner, "Monopoly and the Measurement of Monopoly Power," *Review of Economic Studies*, vol. 1, pp. 157-175, June 1934.

[Lidgate, 1984] D. Lidgate, and B. M. N. Khalid, "Unit Commitment in a Thermal Generation System with Multiple Pumped-storage Power Stations," *Electric Power Energy System*, Vol. 6, no. 2, April 1984.

[Lin, 1989] M. Lin, A. Breipohl, and F. Lee, "Comparison of Probabilistic Production Cost Simulation Methods" *IEEE Transactions on Power Systems*, Vol. 4, no. 4, p. 1326-1333, October 1989

[Lin, 1995] S. Lin, P. Luh, and C. Larson, "Power System Scheduling with Coupled Transaction," in *Proceedings of the 1995 IEEE Conference on Control Applications,* pp. 19-20, September 1995.

[Lindgren, 1976] Bernard W. Lindgren, Statistical Theory, MacMillan Publishing Co. Inc., New York, New York, 1976.

[Lippmann, 1987] R. P. Lippmann, "An Introduction to Computing with Neural Nets," *IEEE Acoustics, Speech and Signal Processing Magazine*, pp. 4-21, April 1987.

[LOC, 1993] LOC, "Energy Policy Act of 1992: Summary and Implications," *Congressional Research Service*, The Library of Congress February 1, 1993.

[Luce, 1985] R. D. Luce and H. Raiffa, *Games and Decisions: Introduction and Critical Survey*, Dover Publications, New York, 1985.

[Luenberger, 1984] D. Luenberger, *Linear and Non-linear Programming*. Addison-Wesley Publishing Co., Menlo Park, 2 ed., pp. 53-58, 1984.

[Luenberger, 1995] D. G. Luenberger, Microeconomic Theory, McGraw-Hill, New York, New York, 1995.

[Maifeld, 1996] T. Maifeld and G. Sheblé, "Genetic-Based Unit Commitment," *IEEE Trans. on Power Systems,* Vol. 11, no. 3, p. 1359, August 1996.

[Maly, 1995] D. K. Maly and K. S. Kwan, "Charge Scheduling with Dynamic Programming," *IEE Proceedings-Sci. Meas. Technology,* Vol. 142, no. 6, pp. 453-458, November 1995.

[Mamandur, 1978] K. R. C. Mamandur and G. J. Berg, "Economic Shift in Electric Power Generations with Line Flow Constraints," IEEE PAS, Vol-97, no. 5, pp. 1618-1625, Sept/Oct 1978.

[Mamandur, 1982] K. Mamandur, "Emergency Adjustments to VAR Control Variables to Alleviate Over-voltages, Undervoltages, and Generator VAR Limit Violations," *IEEE Trans. on Apparatus and Power Sys*tems, Vol. PAS-101, pp. 1040-1047, May 1982.

[Mangoli, 1993] M. Mangoli K. Lee, and Y. Park, "Optimal Real and Reactive Power Control Using Linear Programming," *Electric Power Systems Research,* Vol. 26, pp. 1-10, January 1993.

[Mann, 1990] J. M. Mann, *Neural Networks: Genetic-Based Learning, Network Architecture, and Applications to Nondestructive Evaluation*, M.S. Thesis, Iowa State University, 1990.

[Marshall, 1989] John F. Marshall, *Futures and Option Contracting, Theory and Practice*, South-Western Publishing Company, Cincinnati, 1989.

[Marsten, 1989] R. Marsten, M. Saltzman, D. Shanno, G. Pierce, and J. Ballintijn, "Implementation of a Dual Affine Interior Point Algorithm for Linear Programming," *ORSA Journal on Computing,* vol. 1, no. 4, 1989, pp. 287-297.

[Martin, 1988] J. Martin and S. Oxman, *Building Expert Systems*, Prentice Hall, Englewood Cliffs, New Jersey, 1988.

[McAfee, 1987] R. McAfee and J. McMillan, "Auctions and biddings," *Journal of Economic Literature*, Vol. 25, pp. 699-738, June 1987.

[McCabe, 1989] K. A. McCabe, S. J. Rassenti and V. L. Smith. Designing 'Smart' Computer-Assisted Markets: An Experimental Auction for Gas Networks. *Journal of Political Economy* 5: 259-283, 1989.

[McCabe, 1990] K. A. McCabe, S. J. Rassenti and V. L. Smith. Auction Design for Composite Goods: The Natural Gas Industry. *Journal of Economic Behavior and Organization* 14: 127-149, 1990.

[McCabe, 1991] K. A. McCabe, S. J. Rassenti and V. L. Smith. Smart Computer-Assisted Markets. *Science* 254 (Oct. 25): 534-538, 1991.

[McCalley, 1994a] J. McCalley and G. B. Sheblé, "Competitive Electric Energy Systems: Reliability of Bulk Transmission and Supply," tutorial paper presented at the Fourth International Conference of Probabilistic Methods Applied to Power Systems, 1994.

[McCalley, 1994b] J. McCalley, A. Fouad, V. Vittal, A. Irizarry-Rivera, R. Farmer, and B. Agarwal, "A Probabilistic Problem in Electric Power System Operation: The Economy-Security Tradeoff for Stability-Limited Systems," *Proceedings of the Third International Workshop on Rough Sets and Soft Computing,* November 10-12, 1994, San Jose, California.

[McCalley, 1994c] J. McCalley and B. Krause, "Rapid Transmission Capacity Margin Determination for Dynamic Security Assessment Using Artificial Neural Networks," in press, *Electric Power Systems Research Journal*, 1994.

[McCalley, 1994d] J. D. McCalley and G. B. Sheblé. Class notes from EE653A: "Evaluation of Transmission Service for Secure Power System Operation in A Less Regulated Utility Environment," Department of Electrical and Computer Engineering, Iowa State University, Ames, Iowa, 1994.

[Megiddo, 1991] N. Megiddo, "On Finding Primal- and Dual-optimal Bases," *ORSA Journal on Computing,* Vol. 3, no. 1, 1991, pp. 63-65.

[Merlin, 1983] A. Merlin and P. Sanden, "A New Method for Unit Commitment at Electricite De France," *IEEE Transactions on Power Apparatus and Systems*, Vol. PAS-102, no. 5, pp. 1218-1225, May 1983.

[Merril, 1990] H. M. Merril and A. J. Wood, "Risk and Uncertainty in Power System Planning," 10th Power System Computation Conference - PSCC, Graz, Austria, August 1990.

[Merrill, 1991] H. M. Merrill, "Have I Ever Got a Deal for You. Economic Principles in Pricing of Services," IEEE SP 91EH0345-9-PWR, pp. 1-8, 1991.

[Mescua, 1985] J. Mescua, "A Decoupled Method for Systematic Adjustments of Phase-Shifting and Tap-Changing Transformers," *IEEE PAS,* Vol-104, no. 9, September 1985, pp. 2315-2321.

[Milgrom, 1982] P. R. Milgrom and R. J. Weber, A Theory of Auctions and Competitive Bidding. *Econometrica* 50 (5): 1089-1122, 1982.

[Milgrom, 1989] P. Milgrom, "Auctions and Bidding: A Primer," Journal of Economic Perspectives, Vol. 3, no. 3, pp. 3-22, 1989.

[Miller, 1977] R. M. Miller, C. R. Plott and V. L. Smith. Intertemporal Competitive Equilibrium: An Empirical Study of Speculation. *Quarterly Journal of Economics* 91 (4), pp. 599-624, 1977.

[Miller, 1991] J. M. Miller, B. M. Balmat, K. N. Morris, et al., "Operating Problems with Parallel Flows," committee report submitted to IEEE PWRS 226-1, 1991.

[Miranda, 1994] V. Miranda, "Power System Planning and Fuzzy Sets: Towards a Comprehensive Model Including all Types Of Uncertainties," *Proceedings of PMAPSí94*, Rio de Janeiro, Brazil, September 1994.

[Miranda, 1994] V. Miranda, J. V. Ranito, L. M. Proenca, "Genetic Algorithms in Optimal Multistage Distribution Network Planning," *IEEE Transactions on Power Systems,* Vol. 9, no. 4, pp. 1927-1933, November 1994.

[Miranda, 1995] V. Miranda, L. M. Proença, "A General Methodology for Distribution Planning Under Uncertainty, Including Genetic Algorithms and Fuzzy Models in a Multi-criteria Environment," *Proceedings of Stockholm Power Tech, SPT'95,* Stockholm, Sweden, June 18-22, pp. 832-837, 1995.

[Momoh, 1994] J. Momoh, S. Guo, E. Ogbuobiri and R. Adapa, "Quadratic Interior Point Method Solving Power System Optimization Problems," *IEEE Transactions on Power Systems,* Vol. 9, pp. 1327-1336, August 1994.

[Momoh, 1995] J. Momoh, L. Dias, S. Guo, and R. Adapa, "Economic Operation and Planning of Multi-area Interconnected Power Systems," *IEEE Trans. on Power Systems,* Vol. 10, no. 2, May 1995, pp. 1044-1053.

[Mortensen, 1990] R. E. Mortensen, and K. P. Haggerty, "Dynamics of Heating and Cooling Loads: Models, Simulation, and Actual Utility Data," *IEEE Transactions on Power Systems*, Vol. 5, no. 1, pp. 253-248, 1990.

[Mukherjee, 1992] S. Mukherjee, A. Recio and C. Douligeris, "Optimal Power Flow by Linear Programming Based Optimization," in *Proceedings of the IEEE SOUTHEASTCON '92,* pp. 149-162, April 1992.

[NERC, 1998] NERC, "NERC Control Performance Criteria," *Supplement to NERC Operating Manual,* North American Reliability Council, Princeton, NJ.

[Ng, 1998] K.-H. Ng, and G. B. Sheblé, "Direct Load Control – A Profit-based Load Management Using Linear Programming," *IEEE Transactions on Power Systems*, Vol. 13, no. 2, May 1998.

[Ng, 1997] K.-H. Ng, *Reformulating Load Management Under Deregulation,* Master's Thesis, Iowa State University, Ames, May 1997.

[NU, 1990] NU Files Plan to Hurry Merger. *Electrical World,* 204 (4): 33, 1990.

[NYMEX, 1995] NYMEX, "Financial Options: Flexibility, Control & Opportunity," *New York Mercantile Exchange,* Publication Department, New York, NY, 1995.

[Olofsson, 1995a] M. Olofsson, G. Andersson and L. Soder, "Optimal Operation of the Swedish Railway Electrical System," in *Proceedings of the International Conference on Electric Railways in a United* Europe, pp. 64-68, March 1995.

[Olofsson, 1995b] M. Olofsson, G. Andersson and L. Soder, "Linear Programming Based Optimal Power Flow Using Second Order Sensitivities," *IEEE/PES Winter Meeting,* January 1995, New York.

[O'Neill, 1994] R. P. O'Neill and C. S. Whitmore, "Network Oligopoly Regulation: An Approach to Electric Federalism," *Electricity and Federalism Symposium,* June 24, 1993 (Revised March 16, 1994).

[Open Access, 1990] Open Access: Look Before You Leap. *Electrical World,* Vol. 204, no. 5, pp. 33-34, 1990.

[Oren, 1994] S. S. Oren, P. Spiller, P. Variya and F. Wu, "Nodal Prices and Transmission Rights: A Critical Appraisal," University of California at Berkeley Research Report, December 1994.

[Oren, 1997] S. S. Oren, "Economic Inefficiency of Passive Transmission Rights in Congested Electricity Systems with Competitive Generation," *The Energy Journal,* Vol. 18, no. 1, pp. 63-83, 1997.

[Outhred, 1993] H. R. Outhred, Principles of a Market-Based Electricity Industry and Possible Steps Toward Implementation in Australia. *International Conference on Advances in Power System Control, Operation and Management.* Hong Kong, Dec. 7-10, 1993.

[Papalexopoulos, 1989] A. D. Papalexopoulos, C. F. Imparato and F. F. Wu, "Large Scale Optimal Power Flow: Effects of Initialization, Decoupling and Discretization," IEEE PWRS Vol. 4, no. 2, pp. 748-759, 1989.

[Parker, 1989] B. J. Parker, E. Denzinger, B. Porretta, G. J. Anders and M. S. Mirsky, "Optimal Economic Power Transfers," *IEEE Transactions on Power Systems,* Vol. 4, no. 3, pp. 1167-1175, 1989.

[Parker, 1996] C. Parker and J. Stremel, "A Smart Monte Carlo Procedure for Production Costing and Uncertainty Analysis," *Proceedings of the American Power Conference*, Vol. 58, no. II, pp. 897-900, 1996.

[Pellegrino, 1996] F. Pellegrino, A. Renaud, and T. Socroun, "Bundle and Augmented LaGrangian Methods for Short-Term Unit Commitment," *Proceedings of the 12th Power Systems* Computation *Conference (PSCC),* Dresden, vol.2, August 1996, pp.730-739.

[Pereira, 1982] M. Pereira and L. Pinto, "Decomposition Approach to the Economic Dispatch of Hydrothermal Systems," *IEEE Transactions on Apparatus and Power Systems,* Vol. 10, pp. 3851-3860, October 1982.

[Pereira, 1992] V. Pereira B. G. Gorenstin and Morozowski Fo, "Chronological Probabilistic Production Costing and Wheeling Calculations with Transmission Network Modeling," *IEEE Transactions on Power Systems*, Vol. 7, no. 2, pp. 885-891, May 1992.

[Perl, 1996] L. J. Perl, "Measuring Market Power in Electrical Generation," *Antitrust Law Journal*, vol. 64, pp. 311-321, 1996.

[Peterson, 1995a] W. L. Peterson, and S. R. Brammer, "A Capacity Based LaGrangian Relaxation Unit Commitment with Ramp Rate Constraints," *IEEE Trans.* on *Power Systems,* vol. 10, no. 2, May 1995, pp. 1077-1084.

[Peterson, 1995b] W. Peterson and S. Brammer, "Crew Constraints in LaGrangian Relaxation Unit Commitment," in *Proceedings of the IEEE Southeastcon '95 Conference,* pp. 381-384, March 1995.

[Philips, 1988] L. Philips, *The Economics of Imperfect Information.* New York: Cambridge University Press, 1988.

[Phillips, 1990] R. T. Phillips, The Future of Competitive Power Generation. *Public Utilities Fortnightly,* Vol. 125, no. 6, pp. 13-16, 1990.

[Pierre, 1975] Donald A. Pierre and Michael J. Lowe, "Mathematical Programming Via Augmented LaGrangians," Addison-Wesley Publishing Company, Boston, 1975.

[Plott, 1988] C. Plott, Research on Pricing in a Gas Transportation Network. Technical Report 88-2, FERC Office of Economic Policy, July 1988.

[Ponnambalam, 1992] K. Ponnambalam, V. Quintana, and A. Vannelli, "A Fast Algorithm for Power System Optimization Problems Using an Interior Point Method," *IEEE Trans. on Power Systems,* Vol. 7, no. 2, May 1992, pp. 892-899.

[Post, 1994] D. Post, *Electric Power Interchange Transaction Analysis and Selection.* Master's thesis, Iowa State University, Ames, Iowa, 1994.

[Post, 1995] D. Post, S. Coppinger, and G. Sheblé, "Application of Auctions as a Pricing Mechanism for the Interchange of Electric Power," *IEEE Trans.* on *Power Systems,* vol. 10, no. 3, August 1995, pp. 1580-1584.

[Powers, 1984] Mark Powers and David Vogel, *Inside the Financial Futures Markets,* John Wiley & Sons, January 1984.

[Prasannanm, 1996] B. Prasannanm, P. B. Luh, H. Yan, J. A. Palmberg, and L. Zhang, "Optimization-based Trasnactions and Hydrothermal Scheduling," *IEEE Transactions on Power Systems,* Vol. 11, no. 2, May 1996.

[Prasannan, 1995] B. Prasannan, P. Luh, H. Yan, J. Palmberg, and L. Zhang, "Optimization-based Sale Transactions and Hydrothermal Scheduling," in *Proceedings of the 1995 IEEE Power Industry Computer Application* Conference, pp. 137-142, May 1995.

[Qiu, 1987] J. Qiu and S. Shahidehpour, "New Approach for Minimizing Power Losses and Improving Voltage Profile," *IEEE Transactions on Power Systems,* Vol. PWRS-2, pp. 287-295, May 1987.

[Raiffa, 1968] H. Raiffa, "Decision Analysis," Addison-Wesley 1968.

[Rajan, 1997] S. Rajan, Electric System Operating Strategies in an Energy Brokerage Environment. Ph.D. dissertation, Iowa State University, Ames, Iowa, 1997.

[Rakic, 1994] M. V. Rakic, and Z. M. Markovic, "Short Term Operation and Power Exchange Planning of Hydro-thermal Power Systems," *IEEE Transactions on Power Systems*, Vol. 9, no. 1, February 1994.

[Ramanathan, 1986] R. Ramanathan, Real-Time Wheeling Losses Computation Techniques for Energy Management Systems. *IEEE Transactions on Power Systems,* Vol. 1, no. 3, 314-320, 1986.

[Rassenti, 1982] S. J. Rassenti, V. L. Smith and R. L. Bulfin. A Combinatorial Auction Mechanism for Airport Time Slot Allocation. *Bell Journal of Economics,* Vol. 13, pp. 402-417, 1982.

[Rassenti, 1992] S. J. Rassenti, S. S. Reynolds and V. L. Smith. "Cotenancy and Competition in an Experimental Auction Market for Natural Gas Pipeline Networks," *Economic Theory,* pp. 1-51, February 1992.

[Rau, 1989] N. S. Rau, Certain Considerations in the Pricing of Transmission Service. *IEEE Transactions on Power Systems,* Vol. 4, no. 3, pp. 1133-1139, 1989.

[Reichelt, 1991] D. Reichelt and H. Glavitsch, "Features of a Hybrid Expert System for Security Enhancement," *Proceedings of PICA Conference*, pp. 330-336, May 1991.

[Resource Planning Associates, Inc., 1979] Resource Planning Associates, Inc. "The Florida Electric Power Coordinating Group: An Evolving Power Pool," *Report for U.S. Department of Energy*, October 1979.

[Richter, 1997a] C. Richter and G. Sheblé, "Genetic Algorithm Evolution of Utility Bidding Strategies for the Competitive Marketplace," 1997 IEEE/PES Summer Meeting, PE-752-PWRS-1-05-1997, New York: IEEE, 1997.

[Richter, 1997b] C. Richter and G. Sheblé, "Building Fuzzy Bidding Strategies for the Competitive Generator," in *Proceedings of the 1997 North American Power Symposium,* 1997.

[Richter, 1998a] C. Richter and G. Sheblé, "Bidding Strategies that Minimize Risk with Options and Futures Contracts," in *Proceedings of the 1998 American Power Conference, Session 25, Open Access II-Power Marketing, Paper C,* 1998.

[Richter, 1998b] C. Richter, D. Ashlock and G. Sheblé, "Effects of Tree Size and State Number on GP-Automata Bidding Strategies," *Proceedings of the GP-98 Genetic Programming Conference*, 1998.

[Richter, 1995] C. W. Richter, Jr., T. T. Maifeld, and G. B. Sheblé. "Genetic Algorithm Development of a Healthcare Expert System." *Proceedings of the 4th Annual Midwest Electro-technology Conference,* pp. 35-38. Ames, IA, 1995.

[Roy, 1993] S. Roy, "Goal-programming Approach to Optimal Price Determination for Inter-area Energy Trading," *International Journal Energy Research,* Vol. 17, pp. 847-862, December 1993.

[Rupanagunta, 1995] P. Rupanagunta, M. L. Baughman, and J. W. Jones, "Scheduling of Cool Storage Using Non-linear Programming Techniques," *IEEE Transactions on Power Systems,* Vol. 10, no. 3, August 1995.

[Russel, 1992] T. Russel, "Working with an Independent Grid in the UK - A Generator's View," *Proceedings of the 24th Annual North American Power Symposium*, Manhattan, Kansas, pp. 270-275, September 1992.

[Russel, 1995] T. Russel and L. Mogridge. "Some Operational Issues Between Independent Generators and an Independent Grid," *Proceedings of the 27th Annual North American Power Symposium*, Manhattan, Kansas, pp. 1183-1188, September 1995.

[Ruzic, 1996a] S. Ruzic, N. Rajakovic and A. Vuakovic, "Flexible Approach to Short-term Hydro-thermal Coordination Part I: Problem Formation," *IEEE Transactions on Power Systems,* Vol. 11, pp. 1564-1571, August 1996.

[Ruzic, 1996b] S. Ruzic, N. Rajakovic and A. Vuakovic, "Flexible Approach to Short-term Hydro-thermal Coordination Part II: Dual Problem Solution Procedure," *IEEE Transactions on Power Systems,* Vol. 11, pp. 1572-1578, August 1996.

[Santos, 1988] A. Santos, S. Deckman and S. Soares, "A Dual Augmented LaGrangian Approach for Optimal Power Flow," *IEEE Transactions on Power Systems*, Vol-3, no. 3, pp. 1020-1025, August 1988.

[Schulte, 1995] R. P. Schulte, "An Automatic Generation Control Modification for Present Demands on Interconnected Power Systems," presented at IEEE PES Summer Meeting, 95 SM 187-7 PWRS, 1995.

[Schweppe, 1985] F. C. Schweppe, M. C. Caramanis and R. D. Tabors, Evaluation of Spot Price Based Electricity Rates. *IEEE Transactions on Power Apparatus and Systems,* Vol. 104, no. 7, pp. 1644-1655, 1985.

[Schweppe, 1988] F. C. Schweppe, M. C. Caramanis, R. D. Tabors, and R. E. Bohn, *Spot Pricing of Electricity*, Kluwer Academic Publishers, Boston, 1988.

[Sheblé, 1985] G. Sheblé, "Unit Commitment for Operations," Ph.D. Dissertation, Virginia Polytechnic Institute and State University, March 1985.

[Sheblé, 1989] G. Sheblé, "Real-time Economic Dispatch and Reserve Allocation Using Merit Order Loading and Linear Programming Rules," *IEEE Transactions on Power Systems,* Vol. 4, pp. 1414-1420, November 1989.

[Sheblé, 1992a] G. B. Sheblé and J. W. Lamont, Class notes from EE553: " Steady State Analysis of Power Systems," Department of Electrical and Computer Engineering, Iowa State University, Ames , Iowa, 1992.

[Sheblé, 1992b] G. B. Sheblé and G. N. Fahd, "LaGrangian Relaxation," Class Notes, June 1992.

[Sheblé, 1993a] G. B. Sheblé, Class notes from EE653E: "Power System Optimization," Department of Electrical and Computer Engineering, Iowa State University, Ames, Iowa, 1993.

[Sheblé, 1993b] G. B. Sheblé, "Energy As A Commodity," Power Seminar, Iowa State University, Ames, Iowa, April 6, 1993.

[Sheblé, 1994a] G. B. Sheblé and J. D. McCalley, Class notes form EE653A: *Revaluation of Transmission Service for Secure Power System Operation in a Less Regulated Utility Environment*, Iowa State University, Ames, Iowa, 1994.

[Sheblé, 1994b] G. B. Sheblé, "Electric Energy in a Fully Evolved Marketplace," *Proceedings of the 26th Annual North American Power Symposium*, Manhattan, Kansas, pp. 81-90, September 1994.

[Sheblé, 1994c] G. B. Sheblé, "Simulation of Discrete Auction Systems for Power System Risk Management," *Proceedings of the 27th Annual Frontiers of Power Conference*, Oklahoma State University, Stillwater, Oklahoma, pp. I.1-I.9, 1994.

[Sheblé, 1994d] G. Sheblé and G. Fahd, "Unit Commitment Literature Synopsis," *IEEE Transactions on Power Systems,* Vol. 9, pp. 128-135, February 1994.

[Sheblé, 1994e] G. Sheblé and J. McCalley, "Discrete Auction Systems for Power System Management." Presented at the 1994 National Science Foundation Workshop, Pullman, WA, 1994.

[Sheblé, 1996a] G. B. Sheblé, M. Ilic, B. F. Wollenberg, and F. Wu, Lecture Notes from: Engineering Strategies for Open Access Transmission Systems, A two-day Short Course Presentation, San Francisco, CA., December 5 and 6, 1996.

[Sheblé, 1996b] G. Sheblé, "Priced Based Operation in an Auction Market Structure," *IEEE Trans.* on *Power Systems,* Vol. 11, no. 4, November 1996, pp. 1770-1777.

[Shirmohammadi, 1989] D. Shirmohammadi, P. R. Gribik, T. K. Law, J. H. Malinowski and R. E. O'Donnell, "Evaluation of Transmission Network Capacity Use for Wheeling Transactions," *IEEE Transactions on Power Systems,* Vol. 4, no. 4, pp. 1405-1413, 1989.

[Shirmohammadi, 1991] D. Shirmohammadi, C. Rajagopalan, E. R. Alward and C. L. Thomas, Cost of Transmission Transactions: An Introduction, *IEEE Transactions on Power Systems,* Vol. 6, no. 3, pp. 1006-1016, 1991.

[Shy, 1996] O. Shy, *Industrial Organization: Theory and Application,* The MIT Press, London, England, 1996.

[Simmons, 1975] D. M. Simmons, *Nonlinear Programming for Operations Research*, Englewood Cliffs, New Jersey: Prentice-Hall Inc., 1975.

[Skeer, 1991] J. Skeer, "Highlights of the International Energy Agency Conference on Advanced Technologies for Electric Demand-side Management," in *Proceedings of Advanced Technologies for Electric Demand-side Management,* Sorrento, Italy: International Energy Agency, 1991.

[Slawaji, 1994] S. Slawaji and K. Lo, "New Approach for Solving the Problem of Unit Commitment," in *Proceedings of the 2nd International Conference on Advances in Power System Control, Operation & Management*, Vol. 2, pp. 571-576, December 1994.

[Smith, 1965] V. L. Smith, "Experimental Auction Markets and the Walrasian Hypothesis," *Journal of Political Economy,* Vol. 73, pp. 387-393, 1965.

[Smith, 1967] V. L. Smith, "Experimental Studies of Discrimination Versus Competition in Sealed-Bid Auction Markets," *Journal of Business,* Vol. 40, pp. 56-84, 1967.

[Smith, 1974] V. L. Smith, "Bidding and Auctioning Institutions: Experimental Results," Division of Humanities and Social Sciences, California Institute of Technology, Pasadena, California, Social Science Working Paper, no. 71, 1974.

[Smith, 1982] V. L. Smith, Microeconomic Systems as an Experimental Science. *American Economic Review,* Vol. 72, no. 5, 923-55, 1982.

[Smith, 1988] V. L. Smith, "Electric Power Deregulation: Background and Prospects," *Contemporary Policy Issues*, Vol. 6, pp. 14-24, July 1988.

[Smith, 1993] S. Smith, "Linear Programming Model for Real-time Pricing of Electric Power Service," *Operations Research,* Vol. 41, pp. 470-483, May-June 1993.

[Soucek, 1992] Branko Soucek, *Dynamic, Genetic, and Chaotic Programming,* John Wiley and Sons, Inc., New York, 1992.

[Stott, 1974] B. Stott and O. Alsac, "Fast Decoupled Load Flow," *IEEE Transactions on PAS,* Vol. 93, no. 3, pp. 859-867, 1974.

[Stott, 1978a] B. Stott and E. Hobson, "Power System Security Control Calculation Using Linear Programming, Part I," *IEEE Transactions on PAS*, Vol. 97, no. 5, pp.1713-1720, September 1978.

[Stott, 1978b] B. Stott and E. Hobson. "Power System Security Control Calculation Using Linear Programming, Part II," *IEEE Transactions on PAS*, Vol. 97, no. 5, pp.1721-1731, September 1978.

[Stott, 1979] B. Stott and J. L. Marinho, "Linear Programming for Power System Network Security Applications," *IEEE Transactions on PAS,* Vol. PAS-98, no. 3, May/June 1979.

[Stott, 1987] B. Stott, O. Alsac and A. J. Monticelli, "Security Analysis and Optimization," *Proceedings of the IEEE,* Volume 75, no. 12, pp. 1623-1644, December 1987.

[Sullivan, 1977] R. L. Sullivan, "Power System Planning," McGraw-Hill, 1977

[Svoboda, 1994] A. Svoboda and S. Oren, "Integrating Price-based Resources in Short-term Scheduling of Electric Power Systems," *IEEE Transactions on Energy Conversion,* Vol. 9, pp. 760-769, December 1994.

[Swidler, 1991] J. C. Swidler, "An Unthinkably Horrible Situation," *Public Utilities Fortnightly,* September 15, 1991.

[Tabors, 1994] R. D. Tabors, "Transmission System Management and Pricing: New Paradigms and International Comparisons," Paper WM110-7 presented at the IEEE/PES Winter Meeting, T-PWRS, February 1994.

[Takriti, 1997] S. Takriti, B. Krasenbrink, and L. S.-Y. Wu. "Incorporating Fuel Constraints and Electricity Spot Prices into the Stochastic Unit Commitment Problem," IBM Research Report: RC 21066, Mathematical Sciences Department, T.J. Watson Research Center, Yorktown Heights, New York, December 29, 1997.

[Thompson, 1992] G. L. Thompson, and S. Thore, *Computational Economics: Economic Modeling with Optimization Software*, Boyd and Fraser Publishing Company, Denver, 1992.

[Torres, 1996] G. Torres, V. Quintana, and G. Lambert-Torres, "Optimal Power Flow in Rectangular Form Via an Interior Point Method," *Proceedings of the 28th North American Power Symposium,* November 1996, pp. 481-488.

[U. S. Congress, 1988] U. S. Congress. "Notice of Proposed Rulemaking: Docket no. RM88-4, RM88-5, RM88-6," *Congressional Record*, March 16, 1988.

[U. S. Congress, 1990] U.S. Congress, Senate, Summary of S. 1630 "Clean Air Act Amendments of 1990," *Congressional Record,* April 20, 1990.

[U.S. Environmental Protection Agency, 1982] U.S. Environmental Protection Agency, Office of Air Quality Planning and Standards, *National Air Pollution Estimates*, Research Triangle Park, North Carolina, 1982.

[Varian, 1992] H. R. Varian, *Microeconomic Analysis, Third Edition.* New York: W. W. Norton and Company, Inc., 1992.

[Vickrey, 1961] W. Vickrey, Counterspeculation, Auctions, and Competitive Sealed Tenders. *Journal of Finance,* Vol. 16: 8-37, 1961.

[Virmani, 1989] S. Virmani, E. C. Adrian, K. Imhof, and S. Mukherjee, "Implementation of a LaGrangian Relaxation Based Unit Commitment

Problem," *IEEE Trans.* on Power *Systems,* Vol. 4, no. 4, October 1989, pp. 1373-1380.

[Virmani, 1997] S. Virmani, E. C. Adrian, K. Imhof and S. Mukherjee, "Implementation of a LaGrangian Relaxation Based Unit Commitment Problem," J. Jacobs, "Artificial Power Markets and Unintended Consequences," *IEEE Transactions on Power Systems*, Vol. 4, no. 4, pp. 1373-1380, October 1989.

[Vojdani, 1994] A. Vojdani, C. Imparto, N. Saini, B. Wollenberg and H. Happ, "Transmission Access Issues." Presented at the 1995 IEEE/PES Winter Meeting, 95 WM 121-4 PWRS. New York: IEEE, 1994.

[Wade, 1991] B. Wade, Brorsen, "Futures Trading, Transactions Costs, and Stock Market Volatility." *The Journal of Futures Market*, Vol. II, no. 2, pp. 153-164, April 1991.

[Walsh, 1997] M. P. Walsh and M. J. O'Malley, "Augmented Hopfield Network for Unit Commitment and Economic Dispatch," *IEEE Transactions on Power Sytems*, Vol. 12, no. 4, November 1997.

[Wang, 1989] L. Wang, "Approximate Confidence Bounds on Monte Carlo Simulation Results for Energy Production," *IEEE Transactions on Power Systems*, Vol. 4, No. 1, p. 69 - 74, February 1989

[Wang, 1994a] C. Wang and S. Shahidehpour, "Ramp-rate Limits in Unit Commitment and Economic Dispatch Incorporating Rotor Fatigue Effect," *IEEE Transactions on Power Systems,* Vol. 9, pp. 1539-1545, August 1994.

[Wang, 1994b] C. Wang and J. R. McDonald, *Modern Power System Planning,* McGraw-Hill, 1994

[Wang, 1995] C. Wang and S. Shahidehpour, "Optimal Generation Scheduling with Ramping Costs," *IEEE Transactions on Power Systems,* Vol. 10, pp. 60-67, February 1995.

[Wei, 1996] H. Wei, H.Sasaki, and R. Yokoyama, " An Application of Interior Point Quadratic Programming Algorithm to Power System Optimization Problems," *IEEE Trans. on Power Systems,* Vol. 11, February 1996, pp. 260-266.

[Wei, 1995] D. C. Wei, and N. Chen, "Air-conditioner Direct Load Control by Multi-pass Dynamic Programming," *IEEE Transactions on Power Systems,* Vol. 10, no. 1, February 1995.

[Weiss, 1991] L. Weiss and S. A. Spiewak, *The Wheeling and Transmission Manual, Second Edition.* The Fairmont Press, Inc., 1991.

[Werden, 1996] G. J. Werden, "Identifying Market Power in Electric Generation," *Public Utilities Fortnightly*, Vol. 134, no. 4, pp. 16-21, February 15, 1996.

[Williams, 1984] A. W. Williams and V. L. Smith, Cyclical Double-Auction Markets with and Without Speculators. *Journal of Business,* Vol. 57, no. 1, pp. 1-34, 1984.

[Willis, 1996] H. L. Willis, *Spatial Electric Load Forecasting,* pp. 14-17, ed. Marcel Dekker Inc., NY, 1996.

[Wilson, 1985] R. Wilson, "Incentive Efficiency of Double Auctions," *Econometrica*, Vol. 53, pp. 1101-1115, 1985.

[Wilson, 1986] R. Wilson, Double Auctions. In *Information, Incentives and Economic Mechanisms: Essays in Honor of Leonid Hurwicz.* Eds.: T. Groves,

R. Radner and S. Reiter. Minneapolis, Minnesota: University of Minnesota Press, 1986.

[Winston, 1988] W. E. Winston and C. A. Gibson, Geographical Load Shift and its Effect on Interchange Evaluation, *IEEE Transactions on Power Systems,* Vol. 3, no. 3, pp. 865-871, 1988.

[Wismer, 1978] David A. Wismer and R. Chattergy, *Introduction to Nonlinear Optimization*, North-Holland Publishing Company, Amsterdam, 1978.

[Wood, 1996] A. J. Wood, and B. F. Wollenberg, *Power Generation, Operation, and Control*, John Wiley and Sons, 2nd Edition, New York, 1996.

[Wu, 1993] Y. Wu, A. Debs and R. Marsten, "Direct Nonlinear Predictor-corrector Primal-dual Interior Point Algorithm for Optimal Power Flows," in *Proceedings of the 1993 IEEE Power Industry Computer Applications Conference,* Vol. 1, pp. 894-897, May 1993.

[Wu, 1993] Y. Wu, A. Debs, and R. Marsten, "A Direct nonlinear Predictor-corrector Primal-dual Interior Point Algorithm for Optimal Power Flows," *Proceedings of the 1993 IEEE Power Industry Computer Applications Conference,* May 1993, pp. 138-145.

[Wu, 1995a] F. Wu and P. Varaiya, "Coordinated Multi-lateral Trades For Electric Power Networks: Theory and Implementation," University of California at Berkeley Research Report, June 1995.

[Wu, 1995b] F. Wu, S. Oren, P. Spiller, and P. Varaiya "Folk Theorems On Transmission Access: Proofs and Counter Examples," University of California at Berkeley Research Report, February 28, 1995.

[Wu, 1995c] F. Wu and P. Varaiya, "What is Wrong With Hogan's Contract Network," University of California at Berkeley Research Report, June 1995.

[Yan, 1993] H. Yan, P. Luh, X. Guan and P. Rogan, "Scheduling of Hydrothermal Power Systems," *IEEE Transactions on Power Systems,* Vol. 8, pp. 1358-1365, August 1993.

[Yan, 1996] X. Yan, and V. Quintana, "An Efficient Predictor-corrector Interior Point Algorithm for Security-constrained Economic Dispatch," *Proceedings of the 1996 IEEE/PES Summer Meeting,* Paper 96 SM 506-6 PWRS, pp. 1-8.

[Yan, 1996a] X. Yan and V. Quintana, "An Efficient Predictor-corrector Interior Point Algorithm for Security Constrained Economic Dispatch," in *Proceedings of the 1996 IEEE/PES Summer Meeting,* pp. 1-8, July-August 1996.

[Yan, 1996b] Z. Yan, N. Xiang, B. Zhang, S. Wang and T. Chung, "Hybrid Decoupled Approach to Optimal Power Flow," *IEEE Transactions on Power Systems,* Vol. 11, pp. 947-954, May 1996.

[Ye, 1992] Y. Ye, "On the Finite Convergence of Interior-point Algorithms for Linear Programming," *Mathematical Programming,* vol. 57, no. 2, 1992, pp. 325-335.

[Zangwill, 1967] W. I. Zangwill, "Nonlinear Programming via Penalty Functions," *Management Science*, Vol. 13, no. 5, pp. 344-358, 1967.

[Zeleny, 1982] M. Zeleny, *Multiple Criteria Decision Making,* McGraw Hill ed., 1982.

[Zhang, 1984] G. Zhang and A. Brameller, "On-line Security Constrained Economic Dispatch and Reactive Power Control Using Linear Programming," in

*Proceedings of the Eighth Power Systems Computation Conference*, pp. 396-400, 1984.

[Zhuang, 1988] F. Zhuang, and F. D. Galiana, "Towards a More Rigorous and Practical Unit Commitment by LaGrangian Relaxation," *IEEE Trans. on Power Systems,* Vol. 3, no. 2, May 1988, pp. 763-773.

# GLOSSARY

**Energy Auction/Brokerage System** is defined as a central auction mechanism which provides trading opportunities to the market participants and maintains reliability of power system operation by coordinating generation, transmission, and distribution functions.

**Supportive Services** are the services used for supporting a reliable delivery of electric energy in power system operation.

**Spinning Reserve** is defined as the unused capacity of generating units that are on line in operation. Spinning reserve can be called into operation almost instantaneously.

**Ready Reserve** is defined as the unused capacity of generation that is not on line but can be brought on line within 15 minutes in operation.

**Transmission Losses** are defined as the amount of electric energy dissipated in the electrical transmission network during the process of power delivery.

**Load Following** is defined as the amount of electric energy provided to maintain the contracted tie line flow and frequency in power system operation during the trading period. The load following energy is required to respond in 5-10 minutes of time frame.

**An Unbundled Supportive Service** is defined as the service that is not traded with other services nor with the main commodity, energy.

**A Bundled Supportive Service** is defined as the service that is traded with other services nor with the main commodity, energy.

**Industry Cost** is defined as the cost of providing bundled supportive services. The industry cost should be recovered by some acceptable mechanism other than allocation.

**Forward Market** is defined as the trading place where short-term transactions (as defined in traditional utility industry, i.e., hourly or daily transactions) take place.

**Futures Market** is defined as the trading place where long-term transactions (as defined in traditional utility industry, i.e., monthly or yearly transactions) take place.
**Planning Market** is defined as the market that would underwrite the usage of assets (such as transmission lines) to very long-term commitments (15-20 years or more).

**Swap Market** is defined as the clearinghouse that allows contracts to be terminated with an exchange of physical or financial substitution.

**Option Contracts** give the owner the right but not the obligation to buy or sell a specified trading contract.

**Market Clearing Price** is a vector of prices for which all perceived valuations of the commodity are in equilibrium.

**Available Transfer Capability** is a measure of the transfer capability remaining in the physical transmission network for further commercial activity over and above already committed uses. ATC is defined as the Total Transfer Capability (TTC), less the Transmission Reliability Margin (TRM), less the sum of existing transmission commitments (which includes retail customer service) and the Capacity Benefit Margin (CBM).

**Broker** is a third party who establishes a transaction between a buyer and seller for a fee. A broker does not take title to capacity or energy and is non-jurisdictional to the FERC.

**Call Option** is an option that gives the buyer the right, but not the obligation, to buy capacity and/or energy in the future for a specified price within a specified period of time in exchange for a one-time premium payment. It obligates the seller of the option to sell the capacity and/or energy at the designated price if the option is exercised.

**Capacity Benefit Margin** (CBM) is that amount of transmission transfer capability reserved by load serving entities to ensure access to generation from interconnected systems to meet generation reliability requirements.
Reservation of CBM by a load serving entity allows the entity to reduce installed generating capacity below that which may otherwise have been necessary without interconnections to meet its generation reliability requirements.

**Control Area** is an electric system or systems, bound by interconnection metering and telemetry, capable of controlling generation to maintain its interchange schedule with other Control Areas and contributing to frequency regulation of the Interconnection.

**Day-ahead Market** is market involving trading of multi-hour energy blocks for delivery during the following day. Offpeak market involves daily prescheduled energy for the eight off-peak hours of the following day. On-peak market involves daily prescheduled energy for the sixteen on-peak hours of the following day. The trades reported in the trade press are typically for financially firm or physically firm power transactions. Other terms used to refer to this market include prescheduled and next-day market.

**Derivative** is a financial instrument, traded on or off an exchange, the price of which is directly dependent upon (i.e., "derived from") the value of one or more underlying securities, equity indices, debt instruments, commodities, other

derivative instruments, or any agreed upon pricing index or arrangement (e.g., the movement over time of the Consumer Price Index or freight rates).
Derivatives involve the trading of rights or obligations based on the underlying product, but do not directly transfer property. They are used to hedge risk or to exchange a floating rate of return for fixed rate of return.

**Forward Contract** is a supply contract between a buyer and seller, whereby the buyer is obligated to take delivery and the seller is obligated to provide delivery of a fixed amount of a commodity (e.g., electric energy) at a predetermined price on a specified future date. Payment in full is due at the time of, or following, delivery. This differs from a futures contract where settlement is made daily, resulting in partial payment over the life of the contract.

**Futures Contract** is a supply contract between a buyer and seller, whereby the buyer is obligated to take delivery and the seller is obligated to provide delivery of a fixed amount of a commodity (e.g., electric energy) at a predetermined price at a specified location. Futures contracts are traded exclusively on regulated exchanges and are settled daily based on their current value in the marketplace.

**Hedge** is the initiation of a position in a futures or options market that is intended as a temporary substitute for the sale or purchase of the actual commodity. The sale of futures contracts in anticipation of future sales of cash commodities acts as a protection against possible price declines, or the purchase of futures contracts in anticipation of future purchases of cash commodities as a protection against the possibility of increasing costs.

**Hourly Market** is the market in which hourly blocks of energy are traded.

**Liquidated Damages Contract** is any contract with a provision which obligates the seller of power to pay the buyer's replacement energy costs in the event that the seller fails to deliver the contracted for energy.

**Long Position** is the position of a trader in the futures market who has less contracts obligating him to deliver a commodity at some time in the future than contracts obligating others to deliver the commodity to him.

**Operating Reserve** is that generating capability above firm system demand required to provide for regulation, load forecasting error, forced and scheduled equipment outages and local area protection. It consists of spinning and non-spinning reserve.

**Option Contract** is a contract which give the holder the right, but not the obligation, to purchase or to sell a commodity in the future at a specified price within a specified period of time in exchange for a one-time premium payment.

**Power Marketer** is a wholesale power entity approved by the FERC to buy and sell wholesale power from and to each other and other public utilities at market-based prices. In contrast to Brokers, marketers take title to the power in their transactions.

**Price Discovery** is the manner in which market players find out the bid and offer prices of other buyers and sellers. Auction mechanisms range from one-on-one phone calls to formal exchanges with posted bids and offers.

**Put Option** is an option that gives the buyer the right, but not the obligation, to sell capacity and/or energy in the future for a specified price within a specified period of time in exchange for a one-time premium payment. It obligates the seller of the option to buy the capacity and/or energy at the designated price if the option is exercised.

**Security Coordinator** is an entity that provides the security assessment and emergency operations coordination for a group of control areas.

**Short Position** is the position of a trader in the futures market who has more contracts obligating him to deliver a commodity at some time in the future than contracts obligating others to deliver the commodity to him.

**Sleeve** is a third party trader used as an intermediary between two other traders. Sleeves, which are usually large entities with good credit standing, are used for various reasons including to circumvent credit requirements of one of the parties, to enhance the risk profile of one of the traders, or shield the positions of one or both of the traders.

**Spot Market** is a market where goods are traded for immediate delivery.

**Strike Price** is the price at which the underlying options contract is bought and sold in the event the option is exercised. Also called the exercise price.

**Transmission Reliability Margin (TRM)** is that amount of transmission transfer capability necessary to ensure that the interconnected transmission network is secure under a reasonable range of uncertainties in system conditions.

# INDEX

MIX
Papier aus verantwortungsvollen Quellen
Paper from responsible sources
FSC® C105338

If you have any concerns about our products,
you can contact us on
**ProductSafety@springernature.com**

In case Publisher is established outside the EU,
the EU authorized representative is:
**Springer Nature Customer Service Center GmbH**
**Europaplatz 3, 69115 Heidelberg, Germany**

Printed by Libri Plureos GmbH
in Hamburg, Germany